ROUTLEDGE LIBRARY EDITIONS:
GEOLOGY

Volume 18

GROUNDWATER AS A GEOMORPHIC AGENT

GROUNDWATER AS A GEOMORPHIC AGENT

Binghamton Geomorphology Symposium 13

Edited by
R.G. LAFLEUR

Routledge
Taylor & Francis Group
LONDON AND NEW YORK

First published in 1984 by Allen & Unwin Inc.

This edition first published in 2020
by Routledge
2 Park Square, Milton Park, Abingdon, Oxon OX14 4RN

and by Routledge
605 Third Avenue, New York, NY 10017

Routledge is an imprint of the Taylor & Francis Group, an informa business

British Library Cataloguing in Publication Data
A catalogue record for this book is available from the British Library

ISBN: 978-0-367-18559-6 (Set)
ISBN: 978-0-429-19681-2 (Set) (ebk)
ISBN: 978-0-367-46447-9 (Volume 18) (hbk)
ISBN: 978-1-00-302883-3 (Volume 18) (ebk)

Groundwater as a Geomorphic Agent

Edited by
R.G. LaFleur
Rensselaer Polytechnic Institute, Troy, New York

Boston
ALLEN & UNWIN, INC.
London Sydney

Allen & Unwin Inc.,
9 Winchester Terrace, Winchester, Mass. 01890, USA

George Allen & Unwin (Publishers) Ltd,
40 Museum Street, London WC1A 1LU, UK

George Allen & Unwin (Publishers) Ltd,
Park Lane, Hemel Hempstead, Herts HP2 4TE, UK

George Allen & Unwin Australia Pty Ltd,
8 Napier Street, North Sydney, NSW 2060, Australia

First published in 1984

Library of Congress Cataloging in Publication Data

Main entry under title:
Groundwater as a geomorphic agent.
Proceedings of the 13th Annual Geomorphology Symposium, held at Rensselaer Polytechnic Institute, Sept. 1982.
Includes index.
1. Hydrogeology – Congresses. 2. Geomorphology – Congresses. 1. LaFleur, Robert G. II. Geomorphology Symposium (13th : 1982 : Rensselaer Polytechnic Institute)
GB1001.2.G76 1984 551.3'5 83–15651
ISBN 0–04–551069–5

British Library Cataloguing in Publication Data

Groundwater as a geomorphic agent.
1. Water, Underground
I. LaFleur, R.G.
551.49 GB1003.2
ISBN 0–04–551069–5

Set in 10 on 12 point Times by M.C.S. Ltd, Salisbury, Wilts.
Printed in Great Britain by Mackays of Chatham

Preface

This book is the proceedings of the 13th Annual Geomorphology Symposium, an extension of the Binghamton Series, held at Rensselaer Polytechnic Institute, September 1982. As the title implies, this volume has both a geomorphic and a hydrologic message – delivered by authors of diverse backgrounds and experience. That groundwater as a geomorphic agent can be observed first-hand and quantified is a tribute not only to their talents but also their perseverance, often under difficult circumstances.

Of the many ways groundwater can fashion a landscape, its solutional destruction of carbonate rocks is best known. The cavernous subsurface has been spectacularly illustrated for a long time – traditionally, cave formation is identified by many as an important, if not the only, product of groundwater circulation. However, exciting quantitative studies on rates and styles of cavern formation, stratigraphic and structural controls on karst evolution, and groundwater modeling of karstic aquifers are now emerging, in large measure due to the growing number of hydrologists and geomorphologists who have joined the caver ranks. But only half of this book deals with the role of groundwater in carbonate terrains.

Equally important, and heretofore unheralded, is the role groundwater plays in landscapes where carbonate rocks are not present. There, for temporal reasons, the role may be even more occult than in cavern complexes. In many instances, the contribution of groundwater outflow to early landscape evolution is masked by later more easily observed surface-water processes. The first half of this book considers water in several geological settings; as a soil permeant, as an eroding seeping and piping fluid, as a subsurface etch former on granite, and as a precipitator of crusts.

To aid in evaluation of relatively young landscapes, Foss and Segovia provide several quantitative soil-aging parameters – solum thickness, hue, lamellae development in alluvium, and chemical trends in pH, free iron content and extractable magnesium – based on studies in the southeastern United States.

Higgins calls attention to the overlooked processes of sapping and seepage erosion in early dissection of landscape and suggests these operate over a wide range of scale. Ongoing examples are seen in microdrainage nets on beaches during falling tide. Ancient examples are the pectinate networks of the US High Plains and the wadis of the Gilf Kebir Plateau. Martian valley-like features are the largest scaled of all.

Berger and Aghassy approach seepage erosion from the structural viewpoint identifying in the US Gulf Coast discordant surface drainages related to interplay of groundwater and subsurface folds and faults, and cite the use of remote sensing to detect these often subtle relationships on low-relief terrains.

Surface processes that cause soil and landscape evolution can act in a synergistic or antagonistic manner; most operate both in temperate and tropical climates but as Segovia and Foss point out, not at equal levels of effectiveness. Chemical thresholds may be crossed only in a certain environment as may be the case with thresholds that intervene in the formation of bauxite.

In more arid lands, groundwater activity fashions etch forms, at scales ranging from boulders to inselbergs, and from basins and gutters to large low-relief surfaces. Twidale extends emphasis of the arid-land role of subsurface water to the formation of duricrusts of compositions that vary with geologic age, and cites their usefulness as caprocks that preserve ancient paleosurfaces.

Acid rain impacts a wide area of southern Canada and northeastern United States. Shilts notes that in glaciated terrains it is important to understand patterns of variation in labile drift components in order to evaluate capacity of soil and overburden to absorb or buffer excess hydrogen ions.

Mylroie introduces the section dealing with solution of carbonate rocks with a hydrologic classification of karst landforms. The functional role of caves in the hydrologic cycle is important to an understanding of all karst features, so also are the surface and subsurface environments and the interface separating them. Mass transfer of water carrying dissolved and clastic load across the interface is a dominant process in landform production.

Palmer quantifies solutional reduction of karst features and suggests that rate increases with discharge of solvent water, asymptotically approaching a maximum where solution rate is independent of discharge. Cave-wall retreat can approach 1 mm yr^{-1}, and a threshold width of initial rock opening, for caves to form, is slightly less than 0.1 mm. Distinction between vadose and phreatic cave features provides precise measure of past elevations of local water tables and entrenched river levels.

In further quantification of overall karst denudation rates, White combines new rate equations, available in the literature, with mass transfer calculations, on laboratory-determined kenetics, to suggest rationalization of observed 20–80 mm ka^{-1} rates with a calculated upper limit of 200–300 mm ka^{-1}. These calculations also apply to evolution and time scale for closed depression features, interpretation of convex and concave karst slopes, and relative dissolution and denudation rates on limestone and dolostone terrains.

On a worldwide basis, karst denudation rates may not be too dissimilar, and as Drake also notes, there may be elevation control over the degree of stability of some karst forms.

Back, Hanshaw and Van Driel cite the scalloped rock coast of Yucatan, and identify the subsurface fresh–salt water mixing zone as the place where subsaturation with respect to calcite leads to large scale limestone solution, adequate to cause cavern formation and secondary porosity development.

Flow of groundwater through karst aquifers is simulated by applying finite difference numerical methods to non-Darcian models. Cullen and LaFleur propose and test conceptual models for conduit and fracture-controlled flow and show a technique of constructing aquifer maps that applies interactive computer graphics.

Ford addresses two problems of karst evolution in arctic terrain. The first is that of establishing the extent of modern karst groundwater circulation and karst landform genesis in areas that are permafrozen today. The second problem is that of determining whether there may be significant development of karst beneath glaciers in regions that will become permafrozen upon deglaciation.

Two regional karst studies complete this volume. Crawford models landscape evolution along the Cumberland Plateau escarpment of Tennessee, where interbedded impervious strata distort carbonate conduit systems and a terra rossa plain follows the retreating escarpment. In more structurally deformed rocks of the Sequatchie Anticline, breaching of impervious anticlinal mountain caprock reduces the mountain to karst valleys that are then assimilated into the Sequatchie Valley as it advances headward into the Plateau.

Using uplifted plateaus of Texas, Kentucky and New York as structural settings, Kastning relates karst evolution to fracture patterns. Underground flow paths and doline orientation may reflect faulting not readily mapped by conventional means. Linear cave-passage segments correspond closely to fractures and in places indicate which fractures were the most open initially or which were favorably aligned along prevailing hydraulic gradients.

Many people helped make the 13th Geomorphology Symposium a success – the standing committee on geomorphology symposia, and particularly the speakers and authors who so enthusiastically contributed their time and efforts. Thanks are also expressed to Sydney Archer, Dean of Science at Rensselaer Polytechnic Institute for his financial support of the symposium, to Gloria Daniels for numerous typing chores on manuscripts and to Joan LaFleur for her editorial skills.

Robert G. LaFleur
Rensselaer Polytechnic Institute
April 1983

Contents

Tables

Contributors

Jacob Aghassy (deceased)
Department of Geography, University of Pittsburgh, Pittsburgh, Pennsylvania 15217, USA

William Back
United States Geological Survey, Mail Stop 432, National Center, Reston, Virginia 22092, USA

Zeev Berger
Exxon Production Research Co., P.O. Box 2189, Houston, Texas 77001, USA

Nicholas C. Crawford
Department of Geography and Geology, Western Kentucky University Bowling Green, Kentucky 42101, USA

James J. Cullen, IV
Department of Hydrology and Water Resources, University of Arizona, Tucson, Arizona 85721, USA

John J. Drake
Department of Geography, McMaster University, Hamilton, Ontario L8S 4Kl, Canada

Derek C. Ford
Department of Geography, McMaster University, Hamilton, Ontario L8S 4Kl, Canada

John E. Foss
Department of Soil Science, North Dakota State University, Fargo, North Dakota 58102, USA

Bruce B. Hanshaw
United States Geological Survey, Mail Stop 104, National Center, Reston, Virginia 22092, USA

Charles G. Higgins
Department of Geology, University of California, Davis, Davis, California 95616, USA

Ernst H. Kastning, Jr.
Department of Geology and Geophysics, University of Connecticut, Storrs, Connecticut 06268, USA

Robert G. LaFleur
Department of Geology, Rensselaer Polytechnic Institute, Troy, New York 12181, USA

John E. Mylroie
Department of Geosciences, Murray State University, Murray, Kentucky 42071, USA

Arthur N. Palmer
Department of Earth Science, State University of New York, College at Oneonta, Oneonta, New York 13820, USA

Antonio V. Segovia
Department of Geology, University of Maryland, College Park, Maryland 20742, USA

William W. Shilts
Geological Survey of Canada, 601 Booth Street, Ottawa K1A 0E8, Canada

C. Rowland Twidale
Department of Geography, The University of Adelaide, Box 498, G.P.O., Adelaide 5001, South Australia

J. Nicholas Van Driel
United States Geological Survey, Mail Stop 911, National Center, Reston, Virginia 22092, USA

William B. White
Department of Geosciences, The Pennsylvania State University, University Park, Pennsylvania 16802, USA

Groundwater as a Geomorphic Agent

1
Rates of soil formation

John E. Foss and Antonio V. Segovia

Introduction

The rate of soil formation, or "soil age," is of interest to many scientists because of the interpretations that can be made about the general age of landscapes and the various weathering processes occurring during soil development. Specifically, the rates of soil formation and the resulting soil profiles have been used in dating geologic events, interpreting archeological sites, evaluating geomorphic processes, determining past climatic conditions, and establishing tolerance rates (T factors) for agricultural erosion.In many cases, the preceding interpretations have been limited by the complexity of the soil formation process and lack of precise data on soil age for various geologic materials, landscapes and environments. Recently, however, numerous studies have elucidated the relationship between soil morphological, chemical, physical and mineralogical properties and soil age. This paper will describe the general aspects of soil formation and give specific examples of soil age studies on various geologic materials in New Jersey, Maryland, Virginia, Georgia and South Carolina.

General aspects

The formation of soil from geologic material is the result of complex, dynamic processes whereby changes are continually taking place. In soils on old landscapes, the changes may take place very slowly, but in young geological materials (e.g. glacial till or alluvium), changes may occur rapidly as vegetation becomes established and as environmental factors influence weathering of the materials. In most cases, 2000–3000 years are necessary to show significant profile development in humid temperate regions although organic matter accumulation or structural development may take place in 50 years or less.

Figure 1.1 illustrates the general process of soil formation from two types of geologic material. In the first case, soil has developed from limestone with the resulting soil profile showing A, B and C horizons. A soil profile 2 m thick may have required as much as 20 m of limestone to yield the parent material (non-carbonate mineral impurities such as silicates or iron and magnesium oxides, etc.) for this soil. The ratio of limestone to residual soil may vary depending on the impurities (non-carbonates) in the limestone. In a soil developed from loess, the profile shows A, B and C horizons resulting from soil formation processes and little change in volume from the initial material.

The rate of soil development and thickness of the profile will be greatly influenced by the nature of the geologic material. For example, soil developing in granite for 10 000 years in a cool, humid environment may be only a few centimeters thick, whereas a soil developed from loess for the same length of time may be 1 or 2 m thick. Figure 1.2 (Hall *et al.* 1982) shows some examples of soil thickness and length of weathering (soil age).

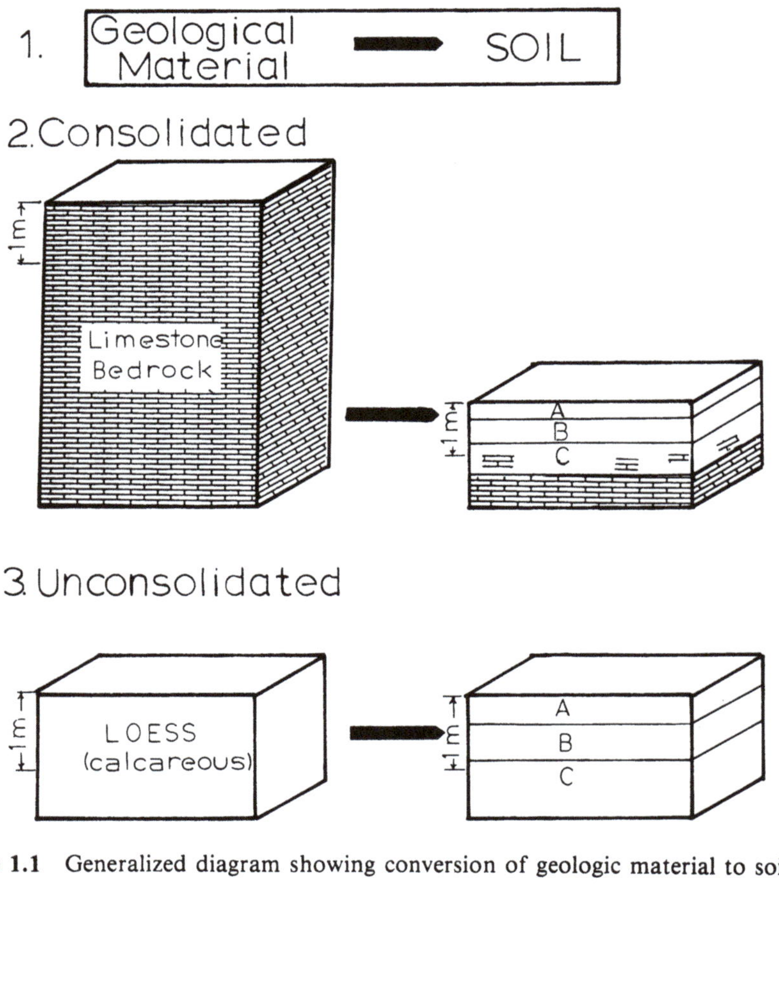

Figure 1.1 Generalized diagram showing conversion of geologic material to soil.

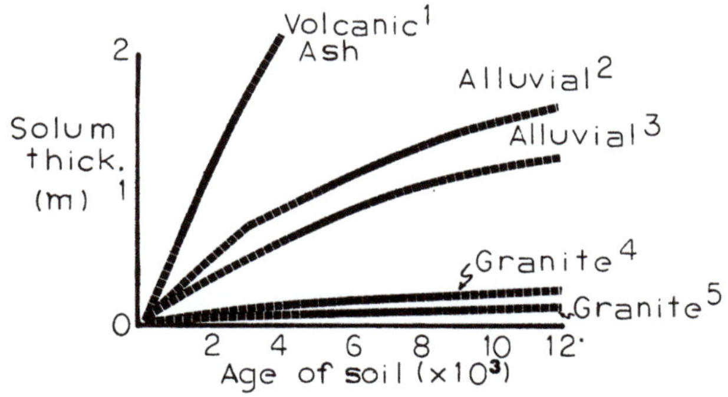

Figure 1.2 Diagram illustrating thickness of solum under different ages of soil and weathering environments (Hall *et al.* 1982). Data compiled from (1) Hay (1960), (2) Ahmed *et al.* (1977), (3) Foss (1974), Foss *et al.*(1978), (4) Leneuf and Rubert (1960), and (5) Owens and Watson (1979).

Although many interrelated processes and factors are involved in soil genesis, researchers have identified specific factors and processes dominating soil development. Figure 1.3 summarizes some of these major factors and processes. The factors of soil formation regulate the overall reactions occurring in the soil system whereas the specific processes (such as additions and losses) result in horizon formation.

Factors of soil formation

The factors of soil formation approach to soil genesis has been useful in explaining variability of soil systems. The equation (after Jenny 1941) for expressing the relationship between a soil or its properties and the soil formation factors is

$$S = f(C, O, G, R, T)$$

where S = an entire soil system or body or a single soil property, f = function of, C = climate, O = organisms, G = geologic or parent material, R = relief or topography and T = length of time for weathering. These factors explain the distribution of certain types of soils (e.g. prairie, strongly weathered or fragipan soils) or the variation in a single soil property such as pH, color or organic matter content.

The importance of time available for weathering is readily apparent in the characteristics of soils developed on different ages of geologic material. For example, soils developed on young alluvial sediments (< 200 years) will have little horizonation and show little effect of pedogenic processes, except for organic additions. In contrast, those soils developed on loess, till or coastal sediments of Pleistocene age in humid environments will generally show good

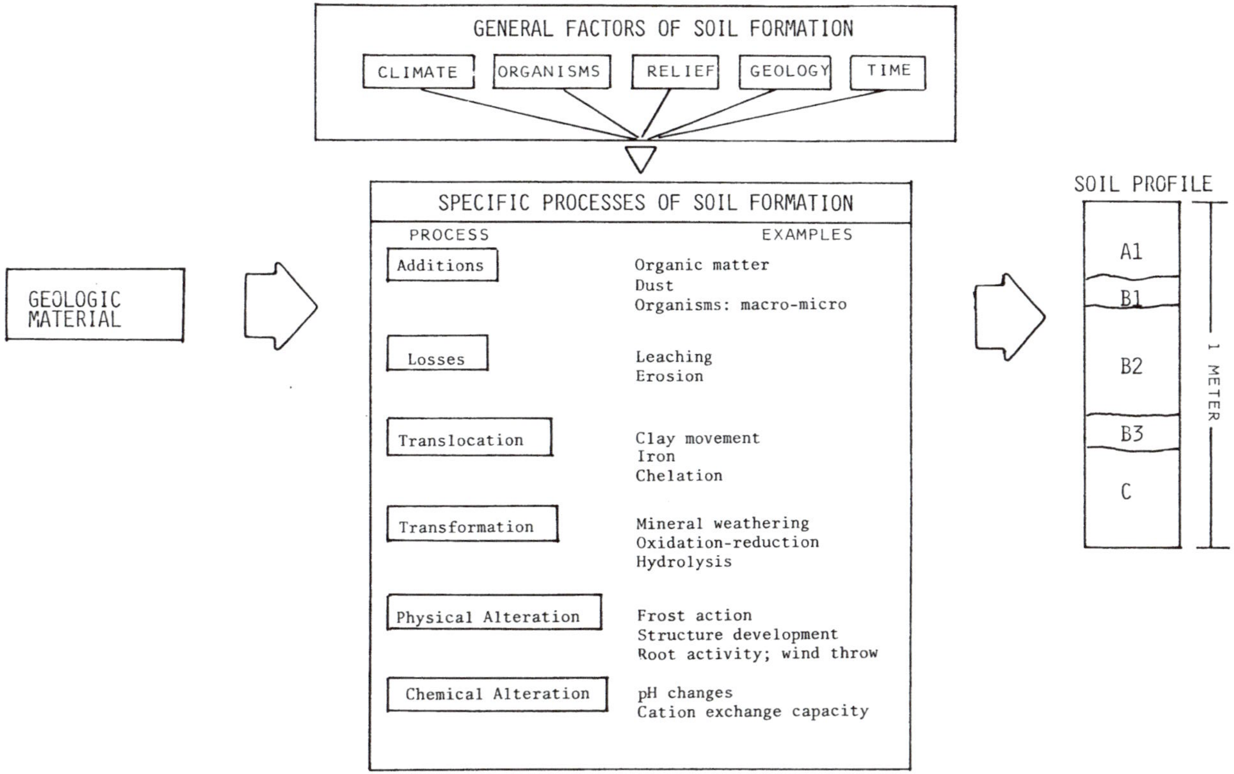

Figure 1.3 Generalized diagram illustrating factors and processes involved in soil genesis. The processes of additions, losses, transformations and translocations have been described previously by Simonson (1959).

horizonation and abundant evidence of pedogenic weathering processes. Certain diagnostic horizons are associated with the length of time for weathering. Weak argillic horizons, e.g., have been found to develop after 4000 years on alluvial sediments near Front Royal, Virginia (Foss 1974). Cambic horizons will develop in 50–100 years on river dredgings in Washington, D.C. (Stein 1978). In Pennsylvania, cambic horizons can develop in <200 years on alluvial sediments (Bilzi & Ciolkosz 1977).

The increase in organic matter content of soils appears to be one of the first recognizable features of soil formation. Hallberg *et al.* (1978) found organic matter accumulation in the upper 10 cm of a 100 year old soil in Iowa to be equivalent to adjacent soils dated at 14000 years. Other studies have also demonstrated the rapid accumulation of organic matter in surface horizons in <60 years (Simonson 1959).

The content of gibbsite and plinthite in soils of the Coastal Plain of North Carolina seems to increase from the youngest to the oldest surface (Daniels *et al.* 1970). The development of petro–calcic horizons in the Southwest is restricted to surfaces 10000 years or older (Ruhe 1967). Oxic horizons in Puerto Rico are generally restricted to old landscapes.

Most soils in the United States are geologically young; i.e. they are generally <1 million years old. Although soils >1 million years in age may be found on the older landscapes, soils are still considered relatively young. For example, the Baltimore gneiss is one of the oldest formations in the Piedmont (1.2 billion years old), and soils developed on this formation in broad upland positions may average 4 m thick. Assuming a uniform rate of soil formation of 5 cm in 10000 years, these soils would be only 800000 years in absolute age. Old geologic formations dating to the Precambrian, Paleozoic or Mesozoic will have much younger soils developed on their respective landscapes than the absolute rock age suggests. Also, soils dating beyond 8000 years in age will usually have undergone a number of major climatic and vegetative cycles. This must be considered in the proper interpretation of soil genesis and distribution. The occurrence of soils with strong argillic horizons in desert areas of the West today are probably relict features from past environments. In humid areas, color mottling caused by iron segregation due to wetness is sometimes found in soils that are now "high and dry"; landscape development with subsequent changes in water–table levels have been associated with this "relict mottling" (Wright 1972).

Processes of soil formation

The processes of soil formation include additions, losses, transformations, translocations, physical alteration and chemical alteration. It is difficult to limit the number of processes in soil formation because of the countless inorganic and organic reactions, physical changes, and additions to the soil system during weathering. The six major processes listed above, however, include those reactions that are commonly associated with development of horizons and subsequent profiles. The older the soil, of course, the greater the influence of these processes on soils.

Table 1.1 lists some examples of the major processes occurring in soils. Those processes generally noted in most soils include organic matter additions, erosion losses, translocation of constituents within soil profiles, and structural changes. As a result of man's activities, some modification and additions have been noted in major processes. In numerous cases, man's influence has been detrimental to soil productivity.

Evaluation of soil formation rates

Soil morphology

In most cases, the evaluation of soil formation is based on a set of soil properties sensitive to weathering processes. For example, Bilzi and Ciolkosz (1977) used a soil morphological rating scale for evaluating profile development. The factors they included in the rating scale were color, texture, structure, consistence, mottling, percent clay films and boundary of horizon.

Table 1.1 Examples of some of the major processes operating in developing soil horizons.

Major process	Examples of process	Examples of man-influenced processes
Additions	Organic matter	S from industrial sources
	Organisms: macro, micro	Pollution: many types of chemicals
	Dust: wind erosion	Salts from irrigation water
	Volcanic; loess	Pesticides
		Sewage sludge
	N from rainfall or micro-organisms	Fertilizer
		Lime
	Deposition of sediment from erosion	Other soil amendments
	Salts from coastal areas or underlying sediments	Increased organic matter under irrigation
Losses	Erosion: wind and water	Accelerated erosion
	Soluble constituents by percolation: Na^+, Ca^{++}, K^+, NO_3^-, Cl^-, $SO_4^=$	Increased leaching from irrigation
		Increased organic matter decomposition by tillage, drainage, clearing and disturbance
	Organic matter decomposition	Nutrient removal by harvesting

Table 1.1 continued

Major process	Examples of process	Examples of man-influenced processes
Translocation	Clay: A to B or C Fe, Al: A to B Humus: A to B Soluble salts Carbonates	Increased translocation of soluble constituents resulting from irrigation Movement of soluble fertilizer constituents Tillage influence – mixes horizons
Mineral & organic transformations	Primary minerals → Secondary (e.g. feldspar → kaolinite) Secondary mineral weathering Illite → montmorillonite Organic Compounds → Synthesized compounds Oxidation–reduction reactions	Drainage effects on oxidation reduction Tillage effects on organic decomposition
Physical alteration	Structure changes; e.g. massive → blocky Bulk density changes: Upper horizons: lower with organic matter Lower horizons: higher with cementing agents Particle size reduction Pedoturbation e.g. wind throw	Puddling soil by improper tillage and added Na Increased aggregation by grass or meadow Subsidence Increased frost action Land forming
Chemical alteration	pH changes Lower: with leaching Higher: movement of bases into horizon Cation exchange capacity Increases with increased organic matter and clay Clay–organic matter interaction Development of "chemical profile"	Increased pH: lime additions Increased P, K, N, micro elements: fertilizer

Table 1.2 shows the morphological characteristics of soils on various landscapes in the Savannah River valley in the Piedmont of Georgia and South Carolina. The ages of the landscapes were estimated by ^{14}C data, diagnostic artifacts, geologic sedimentation rates, and criteria established in other pedologic–archeologic studies (Foss 1974, 1977). Several trends are apparent from the data in the table. For example, with increasing age of soils, relatively unweathered material (C horizon) will transform to a cambic horizon (color B) or a weak argillic horizon (B2t) in < 4000 years. With increasing age, the argillic horizon in well-drained soils becomes stronger, i.e. redder, more clayey and thicker. Erosion, of course, can inhibit this process on sloping land, or deposition of fresh material along levees or terraces will disrupt the B horizon development. Both of these processes are active in the Savannah River valley, especially the latter.

The color of the B horizon of well-drained (perennially oxidized) soils will show the influence of soil age. As noted in Table 1.2, the colors will make a hue progression from 10YR (typical for unweathered alluvium) → 7.5YR → 5YR → 2.5YR as length of weathering time increases. Figure 1.4 shows the free iron content with depth in seven well-drained profiles typical of the major units in the Savannah River valley. As noted in this figure, the free iron contents increase with soil age, and the free iron is associated with soil color as indicated in Table 1.2. The relationships above would not be

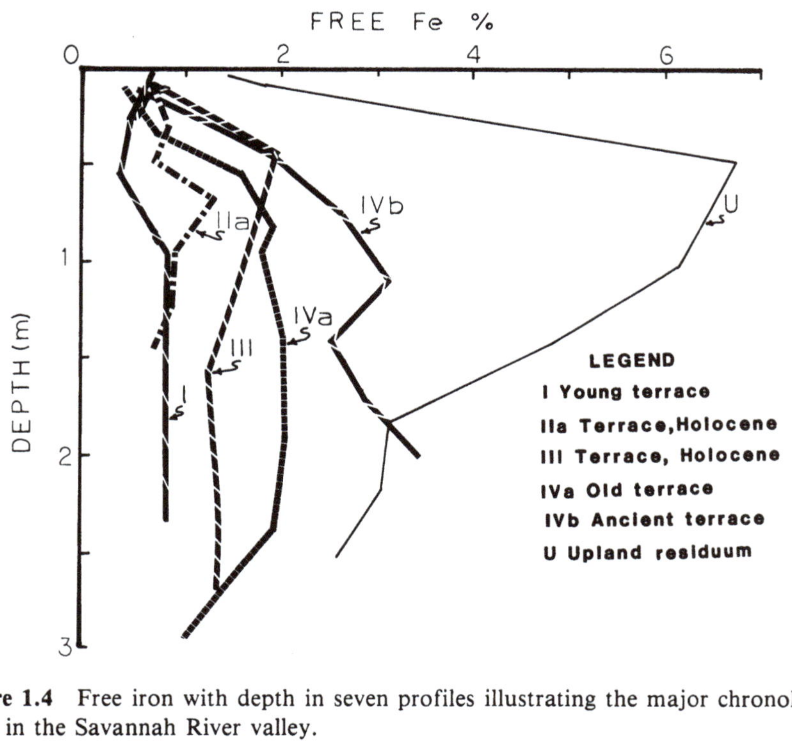

Figure 1.4 Free iron with depth in seven profiles illustrating the major chronologic units in the Savannah River valley.

Table 1.2 Chronology of sediments in the Savannah River valley and associated soil characteristics.

		Diagnostic		Soil characteristics (B horizon)					
Unit	Age years BP	B horizon	Color	Solum thickness	Clay coatings	Structure grade	pH	Extractable Mg pp2m	Extractable Ca pp2m
I	< 250	none or weak cambic	variable	0–0.5 m	none	none to weak	5.5–7.0	< 100	variable
IIa	250–4000	cambic or weak argillic	10YR–7.5YR	0.5–0.8 m	thin, discontinuous	weak	< 6.5	> 150	variable
IIb	4000–6000	weak argillic	7.5YR–10YR	0.8–1.2 m	thin, discontinuous	weak to moderate	< 6.5	> 150	variable
IIc	6000–8000	argillic	7.5YR	0.8–1.2 m	thin, nearly continuous	weak to moderate	< 6.5	> 150	variable
III	8000–10 300	argillic	7.5YR	1.0–1.5 m	thin, continuous	moderate	< 6.5	> 200	300
IVa	10 300–30 000	argillic	5YR–7.5YR	1.0–2.0 m	thin, continuous	moderate	< 6.2	> 200	variable (< 500)
IVb	100 000–250 000	strong argillic	2.5YR–5YR	1.5–2.5 m	thick, continuous	moderate to strong	< 5.6	> 300	< 200
upland residuum U	400 000–1 million +	strong argillic	2.5YR	2.0–5.0+ m	thick, continuous	strong	< 5.0	< 200	< 20

expected in soil developing from red geologic materials; thus, other criteria are needed for age interpretations on these soils. In soils with a high water table or impeded drainage, color or free iron cannot be used to indicate soil age. Because of the reducing environment (low Eh), the resulting gley or mottled condition is not particularly diagnostic. Observations in the Savannah River valley and in other localities have indicated, however, that the intensity and size of the reddish mottles can give an approximation of age. This has especially been noted with mottles of 5YR or 2.5YR colors; these usually indicate an age of > 10000 years. The nature of the geologic material can modify these relationships, however.

The solum (A + B horizon) thickness increases with age of the soil as indicated in Table 1.2. Because of the numerous discontinuities in soils of Units I–III, and in some cases Units IVa and IVb, it is difficult to get accurate thickness values of these soils. In many areas at lower elevations, soils would develop for 1000–2000 years, and would be covered by additional sediment. This would essentially stop development of the B horizon if the sediment exceeded 0.5–1 m in thickness, and the process of soil development would start again with the more recent material. Only at higher elevations that were free from sedimentation would soils continue to develop 10000 years or more with only minor interruptions from additional sediment.

The presence or absence of clay coatings provides a clue to the length of time a soil has been developing. The extent and thickness of the clay coatings on ped (structural unit) surfaces also provide evidence of weathering. Soils in the Savannah River valley show a wide range of clay-coating development. Generally, a soil from the humid East must be 3000–4000 years old before discontinuous clay coatings on ped surfaces are noted, whereas soils 10000 years or more will have continuous clay coatings on peds. Structural grade, or the distinctness of peds, also shows increased development with increased age of soils. Some caution is needed on this interpretation because the texture and the amount of original Ca^{++} of the soil will influence the structural development.

Another form of B horizon is expressed in soils developed in fine sandy sediments; this takes the form of lamellae or layers of finer-textured horizons in a coarse-textured soil profile. In addition to having more clay in the lamellae as compared to the interlamellae area, the lamellae also have higher iron, organic carbon and bases. The major factors necessary to form lamellae in soils developed from alluvium in the Savannah River valley were: (a) > 70% sand in the profile, (b) > 30–35% fine sand, and (c) fine sand : very fine sand ratio of > 1. In Arkansas, soils developed in fine sandy alluvium along the Mississippi River had lamellae only where the soils were acid and well to excessively drained. Soils with a neutral pH and somewhat poorly drained conditions had weakly developed B horizons usually with black concretions.

Figure 1.5 shows a model for lamellae development in soils developed in fine sandy alluvium in the Savannah River valley. As noted on this diagram,

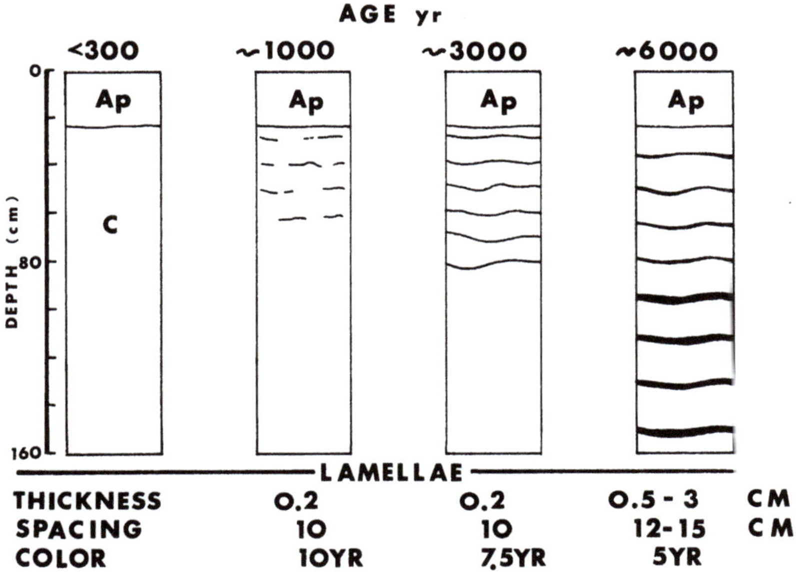

Figure 1.5 Model for lamallae development in soils of the Savannah River valley.

lamellae development may take place in 1000 years, but the lamellae will be thin and broken. Some thin, iron bands were found in the bottom of Ap horizons 200 years in age. In soils 3000 years or more in age, the lamellae will be continuous, and they will become redder than those formed in 1000 years. Soils > 6000 years old will have lamellae with 5YR colors, have 12–15 cm spacing, and be 0.5–3 cm thick. Similar relationships between age and lamellae characteristics were found in soils developed in fine sandy alluvium from the Mississippi River near Leachville, Arkansas. In soils developed in fine sandy sediments near Trenton, New Jersey, soils 9000–10 000 years old showed similar lamellae formation as illustrated for the 6000 year old soil in the Savannah River area. Thus, the lamellae development is sensitive to environmental factors in addition to soil drainage, mineralogy, chemistry and other properties.

Laboratory characteristics

Physical properties Particle-size distribution of soil profiles has been used extensively to study soil age; the clay content has been especially useful for this evaluation. The formation and translocation of clay minerals in soil profiles is well established. The magnitude of the translocation of clay from the A horizon to the B horizon has been useful in the evaluation of the intensity of soil-forming processes. Figure 1.6 shows the clay curves and associated free iron for a chronosequence in the Piedmont physiographic province of Maryland; these soils are formed on micaceous schist. As the soil increases in age, the total clay and free iron increase in both amount and in total depth in the profiles. The Elioak represents a stable land surface for several million years whereas the Glenelg and Manor probably represent soil development after truncation during the late Pleistocene.

The clay distribution curves for a chronosequence in the Savannah River valley are shown in Figure 1.7. Differences in the clay distribution curves of these three contrasting soils are easily detected in the field; thus, this property provides an effective method of delineating these soils. In general, the percent clay and free iron, combined with color, was extremely useful in determining age of well-drained soils on landscapes in the Savannah River valley. Some caution is needed, however, in areas where older soils are transported into colluvial or local alluvial positions with minor modifications. The resulting profile on a young landscape may appear similar to the original *in situ* soil and thus be misinterpreted.

The bulk density of soils is also associated with age. Soils < 1000 years old developed in alluvium along the Shenandoah River in Virginia were found to have low bulk density values (1.1–1.2 g cm^{-3}) as compared to soils 4000–10000 years old (> 1.4g cm^{-3}). The bulk density values above were from B or C horizons.

Chemical characteristics

Chemical properties of soils are related to age, but greater variability is noted in chemical properties than in physical properties. However, Table 1.2 and Figure 1.8 show some generalized relationships between soil age and pH, free iron and extractable magnesium.

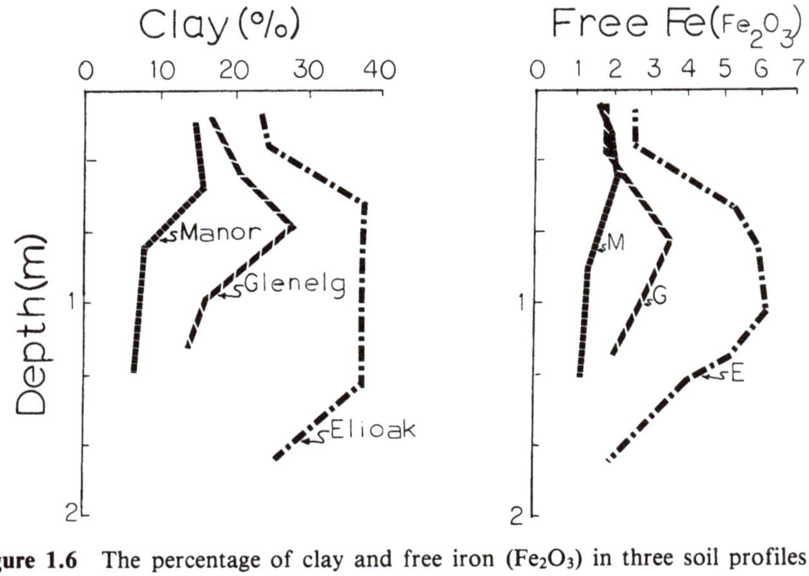

Figure 1.6 The percentage of clay and free iron (Fe_2O_3) in three soil profiles of a developmental sequence from the Piedmont Province area of Maryland.

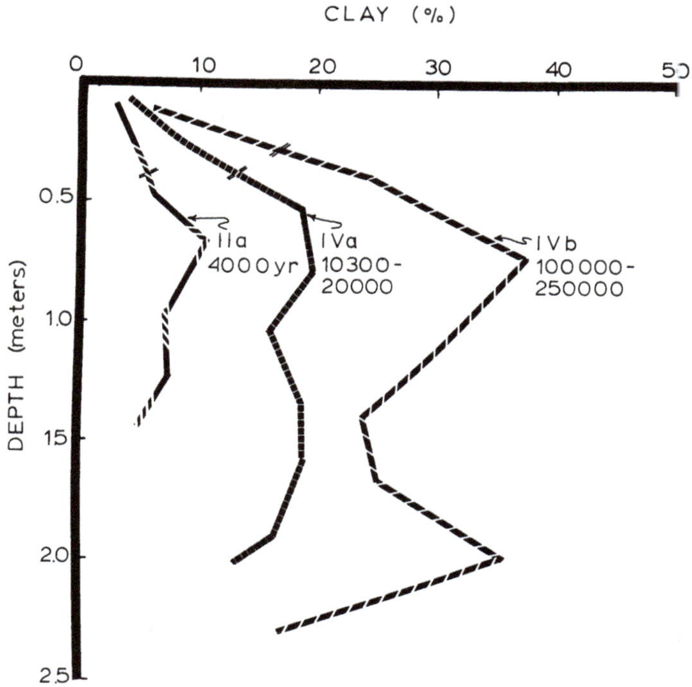

Figure 1.7 Clay distribution curves for a chronosequence in the Savannah River valley in Georgia and South Carolina.

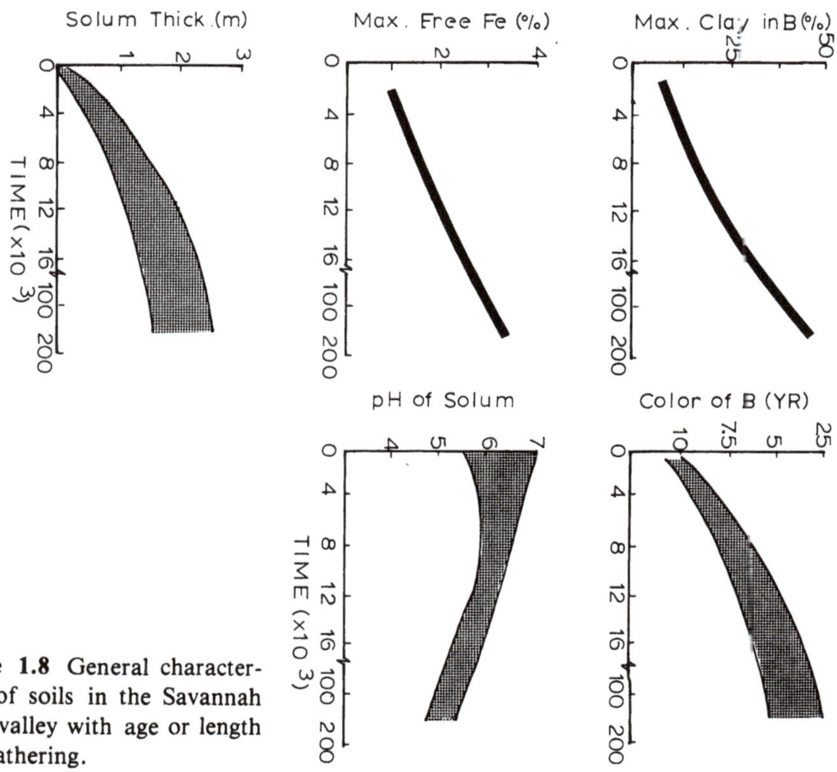

Figure 1.8 General characteristics of soils in the Savannah River valley with age or length of weathering.

Variability of chemical properties is related to parent materials, landscape position, groundwater influence, flooding, vegetation differences and agricultural amendments. Figure 1.8 shows the relationships between several chemical properties (free iron and pH) and other physical and morphological properties in soils of the Savannah River valley. The pH of soils will generally decrease with age, but the younger sediment may be charged with bases or influenced by a water table with substantial basic cations or result from deposition of sediment from very acid ancient soils. Thus, the younger soils (< 3000 years) will have a wide range of pH values in the Savannah River valley.

A chemical analysis technique used in evaluating soil development by Beavers (1960) and Beavers *et al.* (1963) in Illinois employs X-ray spectroscopy of silt-sized particles for various elements. Molar ratios of $CaO:ZrO_2$, $Fe_2O_3:ZrO_2$, $K_2O:ZrO_2$ amd $TiO_2:ZrO_2$ have been used successfully in soil genesis studies. Figure 1.9 shows particle-size distribution, molar ratios and morphology of a soil developed on early Pleistocene, Coastal Plain sediments in southern Maryland (Wright 1972). As noted in this figure, the molar ratios of the fine- and coarse-silt fractions are sensitive to weathering, and they are associated with the clay profile and morphological characteristics. This technique has been used in a number of studies on soil weathering.

Mineralogical characteristics The original mineralogy of geologic materials will change as weathering takes place. The extent of change will be based mainly on the (a) length of time weathering has taken place, (b) original mineralogical composition and (c) the weathering environment. Studies of soil mineralogy as associated with soil age have generally focused on mineral ratios (e.g. quartz : feldspar), index minerals, light and heavy mineral alteration, mineral morphology and clay mineral weathering. Specific studies on mineralogy and soil ages are not numerous. The classic studies by Humbert

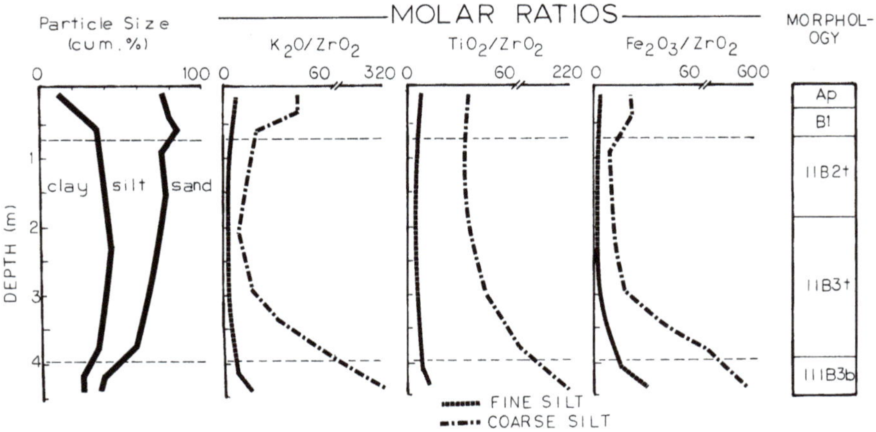

Figure 1.9 Particle size, molar ratios and morphological characteristics of the deeply weathered Magnolia soil in southern Maryland (Wright 1972).

and Marshall (1943), Cady (1960) and Ruhe (1956) have pointed out the value of mineralogical analyses in soil genesis studies.

Confounding factors in soil genesis

Many confounding factors result in confusion about the genesis or rate of soil development on certain landscapes. Some of the major factors that cause difficulty in interpreting soil formation rates on soils > 10 000 years old are the changes in climates, vegetation and rates of weathering processes during geologic time. The climatic changes during the Pleistocene and the late Tertiary were dramatic and widespread. Thus, the older soils on some landscapes may reflect past climatic conditions and weathering processes rather than current ones. Some examples of these anomalies are (a) soils with argillic horizons or lateritic characteristics in desert regions (Stephens 1965) or (b) some underlying paleosols in the northern Coastal Plain in Maryland (Foss *et al.* 1978) or in pre-Wisconsinan glacial till in the Midwest (Thorp 1965) that are not characteristic of soils forming in these areas during the Holocene.

Table 1.3 lists several confounding factors in soil genesis and some typical

Table 1.3 Confounding factors in determining rates of soil formation.

Factor	Examples found in soils
Climatic changes	Changes in precipitation and temperature during the Holocene, Pleistocene and Tertiary
Vegetation changes	Changes associated with climatic fluctuations Plant succession
Landscape evaluation	Drainage changes Erosion–deposition
Deposition of sediment or soil materials	Loess Eolian sands Dust Alluvium Colluvium Volcanic Glacial Man's activities; e.g. along roads, urban areas, fence rows
Erosion and transport of soil materials	Truncation of soil profiles Transfer of soil materials
Disruptive mechanisms	Pedoturbation: roots, animals, tree fall, frost, wetting–drying Earthquake activity Mass movement Man's activities; e.g. plowing, land leveling
Changing rates of soil formation processes	Changing rates with increasing depth of soil, lithologic discontinuities Changes associated with climatic fluctuations

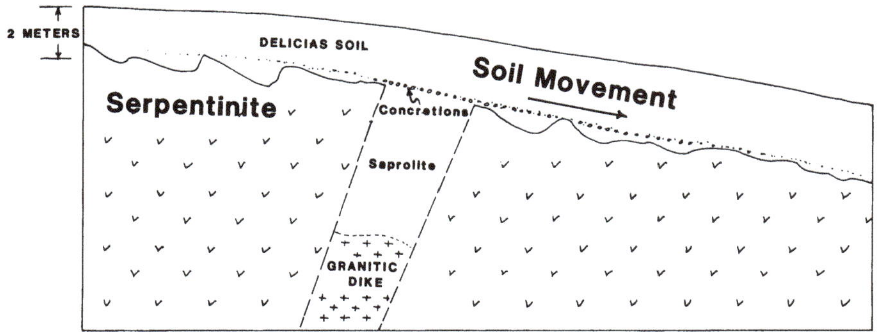

Figure 1.10 Cross section of soil developed on serpentinite in southwestern Puerto Rico. Soil has been transported downslope as evidenced by concretionary layer and serpentinite-derived soils over a granitic dike.

examples found in soils. As observed in recent studies on older landscapes (> 10000 years), many of the soils are polygenetic; i.e. they are formed in more than one weathering environment. During the past few years, a greater degree of local transport of soil materials thought to have developed *in situ* has been found. In Indiana, e.g. Olson *et al.* (1980) found that the red clayey parent material (terra rossa) on limestone is not entirely simple solution products but the result of pediment erosion and deposition. The terra rossa was mainly debris derived from erosion of higher clastic sedimentary rocks. Figure 1.10 shows a landscape in southwestern Puerto Rico on serpentinite. Transport downslope was evident as serpentinite soils (Delicias Series) were found immediately above a granitic dike. A pebble line of concretions was present at the contact between the dike materials and the upper serpentinite soils and also in places between the soil and serpentinite rock.

The other confounding factors listed in Table 1.3 are important in the interpretation of rates of soil formation for particular regions. Few areas of the world are free from "confounding factors." The influence of man's activity (anthropic influence) has always been a factor in the interpretation of soils occurring in agricultural and urban areas (Bidwell & Hole 1965).

References

Ahmad, M., J. Ryan and R.C. Paeth 1977. Soil formation as a function of time in the Punjab River Plains of Pakistan. *Soil Sci. Soc. Am. J.* **41**, 1162–6.

Beavers, A. H. 1960. Use of X-ray spectrographic analysis for the study of soil genesis. *Trans. 7th Int. Congr. Soil Sci.* **1**, 1–9. Madison, Wis.

Beavers, A. H., J. B. Fehrenbacher, P. R. Johnson and R. L. Jones 1963. $CaO-ZrO_2$ molar ratios as an index of weathering. *Proc. Soil Sci. Soc. Am.* **27**, 408–17.

Bidwell, O. W. and F. D. Hole 1965. Man as a factor of soil formation. *Soil Sci.* **99**, 65–72.

Bilzi, A. F. and E. J. Ciolkosz 1977. A field morphology rating scale for evaluating pedological development. *Soil Sci.* **124**, 45–8.

Cady, J. G. 1960. Mineral occurrence in relation to soil profile differentiation. *Trans. 7th Int. Congr. Soil Sci.* **4**, 418–24. Madison, Wis.

Daniels, R. B., E. E. Gamble and J. G. Cady 1970. Some relations among Coastal Plain soils and geomorphic surfaces in North Carolina. *Proc. Soil Sci. Soc. Am.* **34**, 648–53.

Foss, J. E. 1974. Soils of the Thunderbird site and their relationship to cultural occupation and chronology. In *The Flint Run Paleo-Indian Complex,* W. M. Gardner (ed.). Cath. Univ. Am. Occas. Publn. 1, 66–83.

Foss, J. E. 1977. The pedological record at several Palevindian sites in the Northeast. *Ann. NY Acad. Sci.* **288**, 234–44.

Foss, J. E., D. S. Fanning, F. P. Miller and D. P. Wagner 1978. Loess deposits on the Eastern Shore of Maryland. *Soil Sci. Soc. Am. J.* **42**, 329–34.

Hall, G. F., R. B. Daniels and J. E. Foss 1982. *Rate of soil formation and renewal in the USA.* Am. Soc. Agro. Spec. Publn. 45, 23–39.

Hallberg, G. R., N. C. Wollenhaupt and G. A. Miller 1978. A century of soil development in spoil derived from loess in Iowa. *Soil Sci. Soc. Am. J.* **42**, 339–43.

Hay, R. L. 1960. Rate of clay formation and mineral alteration in a 4,000 year old volcanic ash soil on St. Vincent, BWI. *Am. J. Sci.* **258**, 354–68.

Humbert, R. P. and C. E. Marshall 1943. *Mineralogical and chemical studies of soil formation from acid and basic igneous rocks in Missouri.* Univ. Mo. Ag. Exp. Sta. Res. Bull. 359.

Jenny, H. 1941. *Factors of soil formation.* New York: McGraw Hill.

Leneuf, N. and G. Rubert 1960. Calculation of the rate of ferrolitization. *Trans. 7th Int. Congr. Soil Sci.* **4**, 225–8. Madison, Wis.

Olson, C. G., R. V. Ruhe and M. J.Mausbach 1980. The terra rossa limestone contact phenomena in karst, southern Indiana. *Soil Sci. Soc. Am. J.* **44**, 1075–9.

Owens, L. B. and J. P. Watson 1979. Rates of weathering and soil formation on granite in Rhodesia. *Soil Sci. Soc. Am. J.* **43**, 160–6.

Ruhe, R. V. 1956. Geomorphic surfaces and the nature of soils. *Soil Sci.* **82**, 441–55

Ruhe, R. V. 1967. *Geomorphic surfaces and surficial deposits in southern New Mexico.* St. Bur. Mines Min. Resour. New Mexico Inst. Mining Tech. Mem. 8.

Simonson, R. W. 1959. Outline of a generalized theory of soil genesis. *Proc. Soil Sci. Soc. Am.* **23**, 152–6.

Stein, C. E. 1978. *Mapping, classification, and characterization of highly man-influenced soils in the District of Columbia.* MS thesis. University of Maryland.

Stephens, C. G. 1965. Climate as a factor of soil formation through the Quaternary. *Soil Sci.* **99**, 9–14.

Thorp, J. 1965. The nature of the pedological record in the Quaternary. *Soil Sci.* **99**, 1–8.

Wright, W. R. 1972. Pedogenic and geomorphic relationships of associated Paleudults in southern Maryland. PhD dissertation. University of Maryland.

2
Piping and sapping: development of landforms by groundwater outflow

Charles G. Higgins

Introduction

Headward growth of valley heads and the development of drainage networks wholly or partly by the outflow of subsurface water have been among the most neglected and least understood factors of landform genesis. Two separate but related and overlapping processes are involved: erosion by shallow throughflow or outflow of soil water (piping) and erosion by outflow of groundwater (sapping). Although the role of piping in dryland gully development has long been recognized, its importance in humid environments was little appreciated before the late 1960s. Similarly, few authors discussed sapping as a major or contributary process in shaping drainage systems before the mid-1970s, when Mariner 9 imagery of Mars disclosed several sorts of "fluvial" features that seem to have been formed, at least in part, by some process of fluid outflow.

Strictly, sapping refers to the undermining of the base of a cliff, with subsequent failure of the cliff face (Bates & Jackson 1980, p. 556). Such undermining may be accomplished by a variety of processes, including lateral erosion by streams, wave action, artificial excavations, and even by boring molluscs (Vita-Finzi & Cornelius 1973). Sapping is also commonly used to describe the effects of glacial plucking or the action of freeze–thaw and needle ice at the base of cliffs in periglacial environments. Here, however, I shall restrict the term to the effects of groundwater outflow, in the sense of such previously used terms as seepage erosion (Hutchinson 1968, p. 691), artesian sapping (Milton 1973, p. 4042), and spring sapping (Bates & Jackson 1980, p. 604).

First, I shall briefly outline the process and effects of piping, and then discuss various aspects of sapping as they affect both large and small landforms.

Piping

Piping was originally, and still is, a civil-engineering term used for the flushing of sediment from within, under or around the fill or footings of a dam by water seepage or flowage. This commonly creates internal pipe-like openings and sometimes results in catastrophic failure of the structure. In this sense, the term encompasses both piping and groundwater sapping as used here. However in the 1940s, the word was applied by soil scientists and geomorphologists to a form of concentrated soil-water throughflow that creates pipe-like openings in the subsoil, which may later collapse to form a "pseudokarst" or a surface gully system. This process, earlier called tunnel-gulley erosion by Gibbs (1945), was at first believed to be restricted to arid and semi-arid regions.

As described by Buckham and Cockfield (1950), Fletcher *et al.* (1954), Parker (1964) and others, drylands piping works something like this: Overland runoff is intercepted into desiccation cracks, animal burrows or other openings either on hillsides or on relatively level upland or terrace surfaces. The diverted water moves downward until it encounters a particularly porous layer or lateral opening in the subsoil or is blocked by an impermeable layer, by a temporary or perched water table or by the bottom of the vertical opening. Thence, it moves laterally, commonly toward a cliff face, stream bank or gully wall. At the point of emergence, the outflow entrains and removes clay- and silt-sized particles, enlarging the outlet and gradually extending it back into the hillside as a pipe-like or tunnel-like opening. Commonly, sediment is entrained not only at the point of outflow but all along the subsurface flow path, thus forming a continuous underground opening from inlet to outlet. Progressive enlargement and collapse of the tunnel creates sinkholes and/or extends a steep-walled, blunt-ended gully up the slope or into the upland.

Although piping "affects materials ranging in size from clay to gravel" (Mears 1968), it seems to be most effective in unconsolidated to weakly consolidated sediments and soils with a high silt content, such as loess and volcanic ash. Some investigators believe that smectite ("montmorillonite") is a necessary component or that soluble salts or a high exchangeable sodium-ion content are important (e.g. Heede 1971, Parker & Jenne 1966), but it is not clear whether these are necessary conditions or merely local or general associations. Clearly what *is* necessary is a sediment or soil sufficiently friable to be entrained and transported grain-by-grain at fairly low flow velocities but coherent enough to sustain open tubes and tunnels.

Progressive enlargement and collapse of tunnels may create fairly extensive local gully networks (e.g. see Gibbs 1945), or may form sinkholes and other "pseudokarst" features (e.g. Mears 1968, Parker *et al.* 1964, Rathjens 1973). Sinkhole development by piping is likely to be most noticeable on nearly level terrain; Warn (1966) has reported pseudokarst features on a dry Holocene lake bed in the Imperial Valley, California, where the pipes discharge at the bluffs of nearby incised stream valleys. However neither a steep slope nor nearby cliffs or channel walls is prerequisite for piping or sinkhole development. Cavernous openings, one to several meters wide and one to several meters deep, have formed from time to time near the distal edge of alluvial fans of the southwestern Sacramento Valley near Davis, California (Marr 1955).

There, the slope of the fan surfaces is less than $2\,\mathrm{m\,km^{-1}}$, the nearest modern channelways may be several kilometers distant, and the water table is tens of meters below the surface. My own observations near Davis suggest that sinkhole development there is initiated when irrigation water is intercepted by a ground-squirrel burrow or a desiccation crack above a buried distributary channel filled with coarse gravel or cobble alluvium. Such subsurface distributaries can carry large volumes of water and fine sediment, especially during the dry season. Long before they become clogged with fine sediment, erosion and collapse around the inlet may create sizable depressions. This process is different from traditional piping inasmuch as the subsurface conduits have no known outlets and are probably not enlarged by the subsurface flowage. However, the "sinks" that result can be major problems for farmers using furrow – or flood-irrigation. The consequences of piping in drylands can also cause failures of highways and other structures (Parker & Jenne 1966, 1967).

Parker and Jenne (1967) have commented that piping in the drylands of the western United States occurs preferentially under non-vegetated surfaces. Other investigators have noticed a similar association. Parker (1964, 1968) and others have also shown that both the subsurface and surface features of piping develop and change very rapidly. These observations suggest that at least in some areas, piping is a short-lived aspect of "accelerated erosion" that may be triggered by loss of a vegetative cover. If so, its presence may be an indicator that the landscape is unstable and changing.

In humid regions, the erosional effects of piping were relatively unknown before the mid-1960s, although there were occasional reports (e.g. Zeitlinger 1959). Studies of piping in such regions did not begin in earnest until Kirkby and Chorley (1967) presented a model of soil-water throughflow and saturated overland flow as an alternative to Horton overland flow on vegetated slopes. Once investigators began to look for piping and other erosional effects of subsurface throughflow in humid areas, they found numerous examples. Pseudokarst owing to piping has been reported from the humid tropics (e.g. Feininger 1969; Khobzi 1972, in Columbia; Löffler 1974, in New Guinea), and more importantly, several investigators in the United Kingdom and in the eastern United States have recognized that soil-water throughflow with piping or seepage outflow may initiate and extend surface drainage networks (Jones 1971, 1975, 1981, Gilman & Newson 1980, Dunne 1980).

Where the water table is near the surface, as it is likely to be in humid regions, soil-water throughflow merges with shallow groundwater flow, and piping becomes sapping, as the term is used here. Dunne (1980), discussing such a case in Vermont, uses both terms synonymously. Doubtless the two processes do overlap, but throughflow must be an ephemeral phenomenon, occurring for only short intervals after infiltration of rain or irrigation water, whereas groundwater outflow sapping can continue as long as the water table intersects the stream channel or valley headcut. Considering that piping can produce fairly extensive pseudokarst landscapes and gully systems in a wide variety of environments, the more consistent process of sapping should be an even more effective agent of erosion at valley heads, and should be capable of developing much more extensive drainage networks. This process and its effects are discussed in the remainder of this paper.

Sapping

In effluent streams, the outflow of groundwater up through the stream bed may markedly increase the erosibility of the bed, but this has been little studied. In a somewhat different context, Leopold *et al.* (1964, p. 180) wrote: "A static grain bed cannot be sheared without some degree of dilation or dispersion. This dispersion must be upward, against the downward or normal component of the body force pulling the grains toward the bed. Thus all grain flow requires the exertion of some kind of dispersive stress between the sheared grain layers and the bed." The upward lift of effluent groundwater must provide some part of this dispersive stress and thereby help to initiate entrainment of bed particles at flow velocities lower than those required under static or influent conditions. In effect, the flow of groundwater into or out of a channel should alter the competence of the stream. Harrison and Clayton (1970) report that in effluent and influent reaches of a stream on the valley train of Sherman Glacier, Alaska, competence is respectively increased and decreased by about the amounts expected from theoretical considerations. However, they could not duplicate these effects in flume studies; instead, they found that upward or downward seepage caused no change in competence, but altered the bed forms and bed roughness, as Simons and Richardson (1966) had earlier observed. Until the role of groundwater effluent in stream-channel erosion and deposition is better documented, this must remain one of the most tantalizing unknowns in the equations of fluvial hydrology.

Much the same is true of the contribution of groundwater sapping to the maintenance and retreat of some knickpoints and waterfalls. Differential resistance of near-horizontal strata and plunge-pool undercutting by surface flow have long been acknowledged factors in forming falls such as Niagara, but in many falls of this type, such as Burney Falls in northern California, copious outflow of sub-channel water from porous strata at the base of the falls may facilitate undermining by weathering and removal of bedrock and fallen blocks. Unfortunately, the quantitative importance of this is virtually unknowable, given the difficulties of making observations under waterfalls, and doubtless the relative importance of groundwater sapping varies from one site to another.

It is only slightly less difficult to assess the role of sapping in forming headcuts in gullies and streams, and in extending these headward to form drainage systems (Higgins 1983). Fortunately, there are some situations in which this process operates alone, without the obscuring effects of pluvial runoff and erosion. These cases are discussed next.

Model drainage systems on beaches

The clearest examples of drainageways formed entirely by groundwater outflow sapping are little features that develop on the foreshores of some beaches during falling tides. Where I have seen them, they are commonly preceded by diamond-shaped, or "rhombic," surface patterns that are formed by wave backwash and that may indirectly affect the development of later drainages. Much of the following description of these features is only slightly modified from Higgins (1982).

Wave uprush brings a turbulent surge of water and sand up the beach face. During backwash on many beaches, the sheet of water flowing back down the slope becomes thinner and thinner until it breaks into a reticulated diamond pattern that reflects a similar pattern forming under it in the sand of the central part of the swash zone (see Higgins 1982, Fig. 2, Guilcher 1958, p.82).

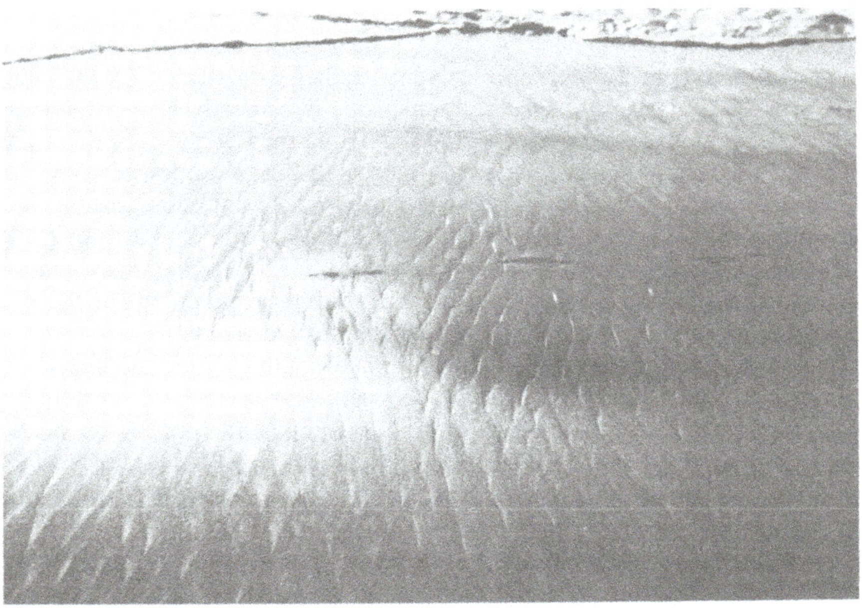

Figure 2.1 Uncommon lobate variety of "rhombic" or reticulate diamond pattern developed in swash zone after wave backwash, Ocean Beach, San Diego, California, March 14, 1971. Surface sand in foreground is drained, in background, is saturated. In mid-distance, near footprints, edges of little lobate bars, or fans (lighter tone), are draining into channels between them (darker tone). Glare from sun at left. Width of view about 4 m.

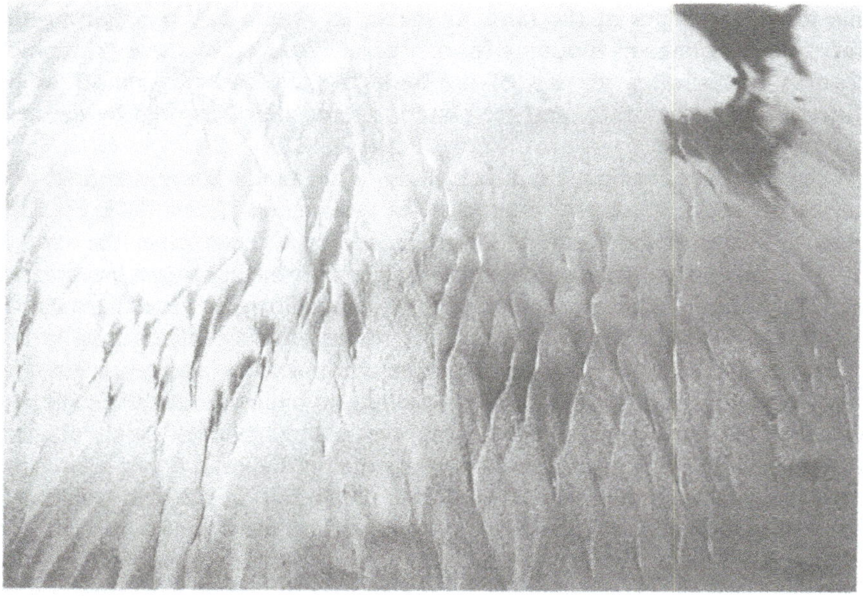

Figure 2.2 Closer view of general area shown in Figure 2.1. After surface backwash has drained away, shallow sapping by outflow drainage of lobate fans may form tiny rills on their edges (not visible here: see Figure 2.3) and may locally extend steep-walled little channels headward between the fans (foreground and left, in sun glare). Under some conditions, these channels may continue to grow into an anastomosing rill network (Fig. 2.4). Small dog in upper right. Width of view about 3 m.

These beach patterns have been variously termed backwash marks, or rhombic, rhomboid, or rhomboidal ripple marks, although they are strictly neither ripples nor truly rhombic. Those shown in Figures 2.1 and 2.2 have curved sides and rounded, lobate ends (also see Komar 1976, Fig. 13-11). The exact manner in which they are initiated is still as obscure today as when they were studied by Woodford (1935) and Demarest (1947). A good review is by Komar (1976).

When first formed, they appear as offset rows of alternating little channels and bars, or fans, the apex of each fan fed by a narrow channel confined between the two fans above it. The remaining backflow becomes concentrated in these channels, deepening them while adding the eroded sediment to the fans below, thus intensifying the reticulate pattern. (In this phase, outflow of water from the saturated fans themselves may extend the inter-fan channels headward, as shown in Figure 2.2, or form tiny tributary rills, only millimeters long, along the edges of the fans, as shown in Figure 2.3, pre-figuring the larger outflow channels that may form later.) Formation of these "rhombic" networks ceases when the last of the backwash dissipates by runoff or by infiltrating the beach face, and the pattern remains until covered by the next wave or tide.

When the tide is falling, the beach water table stands above tide level and intersects the beach face near the top of the swash zone (Grant 1948, Fig. lB). Below this intersection, in what Grant called the effluent zone, the sand is saturated and groundwater seeps to the surface after the backwash has drained way. This surface discharge first emerges in and follows the little channels between the "rhombic" fans. This phase is represented by relict forms in the upper part of Figure 2.11. Where discharge continues long enough, it may, by cross grading micropiracy, integrate these into an angulate–dendritic rill network in which many reaches parallel the sides of the "rhombs" in the original pattern. This phase of development is illustrated in Figure 2.4, and even more clearly in two photographs by H. F. Garner (1974 Figs 5.32 & 9.13). Garner's photos also show that the lower parts of these drainages commonly become broad and braided. This may occur not only on very fine sand beaches such

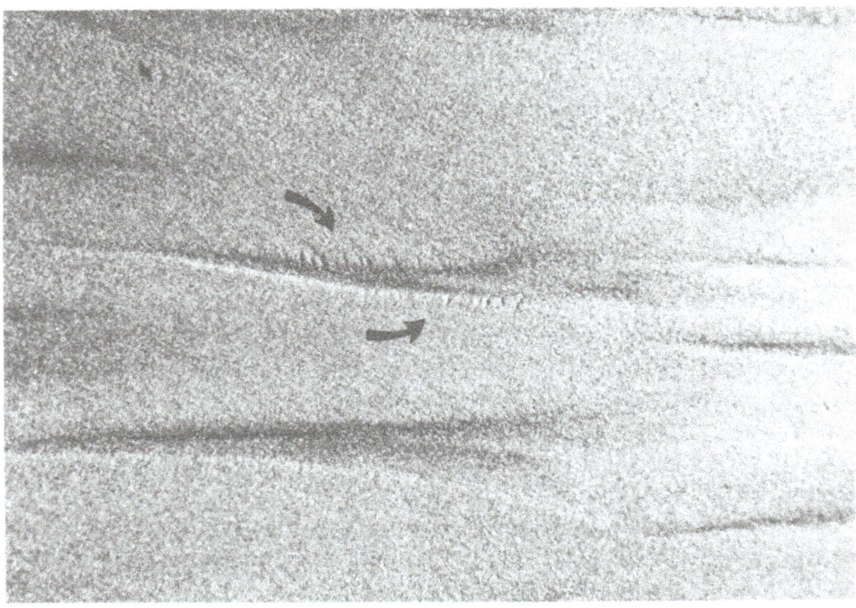

Figure 2.3 Detail of miniature rills (arrows) formed by shallow drainage outflow from the edges of lobate fans such as those shown in Figures 2.1 and 2.2. Darker tone of inter-fan channels owes in part to concentration of black sand. Ocean Beach, March 14, 1971. View parallel to shore; beach slopes toward right. Width of view about 25 cm.

as the one he photographed near Cape May Point, New Jersey, but even on coarse-sand beaches, such as Carmel River Beach, California (Fig. 2.4), or on gravel beaches (see Emery 1962, Fig. 57).

A final phase of development of these foreshore drainage features – and that most analogous to larger-scale drainage systems – represents a continuation of the previous phase of rill network development, and occurs where groundwater is constricted by an impermeable substrate or where for any reason the water table is held at about the same level for an extended period. Then groundwater discharge and sapping begin to develop steep gully heads

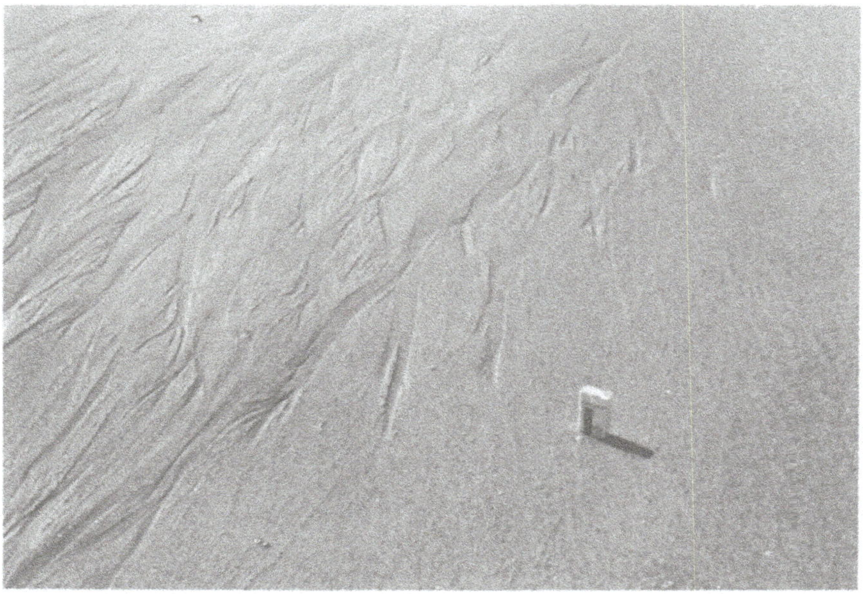

Figure 2.4 Anastomosing shallow rill formed entirely by outflow drainage of groundwater from the upper few millimeters of the swash zone during a falling tide. Steep-sided, flat-bottomed channels, with much cross-grading and piracy, are characteristic. Coarse sand of Carmel River Beach, California, February 1972. Width of view about 1.5 m.

in the rill channels at the upper limit of the saturated zone, all along the intersection of the water table with the beach face (Fig. 2.5). The increased pore pressure, or seepage pressure (Terzaghi 1950, p. 99–100), or dispersive stress caused by the discharge dilates the sand at the gully head and propels the loose grains outward into the channel where they are carried away. The gullies then advance headward up the slope by continued sapping and collapse of the

Figure 2.5 Initiation of small, steep-walled gullies by groundwater outflow sapping along earlier rill channels, Bermuda Avenue Beach, San Diego, California, February 2, 1971. Outflow and sapping were active when photograph was taken, but would not continue much longer; falling water table is indicated by successive terraces in middle gully and at right, and by abandoned highest headscarps. Width of view about 1 m.

Figure 2.6 Active gullying by headward sapping entirely by groundwater outflow in coarse sand where locally steeper slope of the foreshore and constriction of groundwater flow owe to a partially buried boulder (lower left). Low tide at Carmel River Beach, February 1972. Trends of channels closely parallel those of the earlier diamond pattern. Width of view about 1.5 m.

unsaturated sand above (Fig. 2.6). Where outflow is not especially concentrated at the gully heads, basal sapping may cause the gully heads to form broad re-entrants (Fig. 2.7) or the gully sides to retreat more or less uniformly as back-wearing cliffs (Fig. 2.8).

Because the water table slopes less than the beach face and also falls in elevation with time, and because sapping occurs only at or below the exposed edge of the water table, the gully heads grow higher as they advance headward. As the height of gully walls increases, sapping is accompanied by small slumps and block slides (Figs 2.7 & 8). Where the beach face slopes much more steeply than the water table, as is common in coarse-sand beaches, gullies cannot ad-

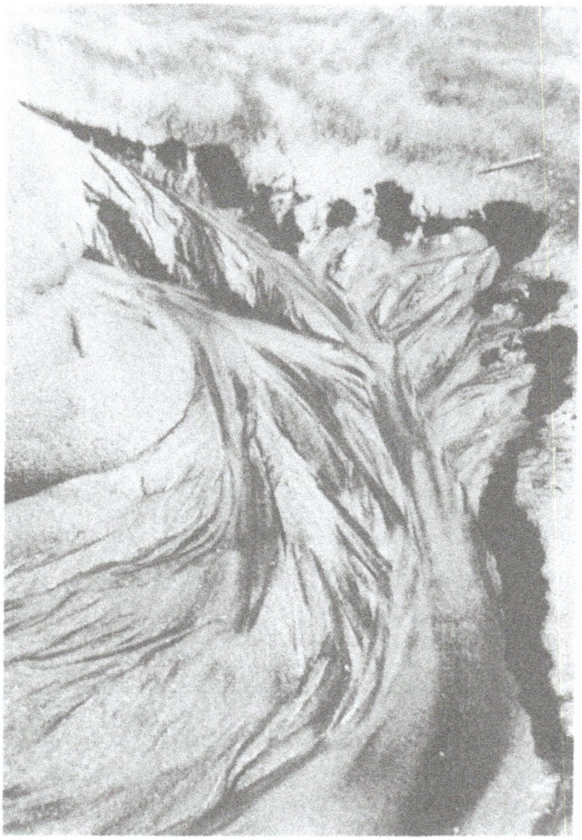

Figure 2.7 Scalloped, embayed "escarpment" formed by relatively nonconcentrated sapping. Groundwater flow is partly obstructed by a boulder (left). Bermuda Avenue Beach, February 2, 1971. Note characteristic braided outflow channels, terraces and abandoned watercourses indicating a falling water table, and fallen masses of sand along the "cliffs." Earlier diamond pattern out-of-focus at top. Width of view about 1 m.

vance very far (Figs 2.5–2.9; see also Komar 1976, Fig. 13-13). However, in fine-sand beaches, the water table may nearly parallel the beach face, and gullies may advance far headward without gaining much in depth (Fig. 2.10).

Although initial outflow may be concentrated in the shallow channels of the earlier-formed "rhombic" pattern, the direction of headward growth of the gullies must be determined almost entirely by the nature of the groundwater flow through the sand at the water table. Because the resulting drainage networks are characteristically angulate, with a preponderance of straight reaches (Figs 2.6, 2.9, 2.10), there must be some kind of directional permeability in

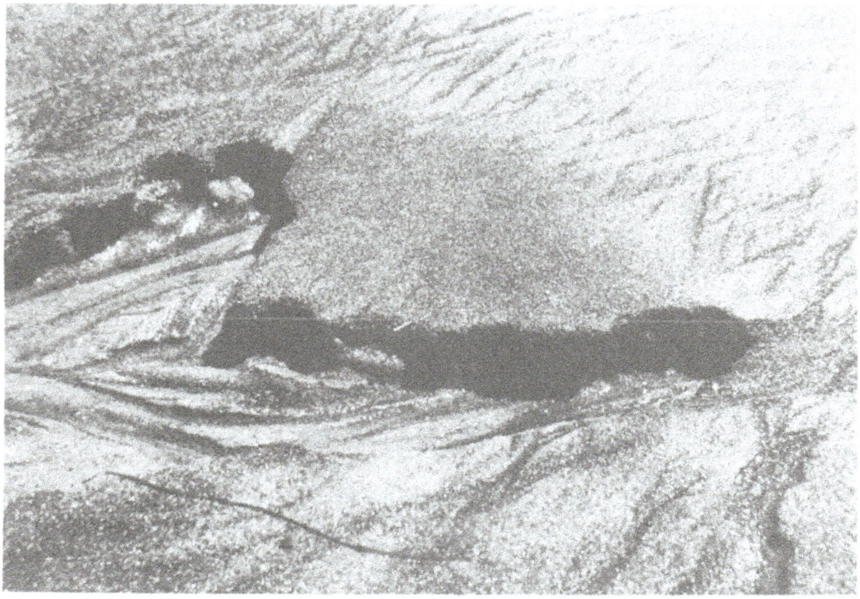

Figure 2.8 Active sapping and small slumps in gully heads and along steep gully slopes, Bermuda Avenue Beach, April 1971. Earlier "rhombic rill" pattern is preserved above the gully heads. Width of view about 0.5 m.

Figure 2.9 Active extensions of gully heads by sapping at Bermuda Avenue Beach where bedrock outcrop (lower right) causes locally steeper slope and obstructs groundwater flow. Low tide, February 2, 1971. Orientations of drainageways roughly parallel earlier diamond pattern. Width of view at center about 4 m.

Figure 2.10 Extensive drainage net developed within an hour or two on relatively gentle slope of Bermuda Avenue Beach, low tide, February 2, 1971. Drainage outflow still active when this photograph was taken. Dark sand is still saturated; light sand along channels and in distance has been drained. Beach slopes toward lower right. Alignments suggesting structure control roughly parallel the orientation of the earlier diamond pattern. Footprints at top give scale; width of view at top about 6 m.

2.9

2.10

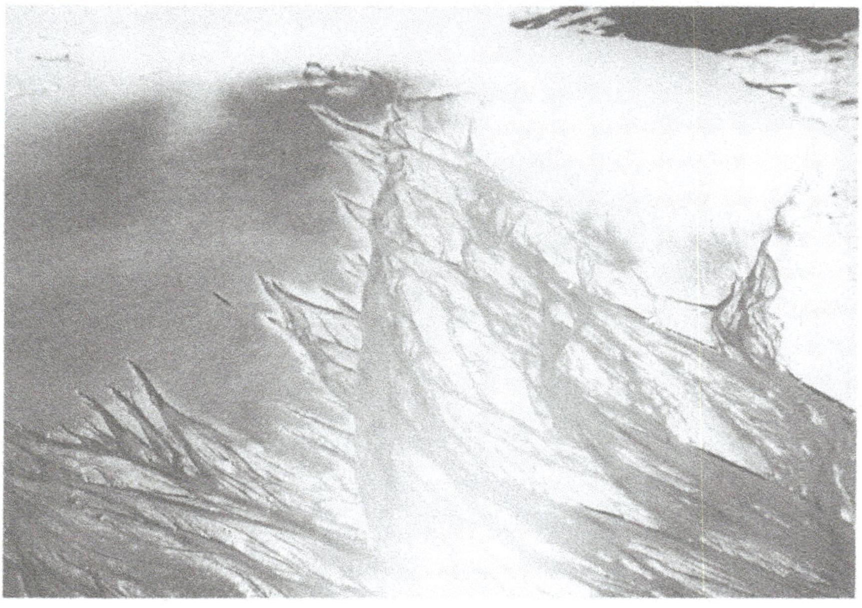

the beach sand. Moreover, because the straight reaches tend to parallel the two directions of the original rhombic pattern, this directional permeability may owe to buried rhombic networks formed during the phase of swash-zone accretion during the preceding rising tide.

Eventually, the water table falls below the base of the gully heads. Outflow and grain movement cease, and the development of the gully system stops. Sometimes a new gully head may be initiated farther down in each drainageway where the water table still intersects the channel, and a new, deeper, little channel may grow headward in the floor of the older one (Fig. 2.11). These secondary channels are likely to be limited and short-lived unless the water table maintains a constant elevation. As this rarely happens, the

Figure 2.11 Secondary active gully heads and channels incised into floor of drained older channel after slight lowering of water table. Earlier "rhombic rill" network partly effaced by footprints. Carmel River Beach, February 1972; width of view about 1 m.

whole drainage system generally remains as a relict feature on the foreshore until it is obliterated by people, animals or the next high tide.

The range in size of these rill-like and gully-like features covers about 1.5 orders of magnitude (compare Figs 2.3 & 2.10). This suggests that the process responsible is to a certain degree scale-independent, so that the features are not only models of drainage systems formed entirely by sapping, but may serve in some ways as analogs of larger-scale drainage systems. In this respect, they have three properties that may be significant in any comparisons that may be made. First and foremost, they are formed entirely by groundwater outflow and sapping. Second, they may develop in preferred directions, apparently controlled by directional permeability or weakness in the material in which they are formed. And third, erosional development is active only when the water table intersects the channel or gully head. Moreover, they exhibit certain distinctive characteristics: they have blunt, steep valley heads and steep valley walls with occasional slumped masses or blocks. Valley floors tend to be aggraded and nearly flat, forming abrupt angles with the valley sides. A smaller version of the main valley may be incised into the valley floor. Still-active valley heads are fed by groundwater outflow. The drainage pattern is generally angulate–pectinate, with relatively few, stubby tributaries. The "upland" into which the valleys are extended may show little or no evidence of contemporary surface runoff, or else pre-existing relict channels there are abruptly cut off by the later, steep-walled valleys. I believe that some, if not all, of these features are distinctive and characteristic of drainage networks of any size that are formed chiefly or entirely by groundwater sapping. Some larger-scale examples are discussed in the following sections.

Drainage systems formed by sapping on Mars

Another place where surface runoff does not obscure the effects of sapping is the surface of the planet Mars, where a variety of features have been attributed to groundwater outflow sapping and collapse. Under present conditions, there is little water in the Martian atmosphere, and it can exist on the surface only briefly in liquid form. However, sources of water for sapping may be (or may once have been) abundant beneath the surface in the form of interstitial ground ice, or permafrost, or even as trapped bodies of liquid.

Speculations about subsurface ice on Mars are not new (e.g. Baumann 1909). Until the Mariner spacecraft flights, however, such speculations were mostly concerned with possible sources of water for organisms or explanations for variable albedo effects, and it was generally assumed that the permafrost was a thin or patchy relict from an ancient water-rich atmosphere. Then in 1969, Mariner 6 sent back three B-camera images that indistinctly show Martian "chaotic terrain." Leighton *et al.* (1969), Sharp *et al.* (1971) and Belcher *et al.* (1971) cautiously ascribed the jumble of blocks and basins to removal of support by thaw and collapse of ground ice as a result of internal heating or climatic warming. They did not speculate on the origin of the ice, nor its thickness, although the dimensions of the chaotic terrain require such ice to have been very thick and to have melted rapidly.

Meanwhile, through some other, independent studies of Mars by Wade and DeWys (1968) and Smoluchowski (1968) and of the Moon by Gold (1966) and Lingenfelter *et al.* (1968; Peale *et al.* 1968, Schubert *et al.* 1970), there emerged a reasonable explanation for a thick zone of ground ice that would not have depended on a denser atmosphere and warmer climate. This hypothesis holds that water was outgassed during differentiation of the planet's interior but was cold-trapped as ice in the sub-freezing, porous regolith before it could reach the surface. Smoluchowski assumed that the thickness of this permafrost zone was only about 10 m, but Gold estimated a thickness of about 1 km for the Moon, assuming a thermal gradient like Earth's. Subsequent observers of Mars have found reasons to believe that ground ice is (or was) widely distributed there, possibly "to depths of kilometers" (Carr & Schaber 1977, also Carr 1981, Fanale 1976, Boyce 1979).

However, concepts of simple thawing of this ground ice were soon found to be insufficient. In November 1971, Mariner 9 reached Mars and began to send back much clearer pictures that eventually provided a synoptic view of the whole planet. Some of the early images show huge outflow channels heading at the chaotic terrain. The fluvial origin of these channels has been discussed by many investigators (e.g. Baker 1979, 1981, 1982, Baker & Kochel 1979), and they have been compared with the Channeled Scabland of eastern Washington (Milton 1973, Baker & Milton 1974, Baker 1978) and with terrestrial deep-sea channels (Komar 1979). Clearly most of them are related to the chaotic terrain in which they originate, and huge amounts of water must have been released very rapidly. This is consistent not only with the vast size of the channels but also with the uniform morphology of the chaotic terrain, which suggests that the subsurface support there must have failed rapidly and successively rather than gradually or piecemeal.

Milton reasoned that to thaw so much ground ice so suddenly "would seem to require the supply of unreasonably large quantities of heat over wide areas in very short periods of time" (Milton 1974, p.655), so he suggested that the Martian ground ice may consist largely of carbon dioxide hydrate, or clathrate. However, Peale *et al.* (1975) pointed out that it takes more heat to release the same amount of water from clathrate as from water ice, and proposed a modification of their hypothesis for the formation of sinuous lunar rilles by running water. They suggested that at the base of the permafrost, where pressure and the thermal gradient allow water to be liquid, a reservoir of interstitial water would be trapped beneath the permafrost, awaiting a meteorite impact or tectonic rupture to release it in a flood. The surface water thus released would rapidly ice over, allowing it to carve a channel before it eventually evaporated or sank into the ground. Peale *et al.*'s lunar hypothesis had earlier been modified for Mars by Maxwell *et al.* (1973), who suggested that the outflow floodwaters might have boiled at such a low rate that an overburden of ice may not have been necessary. Some other explanations that have been proposed include the breakout of artesian water when the permafrost layer is ruptured or when pore pressure exceeds lithostatic pressure (Carr 1979, 1981), gradual melting of the ground ice owing to local subsurface heating, with rapid release "when the outward spreading front of melt water ... intersects a cliff face" (Masursky *et al.* 1977), and spontaneous liquefaction of thawed regolith (McCauley *et al.* 1972, Nummendal 1978). It seems likely that the floods were triggered by some combination of some or all of these events, although my personal preference is for Masursky *et al.*'s scenario because it allows a catastrophic effect from a rather ordinary and reasonable subsurface cause. The condition of the flows that carved the channels is not known, but instead of being completely iced over, they were more likely turbulent floods of water, sediment, frazil slush, ice pans and ice jams (Baker 1979). Clearly, the water did not remain long on the surface, for there is no visible evidence of former ponds, lakes or oceans beyond the ends of the channels (Masursky *et al.* 1977). The water must have evaporated or been absorbed into unsaturated frozen grounds in the lowlands.

The extensive studies of permafrost features and outflow channels on Mars, summarized in part above, show, or at least provide circumstantial evidence, that there is (or used to be) a deep zone of ground ice within the regolith over much of the planet, that interstitial liquid water may be (or may once have been) trapped beneath the ice layer, and that something happened to melt the ice or release water at the surface. Details remain uncertain, but one point is reasonably clear: there has been a subsurface source of water potentially available to develop surface features by groundwater outflow sapping.

Mariner 9 images and the later Viking Orbiter images also show a variety of other channel- and valley-like landforms. Sharp and Malin (1975) attempted to group these into six or seven categories, including outflow channels and features formed primarily by endogenic, or tectonic, processes. They called the others "integrated runoff channels," "dendritic tributaries," "slope gullies" and "fretted channels," and attributed the first three categories to fluvial runoff, largely on the basis of terrestrial features that appear to be morphologically similar.

Where direct observation of formative processes is lacking, hypotheses for the origins of landforms and landscapes must be based either on conjecture or analogy (see Gilbert 1886, p. 287, Sharp *et al.* 1971, p. 334). This is particularly true where landforms are now relict or cannot be studied directly. The channel-like features of Mars are good examples. Most appear to be very old and presumably are not now forming, and virtually all we know about them is what we can interpret from images returned by spacecraft.

The Martian valleys that seem most closely analogous to deeply incised valleys on Earth, or to the model drainage systems on beaches, are Sharp and Malin's "dendritic tributaries," particularly the seemingly structure-controlled, modified dendritic network on the south side of Ius Chasma (Figs

2.12 & 2.13), part of the great Valles Marineris rift valley or graben. The largest of these tributary canyons is ~ 135 km long, ~ 10 km across and > 2 km deep near its mouth, according to the topographic map of the Coprates quadrangle (USGS 1976). In size, it is similar to the Grand Canyon of Arizona, but Grand Canyon has been cut by a through-flowing stream that heads in the Rocky Mountains, whereas the Ius Chasma tributary canyons all head abruptly in the Sinai Planum plateau, with no visible catchment areas. Their blunt headscarps, lack of feeder runoff, and angulate pattern led most early interpreters to attribute them to some kind of sapping – by localized groundwater outflow (McCauley *et al.* 1972, Milton 1973, Maxwell *et al.* 1973, Higgins 1974, Sharp & Malin 1975), or by sublimation or melting of ground ice (Sharp 1973a). After Viking Orbiter images began to become available in 1976, Blasius *et al.* (1977) partially confirmed Sharp and Malin's (1975) speculation that the tributary canyons may be better developed on the south than on the north side of Ius Chasma because the regional northward dip of the strata there would promote stronger groundwater drainage and outflow toward the north.

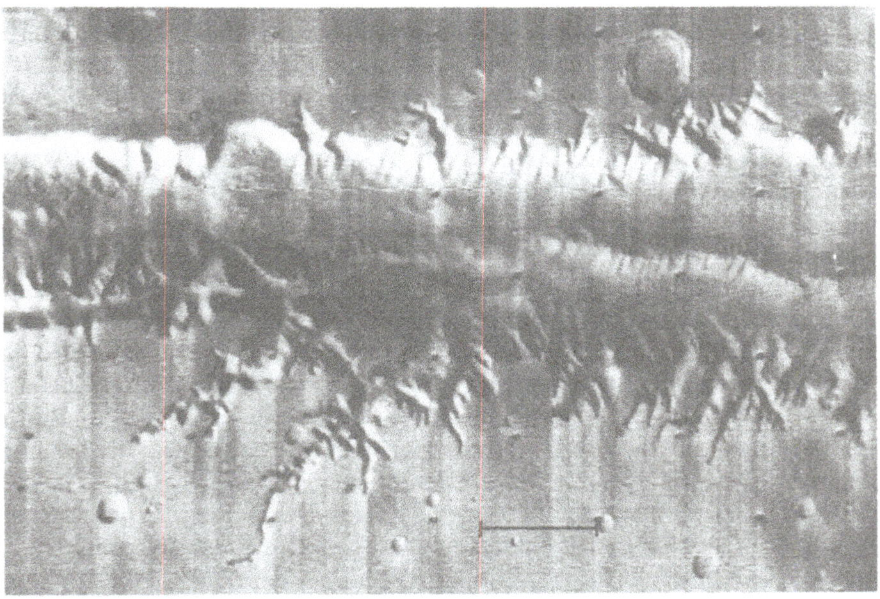

Figure 2.12 Angulate–dendritic tributaries to a portion of Ius Chasma rift graben near the western end of Valles Marineris, Mars. North toward top of picture; bar is about 50 km long, width of view about 350 km. Their blunt headscarps with no visible catchment areas or runoff feeders, angulate patterns, and stubby structure-controlled tributaries are morphologically similar to those of beach drainage systems (Fig. 2.10) or the wadis of the Gilf Kebir (Fig. 2.15), and suggest that they were formed by groundwater outflow sapping. Note large block slide–debris flows on north side of trough, upper left. Part of Mariner 9 image DAS-05851968.

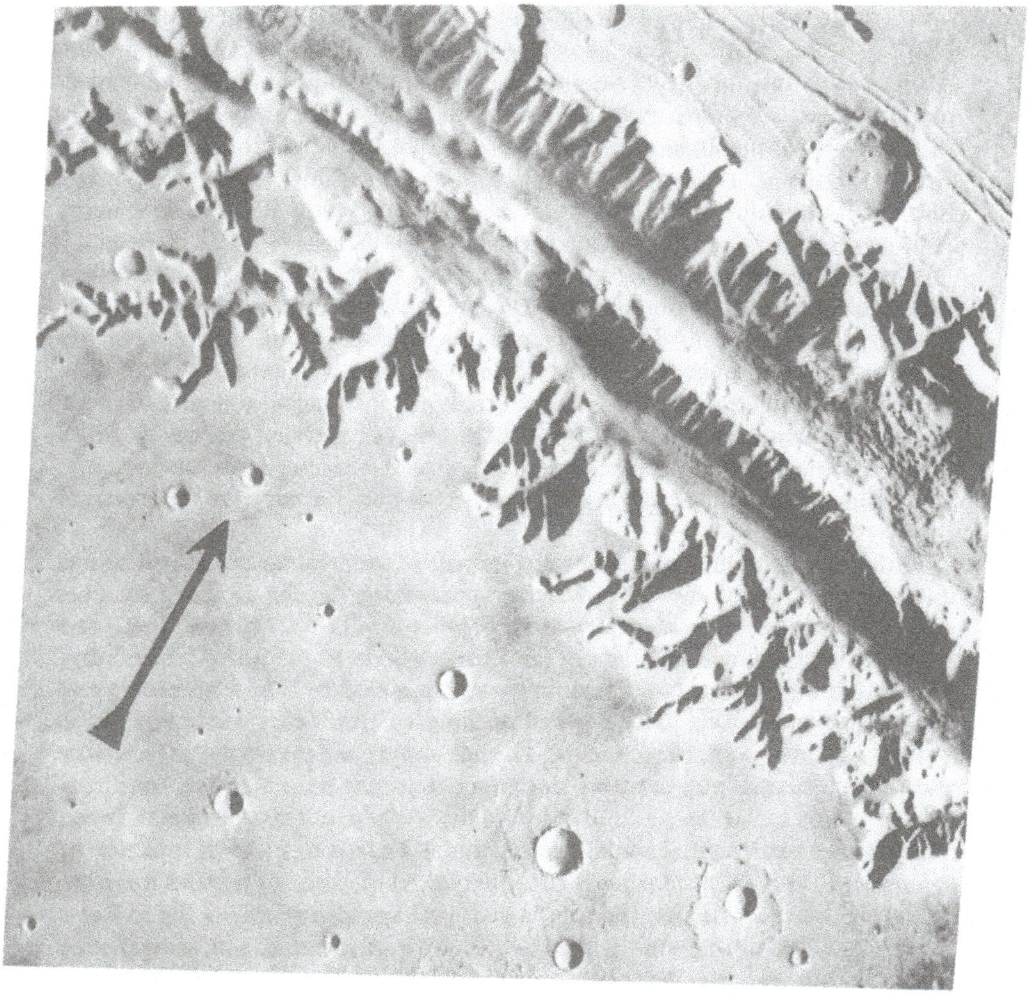

Figure 2.13 Higher resolution image of Ius Chasma tributary valleys shown in Figure 2.12. Specially processed Viking 1 Orbiter orthographic image, no. 645A57. North arrow is 80 km long.

Terrestrial analogs that have been cited for these valleys are discussed in the next section. My own conclusions about the development of the Ius Chasma tributaries by groundwater outflow (Higgins 1974, 1982) were based not on large-scale terrestrial analogs, but on the valleys' many similarities to the little model drainages on California beaches. The distinctive characteristics of the latter are summarized at the end of the previous section, and many of these also characterize the large Martian valleys, as seen in Figure 2.12. There is, however, one notable exception: Viking images (Fig. 2.13) show that the valley cross profiles are V-shaped rather than flat-bottomed and steep-sided as are most of the model drainageways. Maxwell (1979) attributes this to slumping of the valley walls, and Carr and Schaber (1977) suggest that the sloping valley walls are talus. Such talus or scree aprons along the canyon walls could result from weathering and "dry" sapping by sublimation of ground ice, in the manner proposed by Sharp (1973a), some time after excavation of the valleys themselves. Viking images also reveal a narrow strip down the center of each valley. Blasius *et al.* noted that some of these strips show "elongate light and dark markings up to 300 m across ... (suggesting) flow along their length" (Blasius *et al.* 1977, p. 4078). Carr and Schaber (1977) suggest that these might be channel deposits or possibly rock glaciers. As they are not obliterated by the scree aprons, they must be late-stage effects, perhaps contemporary with the scree. According to the topographic map (USGS 1976), gradients of the larger tributaries range from about 2° near their headwaters to 3–3.5° in their lower parts, suggesting a convex longitudinal profile. Such gradients have been measured on solifluction slopes in cold regions, so perhaps the valley floors do owe their present form to some kind of post-incision congeliturbation as Carr and Schaber have suggested. A visit to the site may be necessary to resolve this question.

Initiation of sapping and headward growth of the tributaries of Ius Chasma and elsewhere along the Valles Marineris may have developed along the lines of the following scenario, drawn from Wise *et al.* (1979), Higgins (1982) and others. Subcrustal underplating in the Tharsis region about 3.5–4 billion years ago caused permanent isostatic doming accompanied by lateral spreading and radial graben faulting. The earliest opening of the Valles Marineris "rift" graben system owes to these events. The hot underplate eventually affected the surface, causing a long-lived volcanic stage that may be still active, with continued adjustments along the older faults. When a rising thermal front, generated either by the regional heating under Tharsis or by local injection of magma into the Valles Marineris rift, intersected the zone of ground ice in the plateau bordering the rift, the ice thawed, perhaps also releasing liquid water trapped under or behind it, and water with suspended ash, fragmented volcanic rock and possibly other particles began to flow out into the trough. Where the upper strata of the plateau were thick or very coherent, general sapping along the margin caused great slumps, block slides and debris flows (Fig. 2.12, upper left, also Fig. 2.13, right). Some enormous debris flows farther east in the trough (see Carr 1980, Fig. 21, Carr 1981, Fig. 9.4) suggest that liquefaction or release of water must have been rapid at some sites. At other places, especially where strata dip toward the trough, concentration of outflow along lines of structural weakness caused headward growth of tributary valleys. The time required to form these must have depended largely on the rate of thaw or release of water and the nature and degree of consolidation of the material eroded. The present appearance of the margins of Valles Marineris owes largely to these processes and to subsequent weathering, "dry" sapping and talus accumulation.

Sharp and Malin (1975) also attributed the steep-walled, flat-floored "fretted channels" at least in part to sapping. However, they and most other early interpreters thought that the long "integrated runoff channels," such as Nirgal Vallis (Fig. 2.14), and the relatively small "slope gullies" that occur throughout the old, heavily cratered terrain were products of surface runoff, with concomitant paleoclimatic implications. Recently, however, Baker (1980b, also Carr 1980, Fig. 25) has shown that Nirgal seems to have been formed by outflow sapping, and Pieri (1980a, 1980b), after an exhaustive study of the old, small "slope gullies," or small valley systems, concluded that they are only superficially similar to terrestrial dendritic networks and that they too were formed by sapping. The conditions whereby the permafrost thawed and promoted this sapping relatively early in Mars' history are not

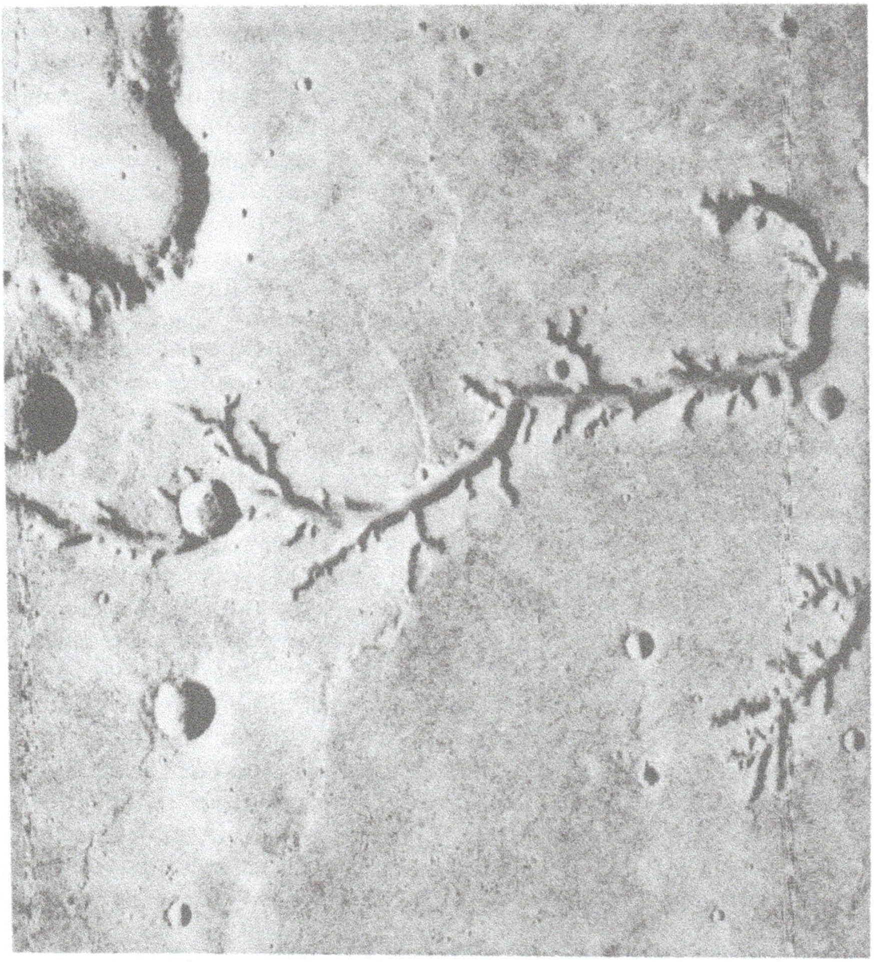

Figure 2.14 High-resolution image of part of the western end of Nirgal Vallis, about 500 km south of the eastern end of Valles Marineris. Nirgal is a valley-like form with a total length of about 700 km, which Baker (1980b) has shown must have been developed by outflow sapping. Specially processed Viking 1 Orbiter orthographic image, no. 466A54. Frame dimensions are 73 km × 80 km.

known. However, it now seems that all of the Martian valley-like and channel-like features, except those owing primarily to crustal fracturing, could have been developed by groundwater outflow sapping rather than by rainfall runoff. Current views on these features are summarized by Baker (1981, 1982).

These discoveries of extensive and varied drainage systems developed by sapping on Mars have led to an intensive search for analogous features on Earth and to a belated recognition that sapping may be a major process of valley development in terrestrial geomorphology.

Drainage systems formed by sapping on Earth

When observers of Mars looked to Earth for analogous terrestrial landforms developed by sapping, they discoverd only a scanty literature. Numerous authors have credited sapping, in a very general way, with the propagation of gully systems, or for the retreat of channel knickpoints (e.g. Leopold *et al.* 1964, p. 445–6, Emmett 1968, Higgins 1983). Until recently, however, it has rarely been considered an important factor in the creation of larger drainage networks, and relatively few examples have been cited of valleys or extensive systems formed chiefly by groundwater outflow.

One reason for this is that in a fully developed, or "mature," stream system there may be no evidence remaining of the manner in which it was formed. It may have formed entirely by rill development and deepening by surface runoff, or it may have formed by headward extension, where sapping played a major role. However, this may no longer be determinable after the valleys have been extended to their maximum possible length and their sideslopes graded by rilling and mass wasting. Another reason is that in humid regions, where water tables are high and sapping is most likely to be effective, its effects are most likely to be obscured by the results of surface runoff and creep. So one must look to incomplete, or "youthful," drainage systems in arid or semi-arid regions for clear examples of development by sapping. Because such incomplete systems are commonly small, one may wrongly conclude that sapping produces only local features, such as the Boçorocas of the State of São Paulo, Brazil, which are small V-shaped canyons, that reach 30–40 m deep and cover areas ≥ 1 km^2 (Terzaghi 1950, p. 102, Pichler 1953). Moreover, many drainage systems are incomplete because they are now relict, and the processes now affecting them are not those that formed them. Indeed, if headward erosion by sapping is active only when the water table intersects the channel and valley head, as the beach models suggest, then drainage systems that were originally formed by sapping but are now relict are themselves witnesses to a climatic or other change that caused the water table to decline.

N. M. Fenneman stated that sapping "is the dominant factor in determining the form of the flat-bottomed, scarp-bordered gully or "draw" of arid regions," with particular reference to drainage systems in unconsolidated Tertiary sediments of the southern High Plains of the United States (Fenneman

1923, p. 127–9). However, the groundwater outflow that formed these features may no longer be as active or effective there as it was formerly when water tables were higher. In many parts of the High Plains, Pleistocene loess is partially dissected by pectinate drainage nets with blunt-ended, steep-sided, wide-bottomed valleys, many of which have straight reaches suggesting some kind of subsurface structural control. The floors of some valleys are incised by smaller, steep-headed channels. These features have generally been attributed to surface runoff, but an aerial photograph published by Gerster (1976, Fig. 132) of valley heads near Waunetta, Nebraska, shows the flow in the incised channels issuing from their headscarps on the valley floors, while the upper slopes are dry. The situation depicted is so similar, except in scale, to that of Figure 2.11 and to that of the dry valleys in the English Chalk, discussed below, as to suggest that the pectinate drainage networks of the High Plains were formed chiefly by groundwater sapping when the water tables there were higher.

A similar relict condition was reported by R. F. Peel (1941) in the Gilf Kebir tableland of southwestern Egypt, part of which is shown in Figure 2.15. This plateau is formed of Mesozoic sandstone that dips gently toward the north and is dissected by long, branching, flat-floored wadis that are as much as 300 m deep in the south and 100 km long in the north (see map in Bagnold *et al.*

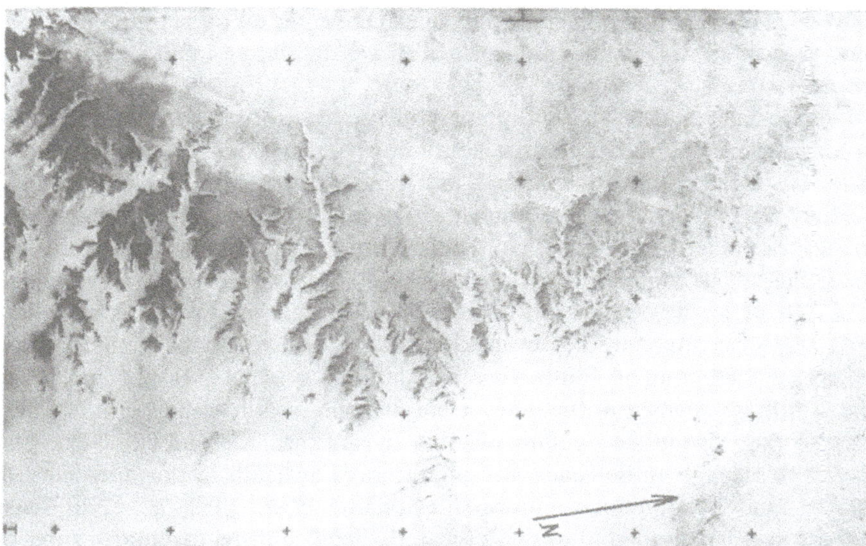

Figure 2.15 Southeastern portion of the Gilf Kebir plateau, southwest Egypt. Part of Landsat 3 RBV image, August 27, 1980. Width of view about 55 km.

1939). These wadis are among the closest terrestrial analogs for the Ius Chasma tributary canyons (Maxwell 1979, El-Baz *et al.* 1980, p. 76–83). The Gilf Kebir is waterless and remote, and few scientists have visited the area (Bagnold *et al.* 1939, El-Baz *et al.* 1980). The geomorphic observations of Peel (1941), made during R. A. Bagnold's remarkable expedition of 1938, and of Maxwell (1982) are the most extensive. The heads and sides of the wadis are cliffs, some with basal nips or caves "showing traces of former spring action" (Peel 1941, p. 18, Fig. 8), and "there is a surprising lack of evidence of water flowing over the plateau surface to drain into the wadi head" (Peel 1941, p. 12). In fact, Peel found "little evidence of water action on the plateau surface at all" (Peel 1941, p. 13). From this, from the angulate drainage pattern, and from the irregular widths of the wadis, Peel concluded that they had been developed chiefly by "basal sapping" by springs. However, he also found that the wadi floors are not only blocked in places by dunes, but are veneered with sediment that locally contains an iron pan and Neolithic artifacts, and that is incised by later channels. Peel thereby concluded that wadi growth is no longer active, but dates back to a time of wetter climate.

A somewhat analogous situation may exist in southern Arnhem Land, Australia, where "Ruined City," with its rectangular box canyons and isolated towers, is etched out of jointed quartz sandstone. Much of the etching and joint-widening there was probably effected by much earlier subsurface weathering in a tropical climate, but removal of the weathered detritus may have owed largely to sapping. During a brief visit to this remote site, Jennings (1970) searched for springs at cliff bases and steep valley heads, but found only signs of former activity. There, too, the effective period of sapping may have been associated with some past episode of wetter climate, and the landforms are now relict.

Even at valley heads where sapping is still active, groundwater outflow may be an intermittent rather than a continuous process. At the heads of box canyons in the "Navajo country" of southwestern United States, H. E. Gregory (1917) found that groundwater seepage or flowage may only appear an hour or more after a storm and then continue for only a few hours or days. Nevertheless, he concluded that weathering and erosion by such seepage and by a film of surface water on the rock face were the major causes of valley-head recession there. In a now-classic study of gully growth in the humid–temperate Piedmont of South Carolina, Ireland *et al.* (1939) also concluded that a film of surface water plays a major role in gully-head recession, but because they found few gullies heading in springs, they added, "The part played by seepage in the headward migration of overfalls in this area is much smaller than is sometimes supposed" (Ireland *et al.* 1939, p. 139). These authors may have failed to observe the gully heads during infrequent times of groundwater outflow, or possibly the effects of such outflow were obscured by the surface runoff that preceded or accompanied it. It seems to me likely that sapping is at least partly responsible for some of these gullies because "many of the gullies of the Piedmont have been developed where formerly there was no channel drainage" (Ireland *et al.* 1939, p. 39). Without a local concentration of surface runoff above a gully head or valley head, sapping is virtually the only process whereby the valley can be extended.

Sapping is most likely to be effective in such humid regions, where water tables are high, but reported examples are rare. Dunne (1980) has studied a small, recently deforested catchment in Vermont in which the channel network has been formed by periodic "spring sapping," and Gilmour and Bonell (1979) infer that subsurface outflow is significant in the Babinda tropical rainforest of northeastern Australia, because the surface soils there have high porosities and "most first order streams in the catchment gather in clearly defined stream head hollows" (Gilmour & Bonell 1979, p. 78, Fig. 5.3). Perhaps the largest drainage network that has been attributed to sapping is that of the central, eastern and southern Netherlands (De Vries 1974, 1976). This region is underlain by permeable Plio–Pleistocene alluvium and Pleistocene eolian sands. The surface is flat or gently rolling, and the infiltration capacity of the soils almost everywhere exceeds the moderate precipitation rates. Consequently, Horton overland flow is rare or nonexistent. De Vries (1976) has proposed a "groundwater outcrop-erosion model" to explain the development of the regional drainage system entirely by groundwater outflow.

The best studied examples of valleys formed by sapping in a humid region are the "coombes," or escarpment valleys, of the English chalk downs. Most of these are now dry, but some are incised by blunt-ended trenches that head at active springs in the valley floor. Sparks and Lewis (1957–8) and Small, in a series of studies (1961, 1962, 1964, 1965), conclude that the valleys were eroded headward by scarp-foot springs when the water table was higher. Small (1978, p. 146) cites a similar origin for escarpment valleys in the Massif Central of France.

Most of the escarpment valleys are short; the largest, such as one near Bratton, Wiltshire, is only 2 mi long and 0.5 mi wide (Small 1964, p. 38). They share with the beach models (and with the wadis of the Gilf Kebir and most of the other sapping-generated systems mentioned above) several or all of the following characteristics. They are relatively steep sided and blunt ended. Valley sides and heads meet the floors in marked angles (in humid climates "steep sided" is a relative term and may mean only that the slopes reach 20–30°; see Sparks & Lewis, 1957–8, p. 26). The valley floors are fairly broad, and their width may vary irregularly along the length. The floors are flat where veneered with sediment, and they may be incised by blunt ended, spring-fed trenches that are miniature versions of the larger valley. Tributary valleys tend to be wide-spaced and stubby. The valleys lack catchment swales or channels above their steep heads and may even head against the slope of the upland. They have markedly angular courses that represent some kind of subsurface control. These characteristics should aid identification of additional examples of drainage systems formed by headward sapping. However, the V-shaped Boçorocas of Brazil, the Vermont catchment studied by Dunne, and the Dutch drainage network studied by De Vries suggest that some systems developed chiefly or in part by groundwater outflow may more closely resemble drainageways developed chiefly by surface runoff.

In their search for terrestrial analogs, investigators of Martian valley-like features have focused on glacial outflow channels of Iceland (Malin 1980), box canyons of the Colorado Plateau in southern Utah (Laity 1980, Laity & Pieri 1980, Laity & Malin in press), and the "theater-headed" valleys of Hawaii (Baker 1980a, 1982). The latter illustrate how groundwater sapping has generally been overlooked or ignored until recently. Hinds (1925) attributed the steep headwalls of the Hawaiian valleys to general differential erosion and sapping by surface agents in weak beds overlain by resistant ones, comparing them with valley heads of tributaries to the Grand Canyon of Arizona. Wentworth (1928) also ascribed the valleys to general sapping, effected by water-table weathering and lateral corrosion by streams. Although Stearns and Vaksvik (1935) and White (1949) noted the springs at the foot of headwalls and cliffs, they did not think that the springs played any part in the development of the valleys. The formative role of groundwater outflow there can be inferred from these older studies, but has only recently been recognized as a major factor in the development of the topography (Baker 1980a, 1982).

Similarly, "theater-headed" valleys or "alcoves" occur in basalt along the Snake River in southern Idaho and are several hundred feet deep and as much as 2 miles long. Johnson (1938–39b, p. 178, 216–7) cited earlier studies by Russell (1902) and Stearns (1936) in attributing the development of these valleys to sapping by the springs that issue from their heads. Johnson (1938–39b, p. 216) also noted that valleys as much as 70 feet deep and 2 miles long in the frontal slope of the Pleistocene Sheyenne River delta in North Dakota were attributed to spring sapping by Hall and Willard (1905), and he cited Sellards and Gunter's (1918) similar explanation for the "steep-heads" in the coastal plain of northwestern Florida, which are 50–100 feet deep and as much as 10 miles long.

Earlier, Henry Stetson (1936) had proposed spring sapping as a formative mechanism for the tributaries to the Grand Canyon of Arizona and for a dissected escarpment near Cameron, Arizona. He used these examples to support an hypothesis, which he credited to Kirk Bryan, that submarine canyons were formed subaerially by groundwater sapping when sea level was much lower. This idea seems to have inspired Johnson's (1938–39, 1939) proposal that submarine canyons were formed below sea level by "artesian sapping" by submarine freshwater springs. Both of these ideas have long been out of favor, chiefly as a consequence of severe contemporary criticism and the predominance of the turbidity-current hypothesis. However, there is now a growing recognition that submarine valleys may have diverse origins, and Belderson (1983), citing Johnson's work, has suggested that some kind of spring-sapping process "may play some part in the overall development of some submarine canyons." Indeed, while Beldersons's Comment was in press, Robb *et al.* (1982) presented some preliminary conclusions, based partly on sidescan-sonar images of the eastern United States continental slope, that some submarine valleys with "amphitheater-shaped heads and cliffed walls" may have been developed in part by groundwater sapping during stages of low sea level, much as Johnson proposed more than four decades ago.

Because subaerial sapping is likely to be obscured by other processes in humid regions or may no longer be active where climatic change or other factors have caused lowering of the water table, it is not surprising that it has largely been overlooked as a valley-forming process. However, now that examples are being looked for, they should be found in abundance.

Exhumation of buried drainage by sapping

Where no pre-existing swales concentrate runoff, control of the direction of headward growth of channels formed by sapping must be determined almost entirely by subsurface permeability and concentration of groundwater flow, as in the beach models. Joints and other zones of structural weakness may concentrate the flow so that the channels grow headward in these zones. Thus, adjustment of surface drainage to subsurface structure in some regions, especially where the initial surface drainage bears little relation to subsurface structure, may owe more to sapping than to surface processes (Small 1965, p. 5–6). In other cases, groundwater flow may be controlled or concentrated by buried topography. For example, a dissected land surface on relatively impermeable rocks may be buried beneath permeable sediments. Later, when a surface stream cuts down below the unconformity, or where the contact is exposed by faulting, groundwater outflow along the valley walls will be concentrated at the lowest exposures of the unconformity – where buried channels have been intersected. Sapping at these sites of concentrated outflow may then extend tributaries headward along the routes of the buried channels, somewhat as illustrated by Dury (1959, p. 21–22, Fig. 10), thus recreating in the modern topography the drainage pattern of the buried topography.

This seems to have occurred in northwestern Sonoma County, California, where an older drainage, developed on melange of the Mesozoic Franciscan Formation, was buried beneath marine standstone in a Pliocene embayment of about 250 km^2. The configuration of the pre-Pliocene contact shows, after allowance is made for post-depositional deformation, that in many parts of the basin the sediments were deposited on fairly level surfaces of Pliocene marine planation, but that in some local areas the buried surface is hilly, with local relief of at least 100 m (Higgins 1960, Fig. 2). Wherever this surface can be inferred in detail, one sees that the courses of many modern streams, including small tributaries, reflect valleys and depressions in the pre-Pliocene topography. One such area is Miller Ridge on the western edge of the basin, 2 km inland from the Pacific coast, 15–25 km northwest of Fort Ross, and 130 km northwest of San Francisco.

Miller Ridge trends northwest and is about 10 km long and 3 km wide. It is bounded on the southwest by the San Andreas fault zone, occupied by the South Fork of the Gualala River, and on the northeast by a subsidiary fault

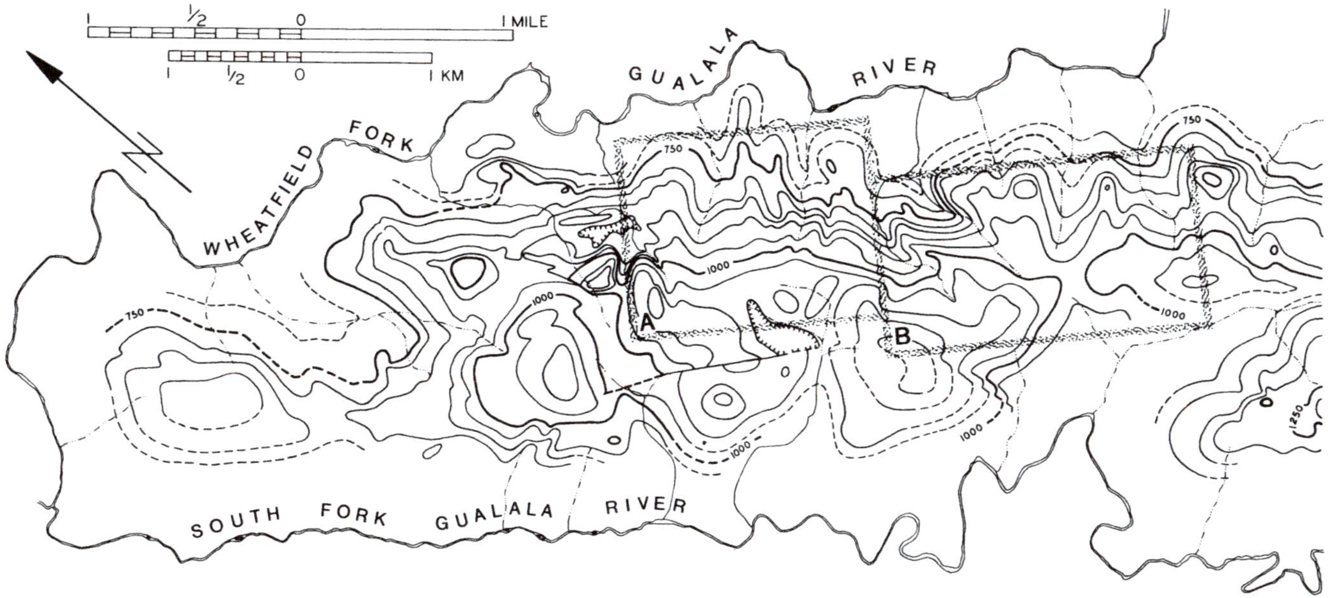

Figure 2.16 Miller Ridge, northwestern Sonoma County, California. Structure contours (50 foot interval; dashed lines where uncertain) show configuration of the unconformable contact, or buried erosion surface, between Pliocene Ohlson Ranch Formation marine sandstone and Mesozoic Franciscan Formation. Modern drainageways indicated by conventional dash–dot and solid lines. Boxed areas are enlarged in Figure 2.17. (Adapted from Higgins 1960, Fig. 2).

zone, occupied by Wheatfield Fork (Fig. 2.16). The ridge has total relief of about 335 m and, except for a few clearings on top, is largely covered with Redwood and Douglas Fir forest. In this area, semi-consolidated Pliocene sandstone, the remaining exposures of which are stippled in Figure 2.17, unconformably overlies a predominantly shaly facies of the Franciscan Formation. The contact between the two units is so irregular along the steep slopes of Miller Ridge that it is possible to draw detailed contours of the buried erosion surface, shown as heavy solid lines (dashed where uncertain) in Figures 2.16 and 2.17. Figure 2.16 shows that the buried surface not only appears to form a ridge under the modern ridge, but that valleys in the slopes of the buried ridge underlie and correspond with the modern drainageways, indicated in Figure 2.16 with dash–dot lines. This is even more apparent in Figure 2.17,

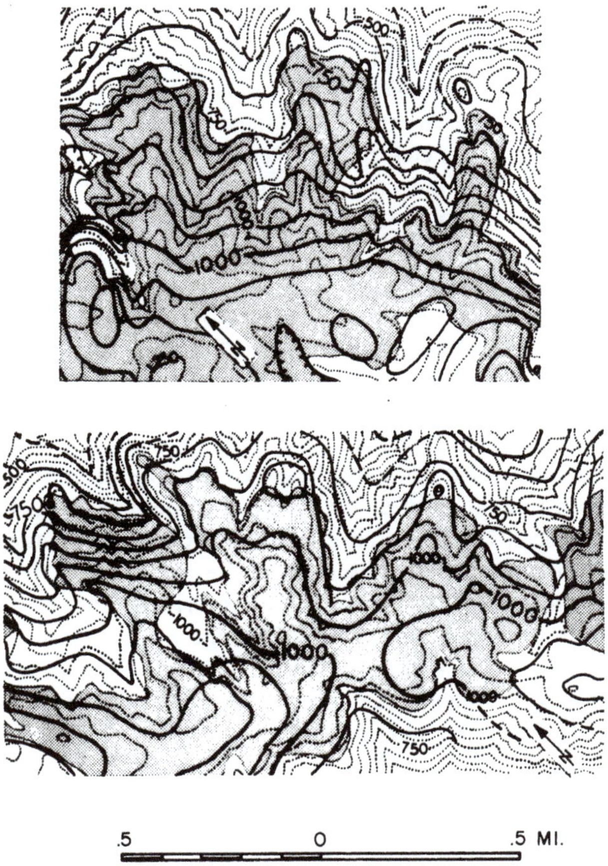

Figure 2.17 Detail of areas boxed in Figure 2.16. Pliocene sandstone stippled. Modern topography indicated by dotted 50 foot contours, from US Geological Survey, Annapolis quadrangle map, 1943. Pre-Pliocene erosion surface indicated by heavy 50 foot contours – dashed where not certain.

which shows enlargements of the two boxed areas of Figure 2.16. In Figure 2.17, contours of the modern surface are indicated by dotted lines. The topographic correspondence is not exact; in the center of box "A," a prominent valley of the buried surface does not seem to be reflected in the modern topography, and at at least three other sites, modern valleys seem to be eroded in or above hills or spurs on the pre-Pliocene surface. Such lack of correspondence may owe in part to overgeneralization of the inferred contours on the buried surface. Throughout the rest of the area, however, the coincidence of the modern with the buried relief is so pronounced as to require an explanation.

Several explanations might be advanced for this apparent remote control of surface drainage by buried topography: control of both pre-Pliocene and modern drainage by an active fault or fracture system; compaction of sediments, thereby outlining pre-Pliocene relief, before or during establishment of the drainage; selective piracy by steams accidentally flowing over pre-Pliocene lows; and headward extension of valleys by sapping. The first two of these are certainly not adequate. The valleys of the South and Wheatfield forks of Gualala River are clearly fault-controlled, but the apparent random orientation of both the modern and ancient tributary valleys suggests that neither is controlled by any kind of regular network of faults or joints. Moreover, a northwest-trending fault (center of Fig. 2.16) that appears to offset the contact seems to have no effect on the modern drainage. This fault and another that has no effect on the modern topography near the south end of the ridge are both sub-parallel to the trend of the San Andreas. No other fault offsets of the Pliocene–Mesozoic contact were detected on Miller Ridge. The irregular subsurface topography on the north slope, just west of box "A" in Figure 2.16, may reflect faulting, slumping, or both. Compaction of the Pliocene sediments is also unlikely. They are of fine to coarse sand, which would have been only slightly compactable. Furthermore, much of the sediment may have been consolidated even before deposition ended. A disconformity in the formation is exposed at an elevation of 950 ft at the Noble Ranch, 3 km southeast of the south end of the ridge. This erosional surface represents an ancient tidal zone, where Pliocene molluscs bored holes in the earlier Pliocene sediments and a whale washed up on the shore. This suggests that the underlying sediments must already have been rather well consolidated to have recorded this erosional episode.

Selective piracy by streams flowing on the pre-Pliocene surface lows is more difficult to evaluate, and may, in fact, account for some adjustment of modern valleys over the sites of ancient ones. However, if this were to explain all of the drainage, one would expect to find some valleys eroded wholly in Pliocene sediments oriented transversely to underlying subsurface valleys, and one would not expect to find many valleys over subsurface valleys that are still completely buried. One does not find the former; one does find the latter.

The remaining explanation, for valley-head retreat by spring sapping, seems most likely. This hypothesis holds that when downcutting of the large, fault-controlled main valleys, which may have been initiated by post-Pliocene deformation, exposed the irregular Pliocene–Mesozoic contact on the slopes of Miller Ridge, groundwater seepage became concentrated at the bottoms of the buried tributary valleys of the pre-Pliocene surface. Then, by subsurface weathering and leaching and by outflow sapping, new valley heads were extended up the courses of the old drainageways. Many of the valleys, especially on the north slope of the ridge, have a near-vertical head scarp, one to several meters high, at the Pliocene–Mesozoic contact, from which seepage or a small spring issues even during the dry summer months. Below this scarp, the valley is steep and V-shaped; above it, the valley is less steep, round-bottomed, and generally lacks a well-defined channel. These upper deep swales seem to owe chiefly to creep and possibly slopewash.

Small (1965), Dunne (1980) and others have discussed groundwater flow models to explain the development of hierarchical drainage networks according to subsurface flow lines, but valley-head sapping and extension may also influence surface runoff and creep simply by creating a local low in the topography. This seems to be the case on Miller Ridge, where swales lie above the headscarps. The few places where the modern tributary valleys are incised into or above subsurface highs must record early drainage that was superimposed onto the buried topography and failed to be adjusted or abandoned after sapping began to develop the tributaries controlled by the buried topography.

Another example of correspondence of modern and buried drainage systems has been reported by Clayton (1960) in the Chelmsford region of southern Essex, England. There, shallow upland stream valleys are consequent on the depositional surface of the Springfield Till of the most recent glaciation, but where the drainage is incised through the till into the underlying interglacial Chelmsford Gravels or still older units, many of the streams "flow along the same courses as pre-glacial valleys ... [and] ... in many cases the old drainage lines were followed again even though they were covered by 100 feet of glacial deposits" (Clayton 1960, p. 63). Clayton postulated that the largest pre-glacial valleys had been reflected by lows in the blanketing drift surface, so that post-glacial consequent drainage was initiated along the same routes although not invariably in the same directions. Many of the tributaries to the largest streams also follow pre-glacial courses, and some of them and their tributaries seem to have eroded headward to capture parts of the original consequent drainage of the till surface (Clayton 1960, p. 65). According to Clayton's maps (Clayton 1960, Figs 2 & 4), almost all of the streams that follow pre-glacial courses or that have extended headward are incised into or through the permeable Chelmsford Gravels, which are sandwiched between two impermeable boulder clays. It thus seems plausible that at least some of the adjustment of the original drainage pattern by re-excavation of pre-glacial drainage lines and capture of consequent streams was effected by valley-head extension by groundwater outflow sapping in the Chelmsford Gravels. Clayton was not looking for sapping and did not mention any springs where valleys head in the Gravel. However, as headward sapping provides such a likely mechanism for re-excavation of the pre-glacial drainage, evidence of present or former valley-head springs could profitably be sought.

In a study of loess-mantled terrain in Nebraska and Iowa, Bariss (1968) reported that the alignment of fourth-, or higher-, order valleys (third-order where the loess is thin) is determined by the configuration of the buried pre-loess landscape. Bariss notes that first-order tributaries are generated by piping and collapse, but he does not suggest a mechanism for the coincidence of the larger valleys with the buried relief. The pre-loess valleys may have been reflected in the depositional surface so that the higher-order streams were consequent, but it is also possible that after the master streams of the region had incised their valleys through the loess, some of their tributaries could begin to grow headward along the courses of the older valleys by sapping resulting from groundwater outflow concentrated in permeable pre-loess channel deposits of the buried drainageways. *Their* tributaries, initiated on or within the loess, do not reflect subsurface control. Until such a developmental sequence can be supported by evidence of present or past sapping, it must remain as conjectural as the sequence suggested above for the Chelmsford region.

I know of no other clear examples of exhumation of buried drainage or relief that might be attributed to sapping. It is difficult to reconstruct a buried or partially exhumed surface without abundant borehole data or exposures of the unconformity, so such cases may be more common than the three cited above might suggest.

On a much smaller scale, both Bunting (1960, 1961) in England and Jutland, and Dunne (1980) in Vermont, have reported "seepage lines" – very shallow damp swales underlain by deeper soil just upslope from spring-fed valley heads. These "seepage lines" may form linear or semi-dendritic patterns on slopes of 10° to < 5°. Bunting attributes these to intensified weathering of the bedrock under concentrated lines of groundwater seepage. How this seepage could initially have become concentrated is not clear. Instead, the circumstances described by Bunting suggest that the bedrock swales pre-date the modern surface, that a shallow pre-glacial drainage system was partly or completely effaced under a thin blanket of periglacial colluvium, and that this system has since been partly exhumed by headward extension of valleys whose courses are determined by groundwater flow concentrated in the buried drainageways. Sapping-generated exhumation of small drainages such as those reported by Bunting and Dunne may be fairly common but has not been widely recognized.

Seepage erosion

Where for any reason groundwater outflow is concentrated, spring sapping may form a headward-growing gully or valley, as discussed in the foregoing sections. But where groundwater emerges uniformly along a slope, the result is a line of water-table seeps. Such seepage outflow has been credited with maintaining the break in slope between geomorphic slope elements by oversteepening the base of the steeper, upper slope. Thus, Schumm (1956), Baker (1980a) and others have attributed pediment-angle maintenance to seepage. Several authors have suggested that seepage is responsible for backwearing of cliff faces – of the English Chalk (Sparks 1960, p. 139; Small 1961), of hamadas in the northwest Sahara (Smith 1978), and elsewhere. McBeth (1961, see also Lewis 1961) even suggests that all tropical slopes owe to vertical leaching and horizontal sapping.

The results of diffuse outflow at the water table should be designated as seepage erosion (Smith 1978) to distinguish it from sapping by concentrated outflow. Hutchinson (1968) and Cook and Doornkamp (1974, p. 155) suggest that seepage erosion occurs when outflow is sufficient to entrain and remove individual soil particles, and that this is most effective in unconsolidated fine sands and silts. However, this mechanism cannot, by itself, account for erosion of consolidated rock, so "seepage erosion" must be a more complex or more varied process or group of processes. These may include intensified chemical weathering, leaching and solution within the seepage zone, coupled with enhanced physical weathering of the periodically damp or saturated rock face by granular disintegration or flaking owing to wetting–drying (Smith 1978), salt-crystal wedging, root-wedging, rainbeat, and, in temperate to cold regions, congelifraction, including needle-ice wedging (Sharp 1976). With sufficient hydraulic gradient, the seepage outflow may at times be forceful enough to entrain weathered particles and carry them outward from the slope, but it is unlikely that such diffuse seepage could transport the grains very far. Removal of the weathered products must then owe to other agents, such as wind, slopewash and channeled streamwash. The combined effect of all these is to sap the base of a cliff so that it retreats by parallel backwasting.

In the southeastern part of the Gilf Kebir (Fig. 2.15), the wadis widen dramatically near their mouths, with no decline in the 35° slopes of their 300 m high walls. Peel interpreted this to mean that not only had the valleys grown headward by sapping but their walls had retreated "by undercutting and basal sapping" (Peel 1941, p. 16). Along the edge of the plateau, such retreat has left isolated mesas and buttes (Fig. 2.15), and beyond these, for tens of kilometers, steep-sided nubbin-like residuals of the same "Nubian" sandstone form a "plains and inselberg" landscape. Peel believed that all of this landscape represents the effects of basal sapping by groundwater seepage and outflow during a wetter climate, aided by sheetwash and later modified by wind action. He further suggested that the planed-off sandstone bedrock surfaces beneath the great "sand sheets" that extend for hundreds of kilometers around the Gilf Kebir represent the end product of complete planation by this process (Peel 1941, p. 21–2), which thus must have accounted for the reduction and partial removal of the "Nubian" sandstone over most of the Libyan Desert.

Similar broad plains bordered by fretted steep cliffs 1–3 km high occur on Mars. Sharp postulated that the plains evolved, "... by recession of a steep bounding escarpment, leaving a smooth lowland floor at a remarkably uniform level. Escarpment recession is speculatively attributed to undermining by evaporation of ground ice exposed within an escarpment face, or, under a different environment, by ground water emerging at its foot. The uniform floor level may reflect the original depth of frozen ground. Removal of debris shed by the receding escarpments could be by eolian deflation ... or by fluvial transport ..." (Sharp 1973b, p. 4073). Later studies (e.g. Carr & Schaber 1977) have not changed this assessment of scarp retreat by some kind of seepage erosion or basal sapping, except to stress the apparent role of a thick zone of ground ice. How much, if any, of the cliff recession owed to seepage outflow of liquid water is not known, although backwasting of the equally high walls of portions of the Kasei Vallis outflow channel, with associated structure-controlled tributary canyons, has been attributed to mass wasting and sapping (Baker & Kochel 1979), most likely by water, judging from the low gradients of the tributary trenches.

Some features of the Martian low plains and their bordering escarpments, isolated residuals and broad channelways resemble the Gilf Kebir. Other analogs that have been sought include the jointed sandstone plateau of Arnhem Land, northern Australia, which Baker describes as "an extensive surface of resistant sandstone surrounded by a steep escarpment from 30 to 330 meters in height" (Baker 1980a, p. 286–7). The plateau is capped by resistant strata overlying weaker rocks at the base. "Spring sapping, cavernous weathering, and rock spalling" undermine the cliffs. The debris is removed by wind or sheetwash, and the escarpment recedes irregularly, with deep embayments "and many sandstone outliers on the broad, flat pediplain that was left by scarp retreat" (Baker 1980a, p. 287). Jennings' (1979) observations in this region, noted in the section on terrestrial drainage systems, suggest that seepage erosion and cliff retreat of the Arnhem Land plateau may no longer be as vigorous as in the past, and that the landscape, like so many that owe to these processes, is now mainly relict.

If these interpretations are correct – and there seems to be no cause to doubt them – then extensive tracts that were once covered by near-horizontal resistant strata and are now bordered by distant lines of residual cliffs, perhaps with scattered residual hills, may owe their denudation more to backwearing by seepage erosion, with its varied attendant processes, than to the conventionally accepted processes of surface runoff, stream incision and downwearing.

Pseudokarst

The surface of Mars also exhibits a variety of huge pseudokarst features, which have been interpreted as thermokarst by analogy with morphologically similar terrain in Siberia and other cold regions (Carr & Schaber 1977). By definition, thermokarst results from "local melting of ground ice and the subsequent settling of the ground" (Bates & Jackson 1980, p. 648), but where the meltwater, with or without entrained sediment, escapes downward or laterally, this process merges with piping or sapping. Such vertical or lateral migration of saturated substrate most likely accounts for the long, structure-controlled lines of coalescing pits (e.g. Tithonia Catena north of Ius Chasma, Coprates Catena south of Coprates Chasma) and large closed depressions, hundreds of kilometers long (e.g. Hebes Chasma, Fig. 2.18, which lies along the continuation of the structural trend of Ganges Catena north of Ophir Chasma) that are associated with the Valles Marineris rift graben system. These pits and depressions doubtless reflect subsurface tension fractures or

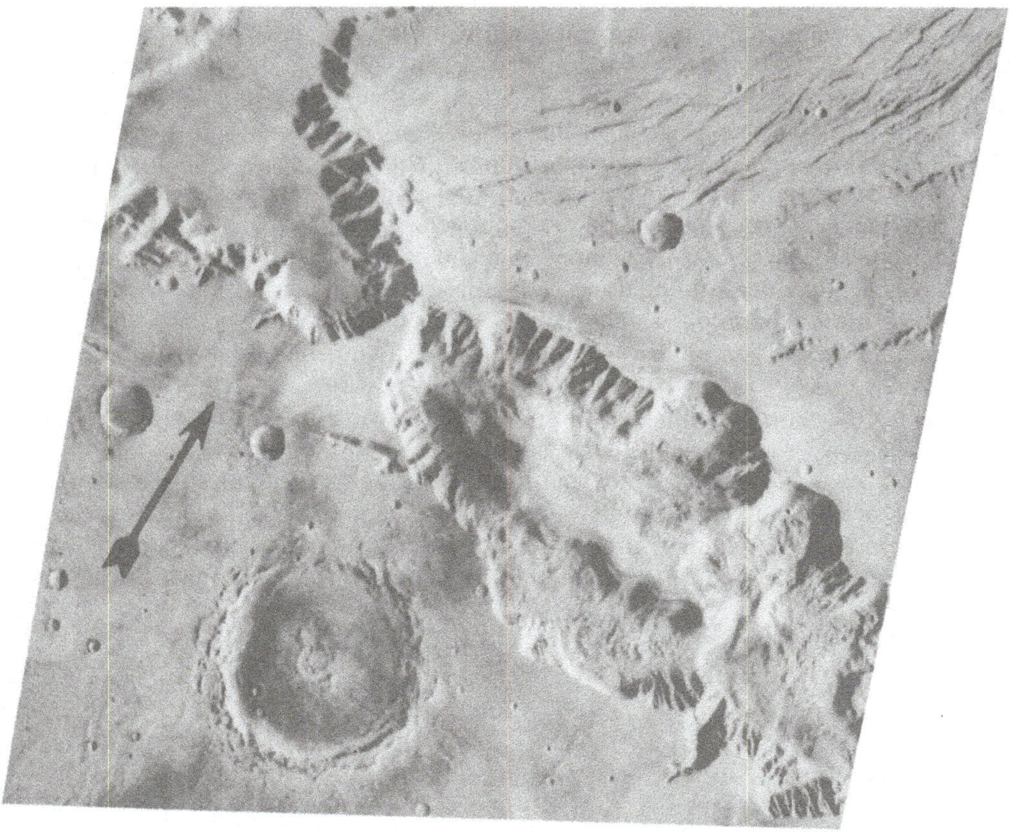

Figure 2.18 Echus (upper left) and Hebes (right) Chasmas, north of the western part of Valles Marineris, Mars. Structure-controlled tributary canyons, especially on the south side of Echus, and scalloped cliffs with block slides and debris flows resemble those of Ius Chasma (Figs 2.12 & 13), and suggest a sapping origin. Material from Echus was removed by breakout northward (upper left) into a great outflow channel. Material from Hebes may have been piped laterally into Echus or removed downward into an underlying east–west trending fault or fracture system, the eastward continuation of which is marked by the Ganges Catena line of pits. Large impact crater at bottom is ~ 80 km in diameter. Specially processed Viking 1 Orbiter orthographic image, no. 645A60.

faults, the opening of which provided downward or lateral egress for thawed ground ice and sediment. It has been suggested that much of the material from Hebes Chasma may have been piped laterally into the adjoining Echus Chasma (Fig. 2.18), which broke out northward to form the abrupt head of a great outflow channel (H. Masursky unpubl.). The inference that the regolith was wet is supported in part by the scalloped and fretted walls of the large, closed basins and by associated stubby, tributary valleys that resemble those of Ius Chasma. Moreover, as Sharp (1973a) has pointed out, the material from these pits and depressions must have been removed downward, since there is no surface process adequate to excavate it.

Terrestrial pseudokarst is a common feature of piping, as noted in the section on piping, but is less commonly associated with sapping, in part because of scale problems. However, Wright (1964) has described an extensive landscape with hundreds of large, shallow depressions in siliceous sandstone on the flat crest of the Chuska Mountains in northwest New Mexico. He attributed these depressions to "collapse of cemented sandstone layers above vacuities that had been produced by piping of uncemented sand out to the steep escarpments of the mountains. The pipes originated at springs or seeps ..." (Wright 1964, p. 589). Because the transporting medium seems to have been groundwater rather than soil water, the process involved should be regarded as sapping rather than piping, as these terms are used here. Wright also noted, "... the process of lake-basin formation seems now to be extinct or dormant. In fact the entire mountain geomorphology is a relic of processes active in the Pleistocene, when the climate was colder, the water table was higher, and the forest cover was restricted" (Wright 1964, p. 589). This reiterates the story of most of the other features described here that owe to sapping or seepage erosion under conditions different from those of the present and that are now relict and undergoing modification by processes different from those that formed them.

Conclusions

Groundwater affects landforms and landscapes in many ways. By solution of soluble minerals, it prepares the development of extensive and varied karst terrains. By subsurface weathering, it may form whole landscapes that are not revealed until the overlying regolith is stripped away. These aspects of the work of groundwater are discussed elsewhere in this volume. Groundwater seepage along sea cliffs may promote solutional nips and notches in permeable calcareous rocks (Wentworth 1939, Higgins 1980) or may aid wave attack on insoluble sediments (Norris 1968). However, the direct role of groundwater in fashioning landscapes and drainage systems in insoluble rocks has largely been neglected or overlooked.

Where the water table intersects the base of an escarpment, groundwater "seepage erosion" – actually a varied group of interacting processes – may undermine the cliffs, promoting their retreat with constant slope angle, and leaving behind a broad pediplain or inselberg landscape. Where this process has denuded large areas of the Earth's surface, such backwasting seems to have predominated over downwearing.

Where groundwater outflow is locally concentrated as springs, extension of valley heads by sapping may develop whole drainage systems. A few such systems have been recognized on Earth and others have been recognized on Mars, where the effects of sapping have not been obscured by rainfall runoff. Most of the terrestrial examples and all of the Martian ones are now relict features, and the processes that formed them are no longer directly observable. For this reason and because the concept of drainage development by groundwater outflow has not been a part of mainstream geomorphic thought, few investigators have recognized the effects of sapping and seepage erosion on terrestrial landscapes. When such effects are looked for, however, the role of groundwater as a geomorphic agent should receive the greater recognition that it justly deserves.

Acknowledgements

Mariner 9 and Viking Orbiter images were kindly provided by the National Space Science Data Center. Drafting is by Ray Roberts, with additions by Ellen Bailey. Field mapping of the Pliocene strata in northwestern Sonoma County and observations of California beaches were supported in part by grants from the Committee on Research, University of California, Davis.

References

Bagnold, R. A., O. H. Myers, R. F. Peel and H. A. Winkler 1939. An expedition to the Gilf Kebir and Uweinet, 1938. *Geogr. J.* **93**, 281–313, map op. 376.

Baker, V. R. 1978. The Spokane Flood controversy and the Martian outflow channels. *Science* **202**, 1249–56.

Baker, V. R. 1979. Erosional processes in channelized water flows on Mars. *J. Geophys. Res.* **84**, 7985–93.

Baker, V. R. 1980a. *Some terrestrial analogs to dry valley systems on Mars.* NASA Tech. Memo. 81776, 286–8.

Baker, V. R. 1980b. *Nirgal Vallis.* NASA Tech. Memo. 82385, 345–7.

Baker, V. R. 1981. The channels and valleys of Mars. *Lun. Planet. Inst. Contrib.* **441**, 12–4.

Baker, V. R. 1982. *The channels of Mars.* Austin, Texas: University of Texas Press.

Baker, V. R. and R. C. Kochel 1979. Martian channel morphology: Maja and Kasei Valles. *J. Geophys. Res.* **84**, 7961–83.

Baker, V. R. and D. J. Milton 1974. Erosion by catastrophic floods on Mars and Earth. *Icarus* **23**, 27–41.

Bariss, N. 1968. A comparative study of selected loess areas in the Missouri River Basin (with remarks on the Carpathian Basin in eastern Europe). In *Loess and related eolian deposits of the world*, C. B. Schultz and J. C. Frye (eds), 81–97. Proc. INQUA VIIth Congr. Vol. 12. Lincoln, Nebraska: University of Nebraska Press.

Bates, R. L. and J. A. Jackson (eds) 1980. *Glossary of Geology*, 2nd edn. Falls Church, Virginia: American Geological Institute.

Baumann, A. 1909. *Mars; erklärung der oberfläche des planeten Mars*. Zürich: Kommissionsverlag von Rascher.

Belcher, D., J. Veverka and C. Sagan 1971. Mariner photography of Mars and aerial photography of Earth: some analogies. *Icarus* **15**, 241–52.

Belderson, R. H. 1983. Comment: "Drainage systems developed by sapping on Earth and Mars". *Geology* **11**, 55.

Blasius, K. R., J. A. Cutts, J. E. Guest and H. Masursky 1977. Geology of the Valles Marineris: first analysis of imaging from the Viking 1 Orbiter primary mission. *J. Geophys. Res.* **82**, 4067–91.

Boyce, J. M. 1979. *Martian subsurface permafrost: evidence from impact craters*. NASA Conf. Publn. 2072, 8–9.

Buckham, A. F., and W. E. Cockfield 1950. Gullies formed by sinking of the ground. *Am. J. Sci.* **248**, 137–41.

Bunting, B. T. 1960. Bedrock corrosion and drainage initiation by seepage moisture on a gritstone escarpment in Derbyshire. *Nature* **185**, 447.

Bunting, B. T. 1961. The role of seepage moisture in soil formation, slope development, and stream initiation. *Am. J. Sci.* **259**, 503–18.

Carr, M. H. 1979. Formation of martian flood features by release of water from confined aquifers. *J. Geophys. Res.* **84**(B6), 2995–3007.

Carr, M. H. 1980. The morphology of the martian surface. *Space Sci. Rev.* **25**, 230–84.

Carr, M. H. 1981. *The surface of Mars*. New Haven: Yale University Press.

Carr, M. H. and G. G. Schaber 1977. Martian permafrost features. *J. Geophys. Res.* **82**, 4039–54.

Clayton, K. M. 1960. The landforms of parts of southern Essex. *Trans. Paps. Inst. Br. Geogs. Publn.* **28**, 55–74.

Cooke, R. U. and J. C. Doornkamp 1974. *Geomorphology in environmental management*. London: Oxford University Press.

Demarest, D. F. 1947. Rhomboid ripple marks and their relationship to beach slope. *J. Sed. Pet.* **17**, 18–22.

De Vries, J. J. 1974. *Groundwater flow systems and stream nets in the Netherlands*. Amsterdam: Editions Rodopi.

De Vries, J. J. 1976. The groundwater outcrop-erosion model: evolution of the stream network in the Netherlands. *J. Hydrol.* **29**, 43–50.

Dunne, T. 1980. Formation and controls of channel networks. *Prog. Phys. Geog. 4*, 211–39.

Dury, G. H. 1959. *The face of the Earth*. Harmondsworth: Penguin.

El-Baz, F., L. Boulos, C. Breed, A. Dardir, H. Dowidar, H. El-Etr, N. Embabi, M. Grolier, V. Haynes, M. Ibrahim, B. Issawi, T. Maxwell, J. McCauley, W. McHugh, A. Moustafa and M. Yousif 1980. Journey to the Gilf Kebir and Uweinat, southwest Egypt, 1978. *Geogr. J.* **146**, 51–93.

Emery, K. O. 1962. *Marine geology of Guam*. US Geol. Surv. Prof. Pap. 403-B.

Emmett, W. W. 1968. Gully erosion. In *Encyclopedia of geomorphology*, R.W. Fairbridge (ed.), 517–9. New York: Reinhold.

Fanale, F. P. 1976. Martian volatiles: their degassing history and geochemical fate. *Icarus* **28**, 179-202.

Feininger, T. 1969. Pseudokarst on quartz diorite, Columbia. *Z. Geomorph.* **13**, 287–96.

Fenneman, N. M. 1923. Physiographic provinces and sections in western Oklahoma and adjacent parts of Texas. *US Geol. Surv. Bull.* 730, 115–34.

Fletcher, J. E., K. Harris, H. B. Peterson and V. N. Chandler 1954. Piping. *Trans. Am. Geophys. Un.* **35**, 258–262.

Garner, H. F. 1974. *The origin of landscapes*. New York: Oxford University Press.

Gerster, G. 1976. *Grand design*. New York: Paddington Press.

Gibbs, H. S. 1945. Tunnel-gulley erosion on the Wither Hills, Marlborough. *NZ J. Sci. Tech.* **27**(2A), 135–46.

Gilbert, G. K. 1886. The inculcation of scientific method by example. *Am. J. Sci.* **31**, 284–99.
Gilman, K. and M. D. Newson 1980. *Soil pipes and pipeflow – a hydrological study in upland Wales*. Br. Geomorphol. Res. Grp Res. Mon. 1.
Gilmour, D. A. and M. Bonell 1979. Runoff processes in tropical rainforests with special reference to a study in north-east Australia. In *Geographical approaches to fluvial processes*, A. F. Pitty, (ed), 73–92. Norwich: Geo Abstracts.
Gold, T. 1966. The Moon's surface.In *The nature of the lunar surface*, W. N. Hess, D. H. Menzel and J. A. O'Keefe (eds), 107–21. Baltimore: Johns Hopkins University Press.
Grant, U.S. 1948. Influence of the water table on beach aggradation and degradation. *J. Marine Res.* **7**, 655–60.
Gregory, H. E. 1917. *Geology of the Navajo country*. US Geol. Surv. Prof. Pap. 93.
Guilcher, A. 1958 *Coastal and submarine morphology*. Translated by B. W. Sparks and R. H. W. Kneese. London: Methuen.
Hall, C. M. and D. E. Willard 1905. *Description of Casselton and Fargo quadrangles*. US Geol. Surv. Folio 117.
Harrison, S. S. and L. Clayton 1970. Effects of groundwater seepage on fluvial processes. *Geol. Soc. Am. Bull.* **81**, 1217–26.
Heede, B. H. 1971. *Characteristics and processes of soil piping in gullies*. US Forest Serv. Res. Pap. RM–68.
Higgins, C. G. 1960. Ohlson Ranch formation, Pliocene, northwestern Sonoma County California. *Univ. Calif. Publns. Geol. Sci.* **36**, 199–232
Higgins, C. G. 1974. Model drainage networks developed by ground-water sapping. *Geol. Soc. Am. Abs. Prog.* **6**, 794-5.
Higgins, C. G. 1980. Nips, notches, and the solution of coastal limestone: an overview of the problem with examples from Greece. *Estuar. Coastal Marine Sci.* **10**, 15–30.
Higgins, C. G. 1982. Drainage systems developed by sapping on Earth and Mars. *Geology* **10**, 147-52.
Higgins, C. G. 1983. Reply: "Drainage systems developed by sapping on Earth and Mars". *Geology* **11**, 55–6.
Hinds, N. E. A. 1925. Amphitheater valley heads. *J. Geol.* **33**, 816–8.
Hutchinson, J. N. 1968. Mass movement. In *Encyclopedia of geomorphology*, R. W. Fairbridge (ed.), 688–95. New York: Reinhold.
Ireland, H. A., C. F. S. Sharpe and D. H. Eargle 1939. *Principles of gully erosion in the Piedmont of South Carolina*. US Dept. Ag. Tech. Bull. 633.
Jennings, J. N. 1979. Arnhem Land city that never was. *Geogr. Mag.* **51**, 822–7.
Johnson, D. W. 1938–39. Origin of submarine canyons. *J. Geomorph.* **1**, 111–29, 230–43, 324–40; **2**, 42–60, 133–58, 213–36.
Johnson, D. W. 1939. *The origin of submarine canyons*. New York: Columbia University Press.
Jones, [J. A.] A. 1971. Soil piping and stream channel initiation. *Water Resour. Res.* **7**, 602–10.
Jones, J. A. A. 1975. *Soil piping and the subsurface initiation of stream channel networks*. PhD dissertation. University of Cambridge.
Jones, J. A. A. 1981. *The nature of soil piping: a review of research*. Br. Geomorphol Res. Grp. Res. Mon. 3.
Khobzi, J. 1972. Erosion chimique et méchanique dans la genèse de dépressions "pseudo-karstiques" souvent endoréiques. *Rev. Géomorph. Dyn.* **21**, 57–70.
Kirkby, M. J. and R. J. Chorley 1967. Throughflow, overland flow and erosion. *Int. Assoc. Scientific Hydrol. Bull.* **12**, 5–21.
Komar, P. D. 1976. *Beach processes and sedimentation*. Englewood Cliffs, NJ: Prentice-Hall.
Komar, P. D. 1979. Comparisons of the hydraulics of water flows in Martian outflow channels with flows of similar scale on Earth. *Icarus* **37**, 156–81.
Laity, J. E. 1980. Sapping processes in Martian and terrestrial valleys. *EOS* **61**, 286–7.
Laity, J. E. and M. C. Malin in press. Sapping processes and the development of theater-headed valley networks on the Colorado Plateau. *Geol Soc. Am. Bull.*
Laity, J. E. and D. C. Pieri 1980. *Sapping processes in tributary valley systems*. NASA Tech. Memo. 81776, 295–7.

Leighton, R. B., N. H. Horowitz, B. C. Murray, R. P. Sharp, A. H. Herriman, A. T. Young, B. A. Smith, M. E. Davies and C. B. Leovy 1969. Mariner 6 and 7 television pictures: preliminary analysis. *Science* **166**, 49–67.

Leopold, L. B., M. G. Wolman and J. P. Miller 1964. *Fluvial processes in geomorphology*. San Francisco: W. H. Freeman.

Lewis, G. M. 1961. F. H. McBeth, Aerial photographic investigation of leaching and sapping as an erosion process. *Geomorph. Abs.* **6**, 6–7.

Lingenfelter, R. E., S. J. Peale and G. Schubert 1968. Lunar rivers. *Science* **161**, 266–9.

Löffler, E. 1974. Piping and pseudokarst features in the tropical lowlands of New Guinea. *Erdkunde* **28**, 13–8.

Malin, M. C. 1980. *Studies of fluvial, eolian, and sapping processes in Iceland*. NASA Tech. Memo. 81776, 300–1.

Marr, J. C. 1955. Sinkholes in irrigated fields. *Calif. Ag.* **9**(11), 6–7.

Masursky, H., J. M. Boyce, A. L. Dial, G. G. Schaber and M. E. Strobell 1977. Classification and time of formation of Martian channels based on Viking data. *J. Geophys. Res.* **82**, 4016–38.

Maxwell, T. A. 1979. *Field investigation of Martian canyonlands in southwestern Egypt*. NASA Conf. Publn 2072, 54.

Maxwell, T. A. 1982. Erosional patterns of the Gilf Kebir plateau and implications for the origin of Martian canyonlands. In *Desert landforms of southwest Egypt*, F. El-Baz and T. A. Maxwell (eds), Chap. 19 p. 281–300. NASA Contr Rep. 3611.

Maxwell, T. A., E. P. Otto, M. D. Picard and R. C. Wilson 1937. Meteorite impact: a suggestion for the origin of some stream channels on Mars. *Geology* **1**, 9–10.

McBeth, F. H. 1961. Aerial photographic investigation of leaching and sapping as an erosion process. *Photogram. Engng* **27**, 154–5.

McCauley, J. F., M. H. Carr, J. A. Cutts, W. K. Hartmann, H. Masursky, D. J. Milton, R. P. Sharp and D. E. Wilhelms 1972. Preliminary Mariner 9 report on the geology of Mars. *Icarus* **17**, 289–327.

Mears, B., Jr. 1968. Piping. In *Encyclopedia of geomorphology*, R. W. Fairbridge (ed.), 849. New York: Reinhold.

Milton, D. J. 1973. Water and processes of degradation in the Martian landscape. *J. Geophys. Res.* **78**, 4037–47.

Milton, D. J. 1974. Carbon dioxide hydrate and floods on Mars. *Science* **183**, 654–6.

Norris, R. M. 1968. Sea cliff retreat near Santa Barbara, California. *Calif. Div. Mines Geol. Min.Inf. Serv.* **21**, 87–91.

Nummedal, D. 1978. *The role of liquefaction in channel development on Mars*. NASA Tech. Memo. 79729, 257–9.

Parker, G. G. 1964. *Piping, a geomorphic agent in landform development of the drylands*. Int. Assoc. Scientific Hydrol. Publn **65**, 103–13.

Parker, G. G. 1968. *Officer's Cave, eastern Oregon, revisited*. Geol. Soc. Am. Spec. Pap. 101, 415.

Parker, G. G. and E. A. Jenne 1966. Piping and collapse structure failures associated with western highways of the United States. *Highway Res. Abs.* **36**(12), p. 115–6.

Parker, G. G. and E. A. Jenne 1967. Structural failure of Western highways caused by piping. *Highway Res. Rec.* **203**, 57–76.

Parker, G. G., L. M. Shown and K. W. Ratzlaff 1964. Officer's Cave, a pseudokarst feature in altered tuff and volcanic ash of the John Day formation in eastern Oregon. *Geol Soc. Am. Bull.* **75**, 393–401.

Peale, S. J., G. Schubert and R. E. Lingenfelter 1968. Distribution of sinuous rilles and water on the Moon. *Nature* **220**, 1222–5.

Peale, S. J., G. Schubert and R. E. Lingenfelter 1975. Origin of Martian channels: clathrates and water. *Science* **187**, 273–4.

Peel, R. F. 1941. Denudational landforms of the central Libyan Desert. *J. Geomorph.* **4**, 3–23.

Pichler, E. 1953. Boçorocas. *Soc. Brasil. Geol. Bol.* **2**, 3–16.

Pieri, D. C. 1980a. Geomorphology of martian valleys. In *Advances in planetary geology*. NASA Tech. Memo. 81979, 1–160.

Pieri, D. C. 1980b. Martian valleys: morphology, distribution, age, and origin. *Science* **210**, 895–7.

Rathjens, C. 1973. Subterrane Abtragung (Piping). *Z. Geomorph. Suppl.* **17**, 168–76.

Robb, J. M., D. W. O'Leary, J. S. Booth and F. A. Kohout 1982. Submarine spring sapping as a geomorphic agent on the East Coast Continental Slope. *Geol Soc. Am. Abs. Prog.* **14**, 600.

Russell, I. C. 1902. *Geology and water resources of the Snake River Plains of Idaho*. US Geol Surv. Bull. 199.

Schubert, G., R. E. Lingenfelter and S. J. Peale 1970. The morphology, distribution, and origin of lunar sinuous rilles. *Rev. Geophys. Space Phys.* **8**, 199–224.

Schumm, S. A. 1956. The role of creep and rainwash on the retreat of badland slopes. *Am. J. Sci.* **254**, 693–706.

Sellards, E. H. and H. Gunter 1918. *Geology between the Apalachicola and Ocklocknee rivers in Florida*. Fla. St. Geol Surv. 10th & 11th An. Reps.

Sharp, J. M., Jr. 1976. Ground-water sapping by freeze-and-thaw. Z. *Geomorph.* **20**, 484–7.

Sharp, R. P. 1973a. Mars: troughed terrain. *J. Geophys. Res.* **78**, 4063–72.

Sharp, R. P. 1973b. Mars: fretted and chaotic terrains. *J. Geophys. Res.* **78**, 4073–83.

Sharp, R. P. and M. C. Malin 1975. Channels on Mars. *Geol Soc. Am. Bull.* **86**, 593–609.

Sharp, R. P., L. A. Soderblom, B. C. Murray and J. A. Cutts 1971. The surface of Mars: 2. uncratered terrains. *J. Geophys. Res.* **76**, 331–42.

Simons, D. B. and E. V. Richardson 1966. *Resistance to flow in alluvial channels*. US Geol Surv. Prof. Pap. 422–J.

Small, R. J. 1961. *The morphology of Chalk escarpments: a critical discussion*. Trans Paps. Inst. Br. Geogs Publn **29**, 71–90.

Small, R. J. 1962. A short note on the origin of the Devil's Dyke, near Brighton. *Proc. Geologs Assoc.* **73**, 187–92.

Small, R. J. 1964. *The escarpment dry valleys of the Wiltshire Chalk*. Trans. Paps Inst. Br. Geogs Publn **34**, 33–52.

Small, R. J. 1965. *The role of spring sapping in the formation of Chalk escarpment valleys*. Southampton Res. Ser. Geog. **1**, 3–29.

Small, R. J. 1978. *The study of landforms*, 2nd edn. Cambridge: Cambridge University Press.

Smith, B. J. 1978. The origin and geomorphic implications of cliff foot recesses and tafoni on limestone hamadas in the northwest Sahara. *Z. Geomorph.* **22**, 21–43.

Smoluchowski, R. 1968. Mars: retention of ice. *Science* **159**, 1348–50.

Sparks, B. W. 1960. *Geomorphology*. New York: Wiley.

Sparks, B. W. and W. V. Lewis 1957–58. Escarpment dry valleys near Pegsdon, Hertfordshire. *Proc. Geologs Assoc.* **68**, 26–38.

Stearns, H. T. 1936. Origin of the large springs and their alcoves along the Snake River in southern Idaho. *J. Geol.* **44**, 429–50.

Stearns, H. T. and K. N. Vaksvik 1935. *Geology and groundwater resources of the island of Oahu, Hawaii*. Hawaii Div. Hydrography Bull. 1.

Stetson, H. C. 1936. Geology and paleontology of the Georges Bank canyons. Part 1. Geology. *Geol Soc. Am. Bull.* **47**, 339–66.

Terzaghi, K. 1950. *Mechanism of landslides*. Geol Soc. Am. Berkey Vol., 83–123.

USGS 1976. *Topographic map of the Coprates quadrangle of Mars*. US Geol Surv. Misc. Invest. Ser. Map I-976 (MC-18).

Vita-Finzi, C. and P. F. Cornelius 1973. Cliff sapping by molluscs in Oman. *J. Sed. Pet.* **43**, 31–2.

Wade, F. A. and J. N. De Wys 1968. Permafrost features on the Martian surface. *Icarus* **9**, 175–85.

Warn, F. 1966. Sinkhole development in the Imperial Valley. In *Engineering geology in Southern California*, R. Lung and R. Proctor (eds), 144–5. Los Angeles Sec. Assoc. Engng Geols Spec. Publn.

Wentworth, C. K. 1928. Principles of stream erosion in Hawaii. *J. Geol.* **36**, 385–410.

Wentworth, C. K. 1939. Marine bench-forming processes: II. solution benching. *J. Geomorph.* **2**, 3–25.
White, S. E. 1949. Processes of erosion on steep slopes of Oahu, Hawaii. *Am. J. Sci.* **247**, 168–86.
Wise, D. U., M. P. Golembek and G. E. McGill 1979. Tharsis Province of Mars: geologic sequence, geometry, and a deformation mechanism. *Icarus* **38**, 456–72.
Woodford, A. O. 1935. Rhomboid ripple mark. *Am. J. Sci.* 5th Ser. **29**, 518–25.
Wright, H. E., Jr. 1964. Origin of the lakes in the Chuska Mountains, northwestern New Mexico. *Geol Soc. Am. Bull.* **75**, 589–98.
Zeitlinger, J. 1959. Beobachtungen über unterirdische Erosion in Verwitterungslehm. *Öster. Geog. Gesell. Mitteil.* **101**, 94–5.

3
Near-surface groundwater and evolution of structurally controlled streams in soft sediments

Zeev Berger and Jacob Aghassy

Introduction

Geomorphic literature includes many papers describing relationships among groundwater, surface water and evolution of the drainage system. Most authors attribute drainage network evolution and initial channel erosion to surface runoff processes. Horton (1945), Chorley (1957), Strahler (1958) and others used quantitative techniques to describe morphology of drainage networks in various climatic regions. They emphasized that initiation of stream channels and the resulting drainage networks are related to relief, vegetation, lithology and other surface elements that control surface runoff.

In the past two decades, however, an increasing number of researchers have shown that drainage networks are initiated by and evolve under the influence of near-surface groundwater flows (Kirkby & Chorley 1967, Dunne 1980). It was demonstrated by Bunting (1961), Small and Lewin (1965) Löffler (1974) and others that the presence and movement of near-surface groundwater can promote continuous corrosion of bedrock and soil and thus create new drainage networks. This type of erosion, commonly called sapping or piping, has been reviewed by Chorley (1978), Dunne (1980) and Higgins (1984).

For many years, geomorphologists engaged in petroleum exploration have considered surface runoff and related erosional processes as the primary cause for the evolution of unique stream patterns that often reveal the presence and configuration of buried structures. They have postulated that buried structures with surface expressions too subtle to be detected by conventional mapping techniques can cause significant changes in surface runoff of low-gradient streams. These changes trigger both an adjustment of existing stream channels and the evolution of new tributaries. These local drainage features are often called "drainage anomalies" or "structurally controlled streams" (Lattman 1959).

In searching for buried structures in low-relief terrains, we have noted that structurally controlled streams are often characterized by deeply entrenched box-shaped valleys and other anomalous erosional features. Such landforms were developed in areas of intense flow of near-surface groundwater and abnormal concentration of ground moisture. We have learned that these conditions are caused by the local influence of buried structures on the otherwise uniform flow of near-surface groundwater, and that in such areas drainage systems often develop by piping and sapping processes resulting from subsurface flow rather than by channel erosion resulting from surface runoff.

The relationships among buried structure, near-surface groundwater, and drainage network evolution are examined in this paper. Examples cited are primarily from the Gulf Coast region of the United States, where structures are buried by a thick cover of recent sediments or soils. The water table in this area is near the surface, and after heavy rains, it merges with shallow subsurface flow, throughflow and surface flow. The term near-surface groundwater is applied to groundwater in the phreatic, saturated aquifer and to the vadose, unsaturated column of sediments and soils that cover the buried structures. Emphasis is placed on multi-altitudinal aerial photographs as a reconnaissance technique for detecting near-surface moisture concentrations and patterns related to buried structures.

Effect of buried structures on near-surface groundwater and related erosional processes

Movement of water in soils and unconsolidated sediments that constitute the unconfined aquifer of a basin takes place in four paths: surface runoff, shallow subsurface flow, throughflow and groundwater flow (Kirkby & Chorley 1967, Dunne 1980). Whereas surface runoff and related erosional processes are predominantly dictated by surface conditions, the other three paths of groundwater and related sapping and piping processes can also be affected by variations in subsurface conditions. Buried structures, such as faults and folds, often disrupt the uniform flow of near-surface groundwater and promote erosion for the following reasons:

(a) Buried structures containing strata of contrasting permeability to underlying sediments may become a path of intensive shallow subsurface flows. These flows promote erosion by piping and sapping, particularly adjacent to cliffs and stream banks where outflow of groundwater occurs (Terzahgi 1931, Dunne 1980) (Fig. 3.1).

(b) Increased fracturing and cross-formational flow over and in the vicinity of buried structures may focus and encourage vertical upward movement of groundwater, which in turn may cause a slight bulge in the water table

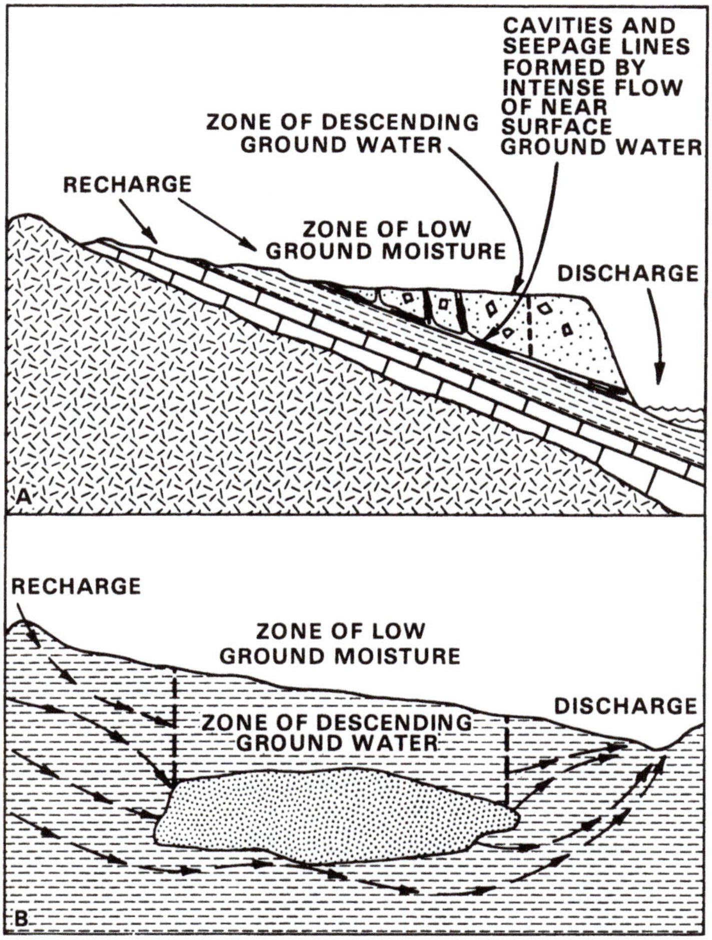

Figure 3.1 Schematic illustrations showing common geological settings, which promote a zone-of-descending-movement of near-surface groundwater and areas of low ground moisture. (A) Shallow groundwater flows along contact between bedrock and underlying sediments and soils. (B) Uniform flow and pressure of near-surface groundwater are disrupted by a lens of highly permeable sand enclosed in matrix of lower permeability (modified from Töth 1980).

(Kudrykov 1974, Töth, 1980) (Fig. 3.2). Excessive supply of moisture weakens the resistance of soils and sediments to weathering, and thus initiates and accelerates erosion by piping and sapping (see discussion by Higgins 1984).

Detection of abnormal moisture conditions related to buried structures

In low-relief terrains where buried structures have little or no topographic expression, the disruption in net movement of near-surface groundwater may produce areas of abnormal ground moisture. A relative deficiency in groundwater is expected in areas where rain water rapidly penetrates the ground

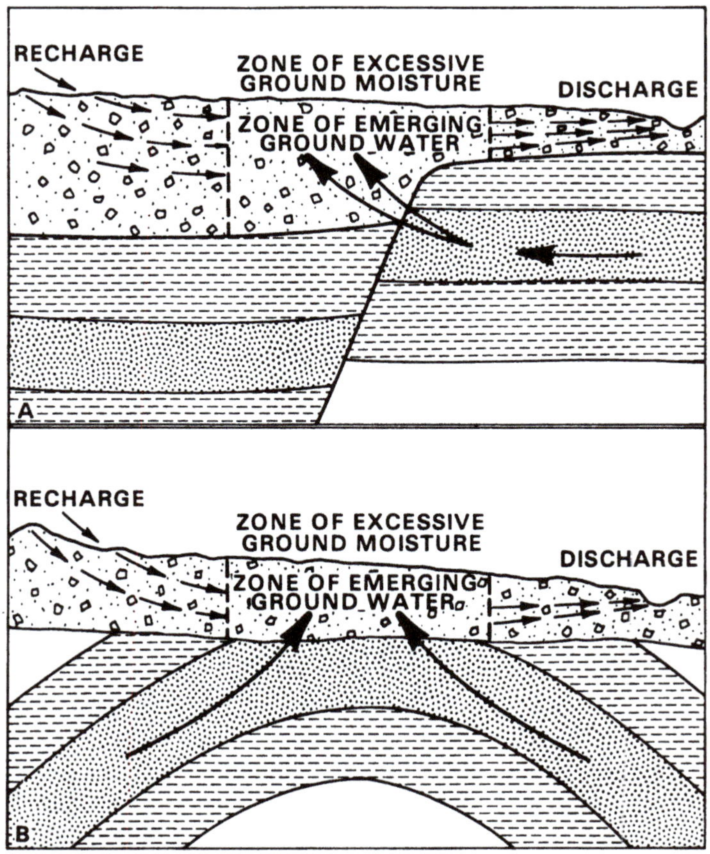

Figure 3.2 Schematic illustration showing common geological setting wherein cross-formational groundwater flows cause zone of emerging movement of groundwater and areas of excessive ground moisture (modified from Kudrykov 1974). (A) Excessive ground moisture occurs along a buried fault scarp. (B) Excessive ground moisture outlines the position of a buried breached anticline.

(A)

(B)

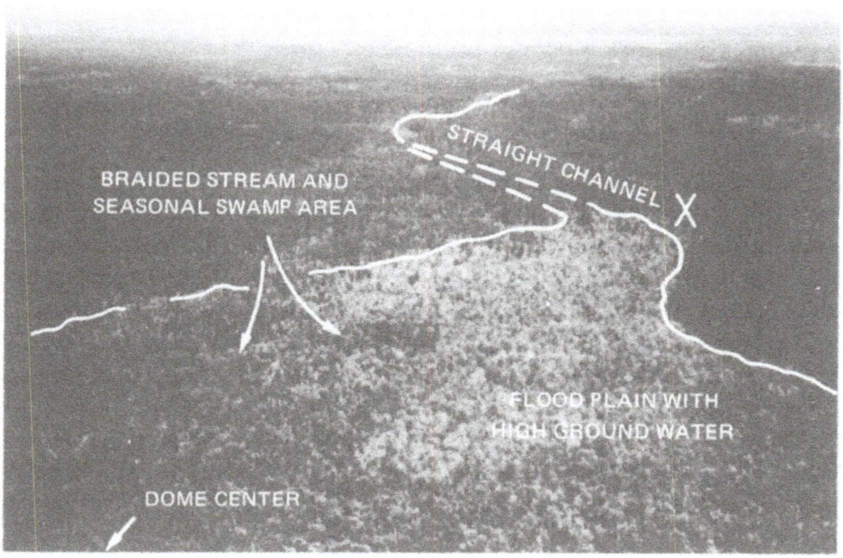

Figure 3.3 Low-altitude color-infrared photographs taken over two salt domes in interior Gulf Coast region, USA. These tectonically stable salt domes are buried under a few hundred feet of sediment and do not produce a clear topographic expression (Aghassy & Berger 1982). (A) Excessive ground moisture over Oakwood Dome, Texas, causing significant changes in channel and floodplain configuration, which are located at X. The excessive moisture conditions are usually depicted on infrared photographs as dark red color. (B) Excessive ground moisture over Cyprus Creek Dome, Mississippi, causing observable contrast in vegetation. The edge of the surface affected by the dome is located at X. The light red and pink vegetation marks the area of poor drainage over the breached dome.

through pipes and seepage lines, whereas excessive ground moisture is expected in areas of emerging water (particularly in humid environments where the water table is near the surface). In both cases, the local variations in near-surface moisture conditions produce observable contrasts in vegetation, soils, accumulations of dissolved minerals, and drainage network characteristics (Töth 1980).

The initial recognition of such conditions over large areas can be best accomplished by various remote-sensing techniques. Aerial photographs, Landsat imagery and other remote-sensing data have been used to define areas whose brightness (or spectral-reflectance) characteristics differ significantly from their surroundings (Sabins 1978). In low-relief terrain, differences in brightness may be caused by changes in surface roughness, but they are most likely caused by local variations in ground moisture and related changes in vegetation, soils, etc. In general, locally bright areas are caused by reduced ground moisture and vegetation whereas darker areas are caused by excessive soil moisture and an increase in vegetation (Figs 3.3 & 4).

Drainage network patterns and valley forms related to piping and sapping

Several authors have described the unique characteristics of drainage networks and valley forms in areas affected by near-surface groundwater and the related sapping and piping processes. Examples range from small-scale features, such as valley forms in the Colorado Plateau (Campbell 1973) to regional drainage networks in Western Highland Rim Peneplain in Tennessee (Stearns 1967). The process of sapping has recently received greater attention because it is proposed as an important mechanism for evolution of large, deep canyons on Mars (Higgins 1982).

Most authors point out that drainage networks and valley forms that develop predominantly by groundwater erosion differ considerably from fluvially eroded drainage systems. Some of these differences can be detected

Figure 3.4 An aerial photograph mosaic and drainage network map showing the relationships between ground moisture and drainage networks as they appear along Pleasant Ridge in Jones County, Mississippi. This long, straight ridge of the Williana deposits, flanked along its length by aligned tributaries and small ponds that suggest the existence of a structure (probably a fault) that is buried under this topographic ridge. (A) Excessive ground moisture is shown by dark tone area. (B) Drainage networks in the area of excessive moisture exhibit angular patterns that are typically formed by piping and sapping.

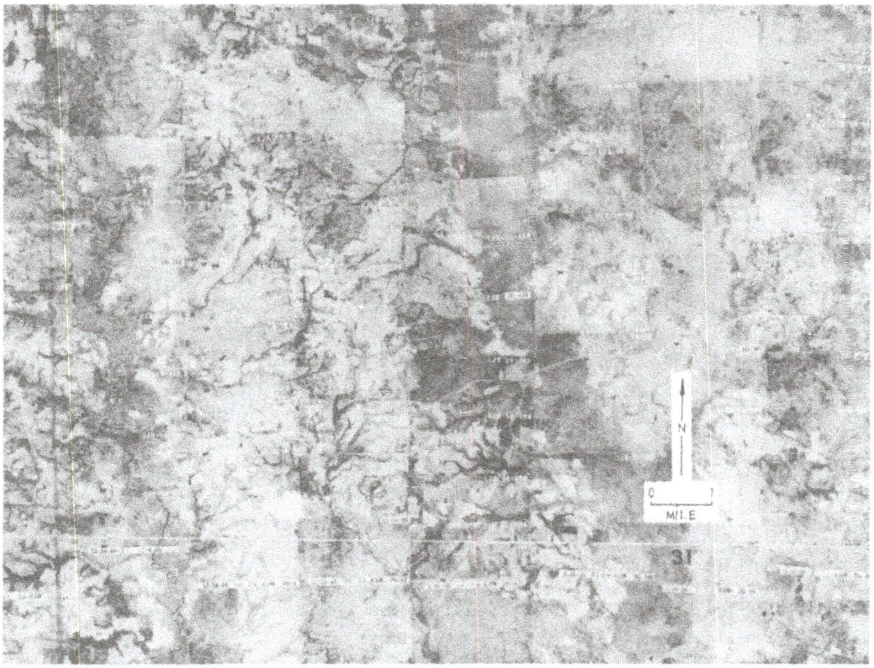
N
0
1
MILE
31

A

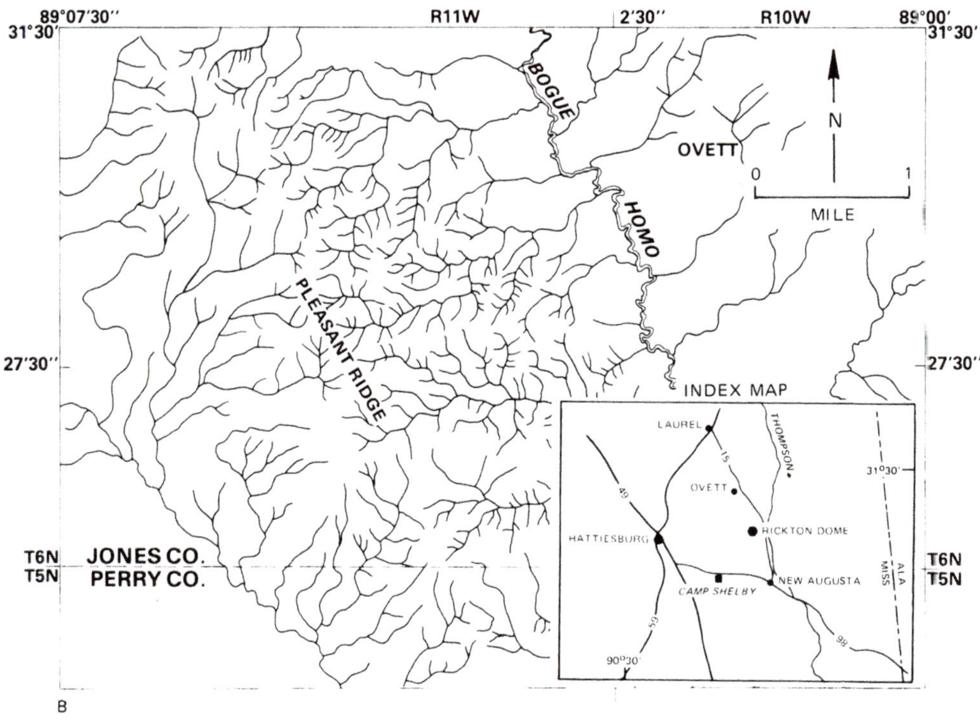
89°07'30"
R11W
2'30"
R10W
89°00'
31°30'
31°30'
BOGUE
HOMO
OVETT
N
0
1
MILE
PLEASANT RIDGE
27'30"
27'30"
INDEX MAP
LAUREL
THOMPSON
15
31°30'
OVETT
49
HATTIESBURG
RICKTON DOME
ALA
MISS
T6N
T5N
JONES CO.
PERRY CO.
T6N
T5N
CAMP SHELBY
NEW AUGUSTA
59
98
90°30'

B

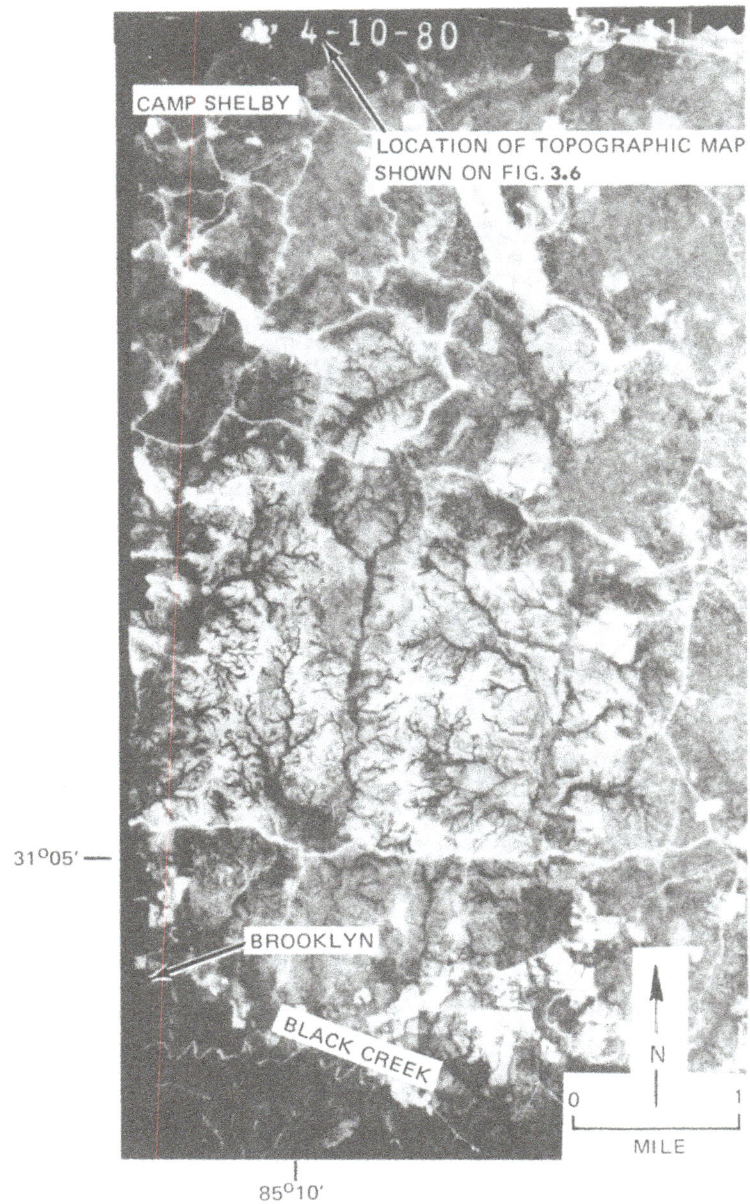

Figure 3.5 An aerial photograph of Camp Shelby Military Reservation in Perry County, Mississippi (see index map, Fig. 3.4). Parts of the camp area are being kept free from forest and vegetation cover to prevent fire during artillery training. Thus, the area provides an opportunity to observe on aerial photographs the unique drainage patterns that develop under the influence of sapping and piping processes.

by various photographic means and from topographic maps (Figs 3.5 & 6). They are summarized as follows:

(a) *Drainage density*: Parts of the basin that are affected by sapping and piping may have drainage networks of a density different from those of their surroundings, even though they are developed over relatively similar surface conditions (Baker 1980).

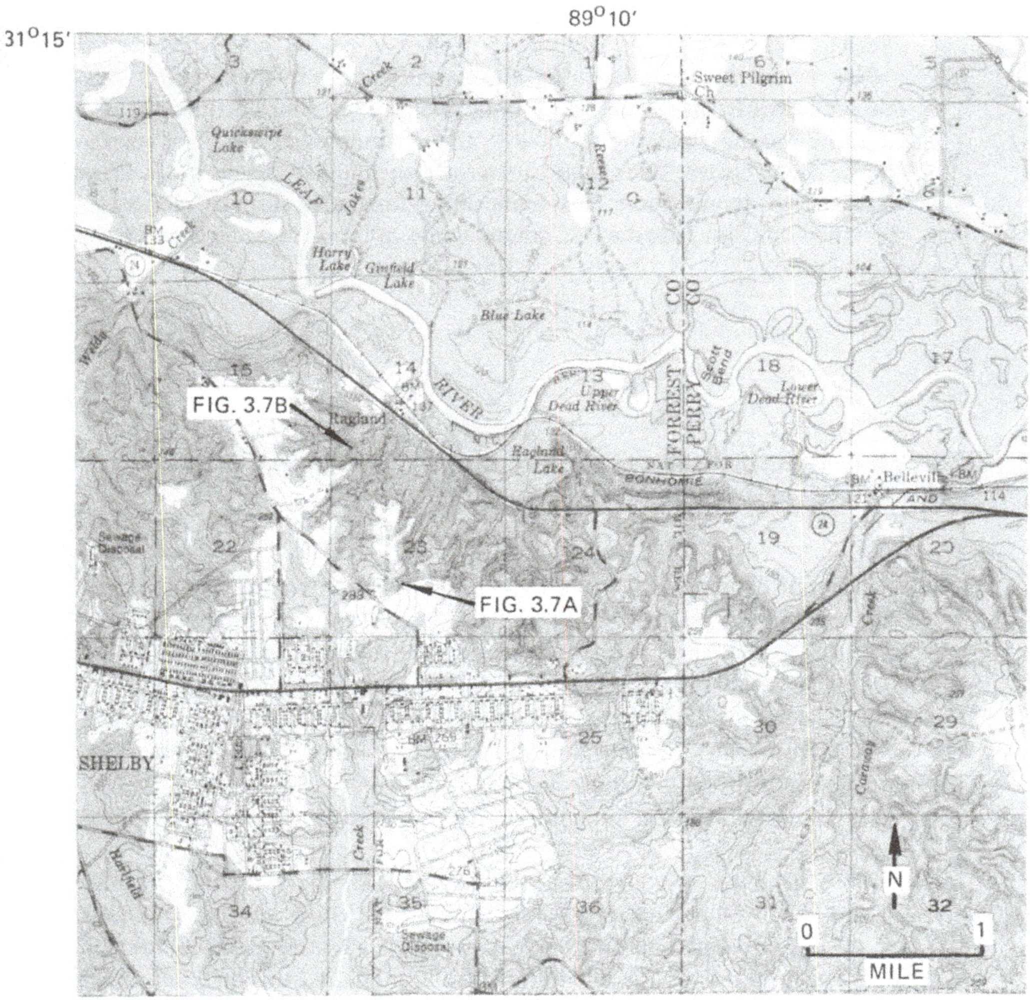

Figure 3.6 A topographic map of Camp Shelby showing an area that is covered by heavy vegetation. The cliff facing the Leaf River is being dissected by dense drainage network that developed primarily by sapping and piping processes. The geological controls that have triggered these erosional processes in Camp Shelby area are not known.

(b) *Drainage network patterns*: The dendritic drainage pattern that is common to low-relief sediments and soil-covered basins is notably absent in areas affected by sapping. More common are angulate drainage patterns. These are characterized by numerous short tributaries, which join the main trunk in high junction angles rather than at low angles (Pieri *et al.* 1980).
(c) *Drainage orientation*: Orientations of major trunks and other tributaries are often dictated by the attitude of the covered bedrock and by major joint systems. Thus, drainage orientation may be in contrast to the present topography and the regional drainage network orientation (Stearns 1967).
(d) *Valley forms*: Incised tributaries and main trunks with steep valley walls characterize areas affected by sapping and piping. Short tributaries join the major trunk in hanging valleys and terminate abruptly in amphitheater valley heads. Such valley forms are particularly developed in areas where active groundwater flow undermines the basal support and causes the valley to extend headward, owing to the collapse of valley-head walls (Small 1964).

Evolution of structurally controlled streams over a buried stream channel at Richton Salt Dome, Louisiana

The Richton Salt Dome in southeast Mississippi forms a positive dome-shaped mass that rises above its surroundings. The topography controls the orientation of drainage, which radiates 360° from the dome's center. The radial drainage pattern, however, is disrupted at the southwest portion of the dome area by drainage patterns that are different in density, orientation and texture from their surroundings (Fig. 3.7). Deeply entrenched, box-shaped valleys and other features that are reminiscent of those described by others as related to sapping and piping are observed on topographic maps (Fig. 3.8).

A shallow geologic cross section that was constructed from resistivity surveys and shallow boring programs shows that these unique drainage features are developed over a buried stream valley. (These paleogeomorphic features are common around salt domes throughout the interior Gulf Coast region; Berger *et al.* 1980.) The valley was formed in clay and sand units of the Miocene-aged Hattiesburg Formation and later buried by a thick blanket of coarse-grained sands and gravel of the Plio–Pliestocene-aged Citronelle Formation (Law 1980). The presence of a buried valley, however, has no apparent influence on the upland topography or distribution of the exposed

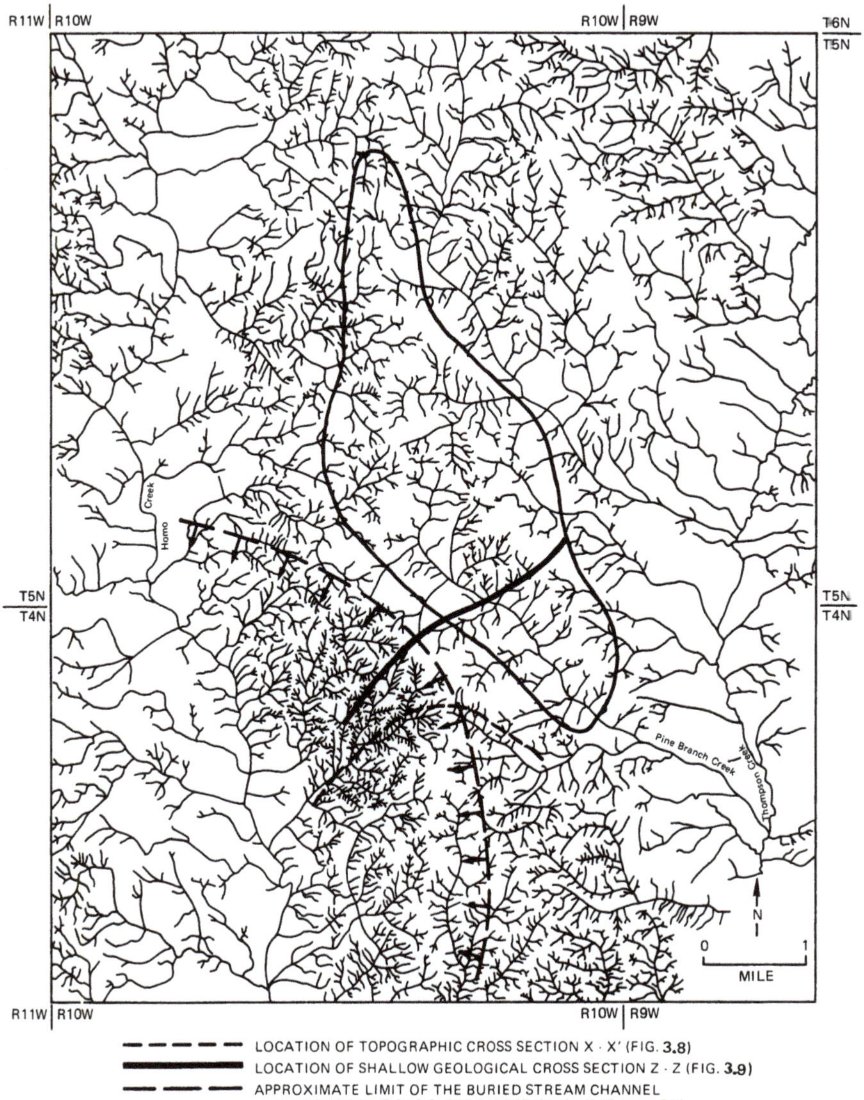

Figure 3.7 Drainage network map of Richton Salt Dome area, showing the radial drainage pattern that surrounds the area affected by the dome and the local "anomalous drainage" that occupies the area of the buried stream channel (after Law 1980).

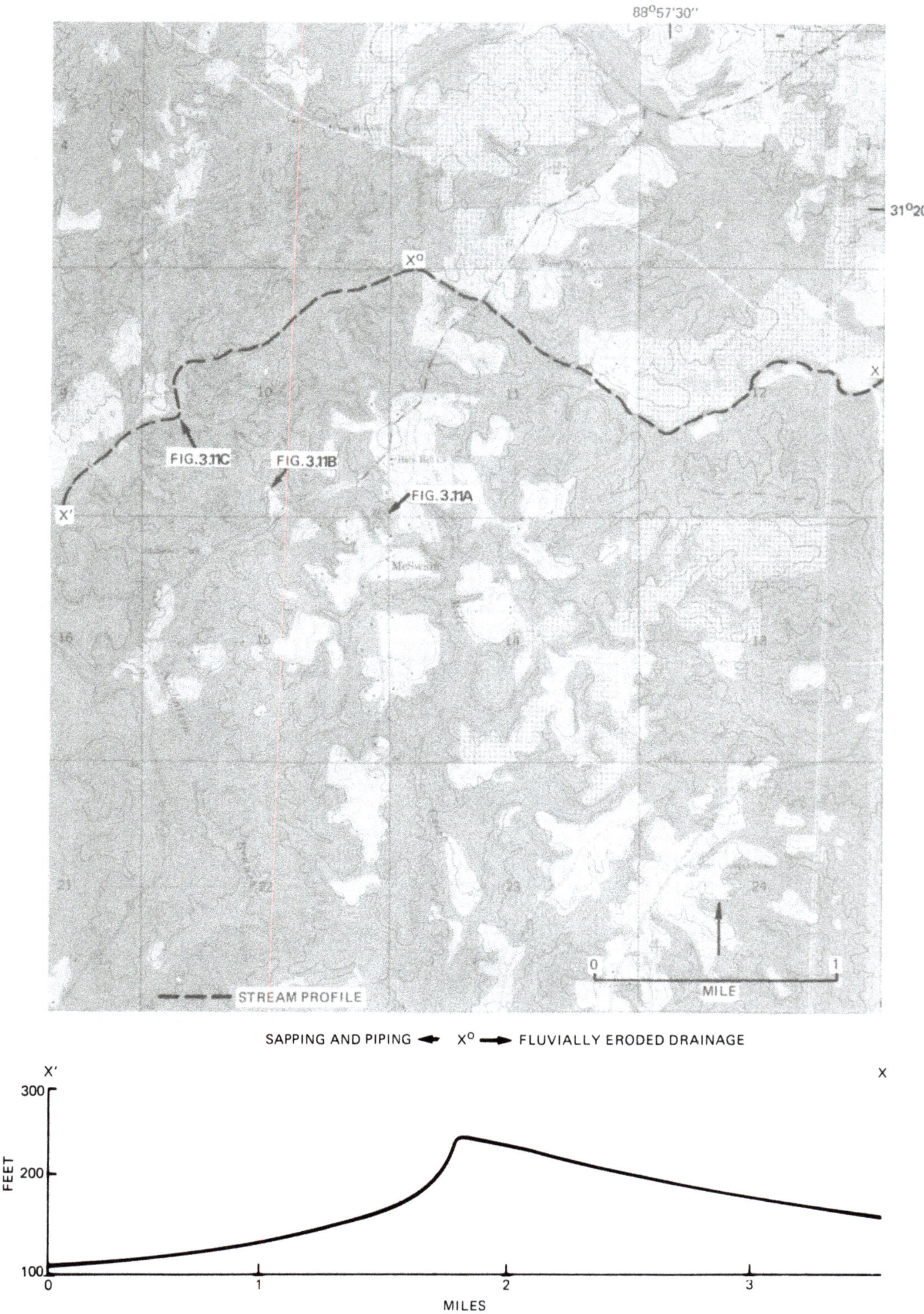

Figure 3.8 A topographic map and a stream profile of the southern part of Richton Dome area showing the distinct differences in topographic expression between fluvially eroded landforms and those that are formed under the influence of sapping and piping.

rock units (Figs 3.9 & 10); thus, surface runoff and related erosional processes cannot be the mechanism that causes the evolution of these unique drainage networks.

A field survey was designed to identify the processes that cause the evolution of anomalous drainage features over the buried stream valley. The survey that was carried out after heavy rains and a snow melt (January 1982) enabled us to trace the movement of water in several tributaries, from the drainage divide to the main stream at the bottomland. We observed that in spite of the excessive supply of rain water, there was little surface runoff in the upper tributaries. Most of the rain water quickly disappeared in seepage lines, pipes and other

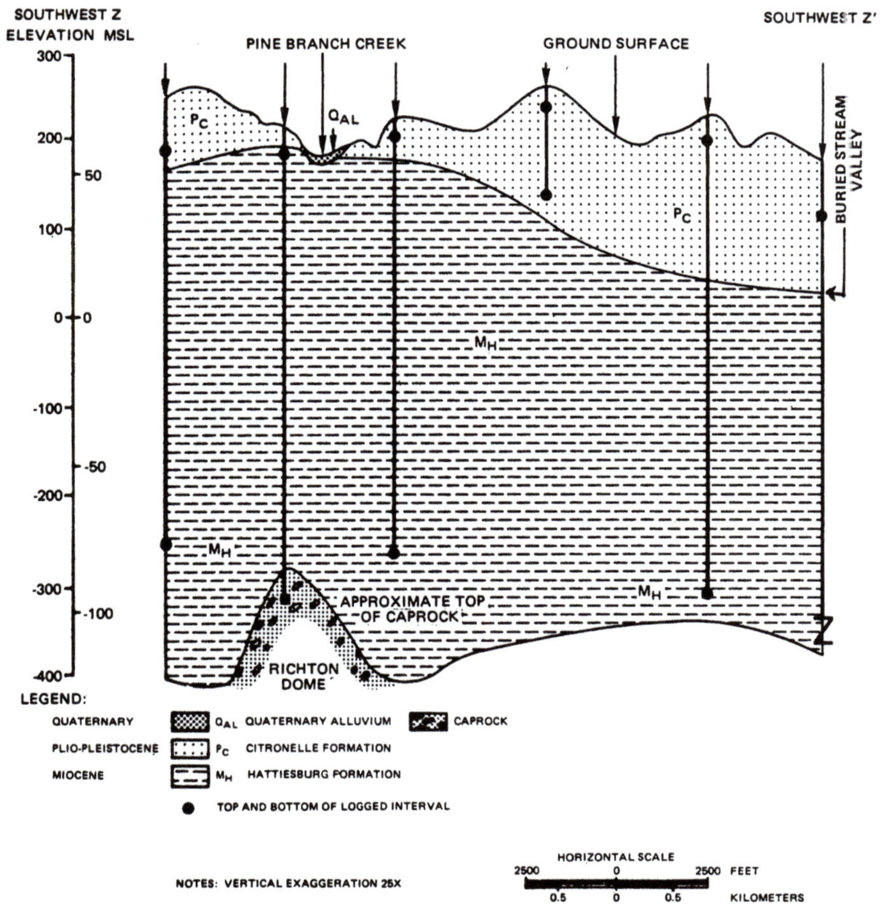

Figure 3.9 A shallow geological section showing the buried stream channel. Note that the upland topography and the exposed rock units are not affected by the buried stream channel (modified from Law 1980).

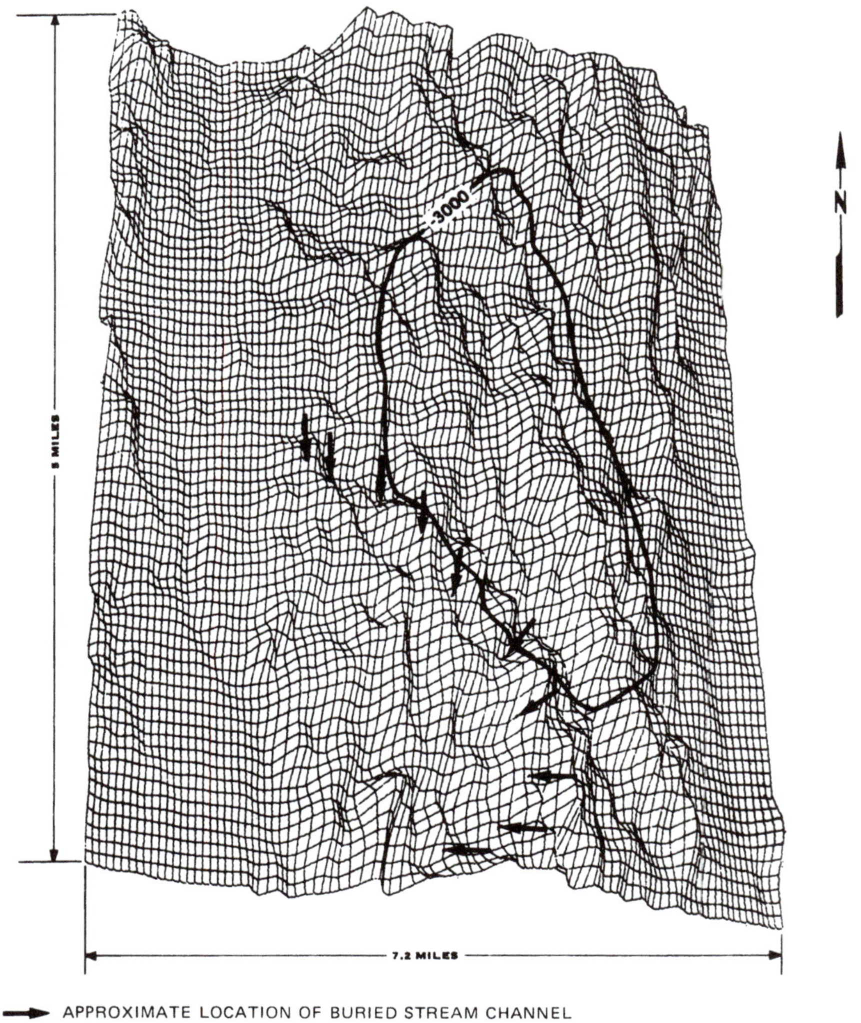

Figure 3.10 A three-dimensional topographic projection of Richton Salt Dome area, showing the positive land mass that was formed over the buried salt dome. The buried stream channel, however, has no apparent influence on the upland topography (after Law 1980).

openings in the ground (Fig. 3.11a). The water flowed downward until it reached the relatively impermeable beds of the buried stream valley and then moved downstream in subsurface tunnels (Fig. 3.11b). It reappeared as small springs or outflows in low areas at exposed contact of the stream valley with the underlying sedimentary fill (Fig. 3.11c).

Concerning erosional processes, we have observed that initiation and predominant extension of the upper tributaries result from collapse of subsur-

(A)

(B)

(C)

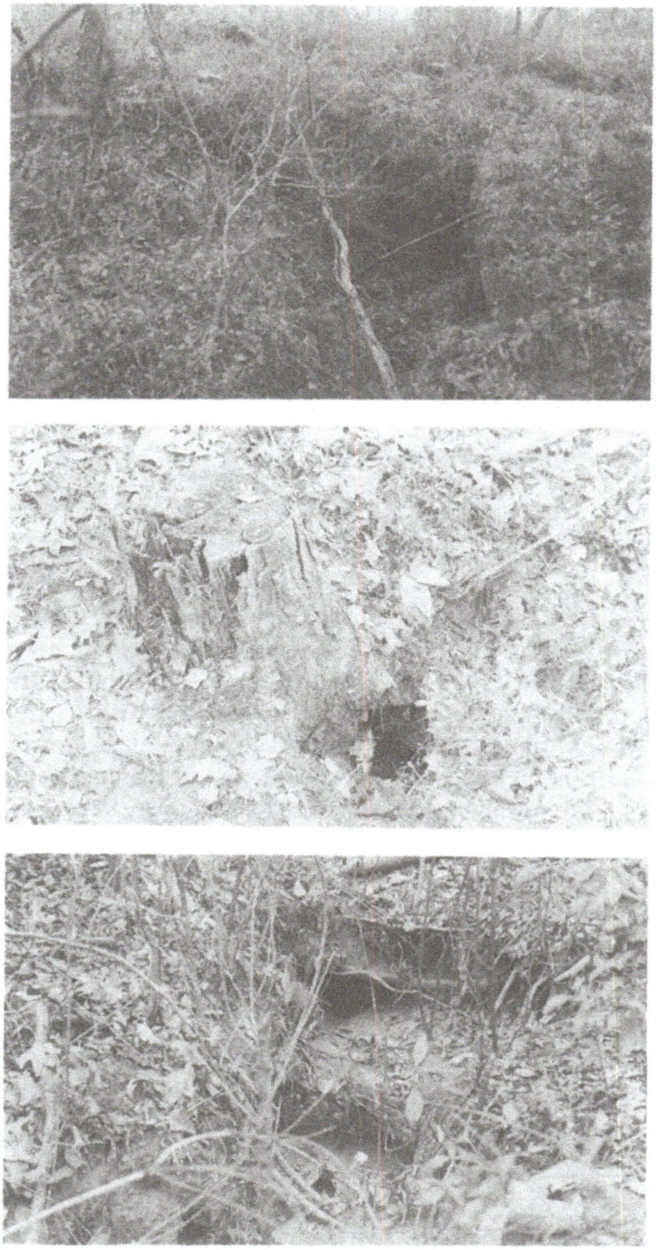

Figure 3.11 Sequential photographs taken from the drainage divide to the bottomland in tributaries that developed over the buried stream channel. (A) Initiation and expansion of upper tributaries occur primarily due to collapse of pipes, subsurface tunnels and other openings in the ground. (B) Surface waters rapidly penetrate the ground and form subsurface tunnels along the less-permeable beds of the buried valley. (C) The water reappears at the bottomland as seeps and springs. Groundwater flow triggers headward erosion of valleys that extend upstream along existing subsurface tunnels.

face tunnels or extension of pipes (Fig. 3.12,A). The intensity and orientation of newly formed tributaries are controlled primarily by the position and configuration of the buried stream valley (Fig. 3.12,B). The most severe erosion takes place in the bottomlands where intense outflow of subsurface water and related sapping processes cause headward growth of the valley. The direction in which the valley expands upstream is largely controlled by the position and configuration of the buried stream valley (Fig. 3.12,C).

We conclude that newly formed channels are created by upward propagation of subsurface erosional features caused by intense flow of near-surface groundwater, rather than by downcutting of stream valleys due to surface runoff. Subsequently, drainage patterns are controlled by, and often mimic, the configuration of the buried structure rather than being controlled by upland topography and/or surface lithology.

Conclusions

Evolution of unique drainage patterns and valley forms over and around buried structures is a subject of long-standing interest in geomorphology (Tator 1958, Lattman 1959, Berger & Aghassy 1980). Buried structures are believed to produce subtle topographic expression, which in turn, affects runoff regimes of low-gradient streams. The topographic expression of buried structures is attributed to differential compaction of the underlying sediments and to reactivation of the buried structures caused by differential loading and renewed regional stress (Norman 1976, Berger 1982).

We believe that the effect of buried structures on near-surface groundwater flow and related erosion by sapping and piping is also an important mechanism that enhances the evolution of structurally controlled streams over and around buried structures. Sapping and piping are particularly important as mechanisms that produce anomalous drainage patterns and valley forms over buried structures that have no clear topographic expression. Such processes are likely to be the most effective in humid and subhumid low-relief terrains where the water table is high and frequently fluctuates during wet and dry seasons or after heavy rains. This phenomenon, however, is not fully recognized because erosional features that are related to sapping and piping are often masked by the results of surface runoff and creep and by heavy vegetation cover.

We recognize that in most cases drainage anomalies over and around buried structures, as well as abnormal concentrations of ground moisture, are related to both topographic controls on surface runoff and subsurface controls on near-surface groundwater flow. Nevertheless, recognition of the two different erosional processes and their resulting landforms should help us understand the mechanisms that develop structurally controlled streams and help us improve geomorphic techniques for detecting buried structures in low-relief terrains.

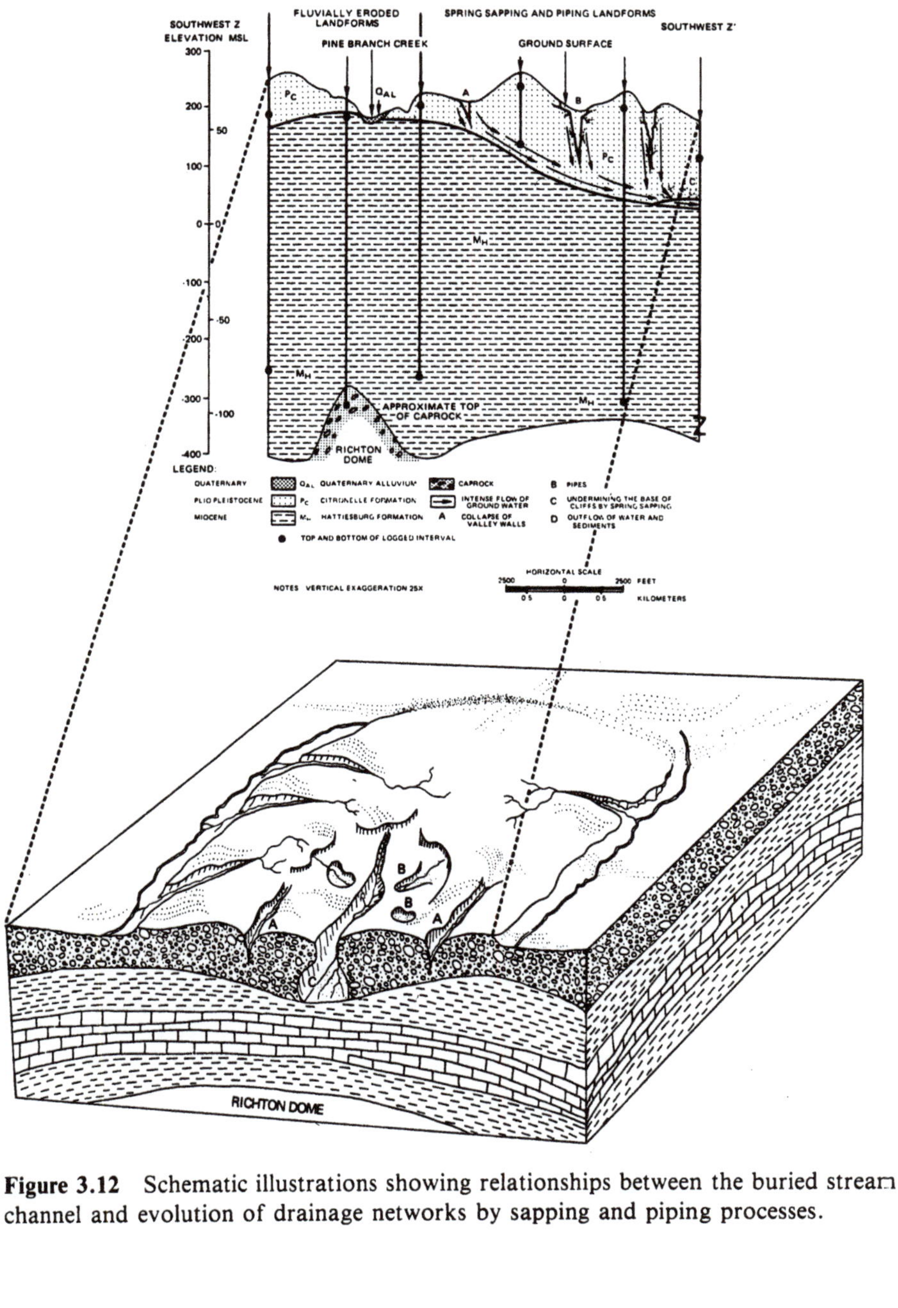

Figure 3.12 Schematic illustrations showing relationships between the buried stream channel and evolution of drainage networks by sapping and piping processes.

Acknowledgements

We thank R. L. Kite, K. E. Green, C. C. Wielchowsky and H. R. Hopkins for discussions and reviews of various stages of this paper, and O. F. Swenson for providing geological data from the Gulf Coast Salt Dome Project.

References

Aghassy, J. and Berger, Z. 1981. An application of side-looking color infrared (photography) for structure detection in subtle topography. *Proc. Fifteenth Int. Symp. Remote Sens. Environ.* 491–8. Ann Arbor, Mich.: Environmental Research Institute of Michigan.

Baker, V. R. 1980 *Some terrestrial analogs to dry valley systems on Mars.* NASA Tech. Memo. TM 81776, 286–8.

Berger, Z. 1982. The use of Landsat data for detection of buried and obscured geological structures in the East Texas basin, U.S.A. *Proc. Second Thematic Conf. Remote Sens. Explor. Geol.,* Ann Arbor, Mich.: Environmental Research Institute of Michigan.

Berger, Z. and J. Aghassy 1980. Geomorphic manifestations of salt dome stability. In *Applied geomorphology*, R. G. Craig and J. L. Craft (eds), 72–84. London: George Allen & Unwin.

Berger, Z., C. Joyce, A. Simcox and J. Sullivan 1980. The influence of differential erosion on the stratigraphic sequence near salt domes in northern Louisiana. *Geol Soc. Am. Abs. Prog.* **12**(7), 386.

Bunting, B. T. 1961. The role of seepage moisture in soil formation, slope development, and stream initiation *Am. J. Sci.* **259**, 503–18.

Campbell, I. A. 1973. Controls of canyon and meander forms by jointing *Area* **5**(4), 291–6.

Chorley, R. J. 1957. Climate and morphometry. *J. Geol.* **65**, 628–38.

Chorley, R. J. 1978. The hillslope hydrological cycle. In *Hillslope hydrology*, M. J. Kirby (ed.), 1–42. Chichester: Wiley.

Dunne, T. 1980. Formation and controls of channel networks. *Prog. Phys. Geog.* **4**, 211–39.

Higgins, C. G. 1982. Drainage systems developed by sapping on Earth and Mars. *Geology* **10**(3), 147–52.

Higgins, C. G. 1984. Piping and sapping: development of landforms by groundwater outflow. In *Groundwater as a geomorphic agent,* R. G. LaFleur (ed.), 18–58. Boston: Allen & Unwin.

Horton, R. E. 1945. Erosional development of streams and their drainage basins; hydrophysical approach to quantitative geomorphology. *Geol Soc. Am. Bull.* **56**, 275–376.

Kirkby, M. J. and R. J. Chorley 1967. Throughflow, overland flow and erosion. *Bull. Int. Assoc. Sci. Hydrol.* **12**, 5–21.

Kudryakov, V. A. 1974. Piezometric minima and their role in the formation and distribution of hydrocarbon accumulations. *Dokladv. Acad. Sci. USSR* **207**, 240–2.

Lattman, L. H. 1959. Geomorphology applied to oil exploration. *Min. Indus.* **28**(6), 1–4.

Law Engineering Testing Co. 1980. *Geological area characterization Gulf Coast Salt Dome Project, Mississippi Area.* Prepared for Batelle Memorial Institute Office of Nuclear Waste Isolation.

Löffler, E. 1974. Piping and pseudokarst features in the tropical lowlands of New Guinea. *Erdkunde* **28**, 13–8.

Norman, J. W. 1976. Photogeological fracture analysis as a subsurface exploration technique. *Trans. Inst. Min. Metal.* **85**, 852–61.

Pieri, D. C., M. C. Malin and J. E. Laity 1980. *Sapping: network structure in terrestrial and Martian valleys.* NASA Tech. Memo. TM-81979, 292–3.

Sabins, F. F. 1978. *Remote sensing principles and interpretation.* San Francisco: W. H. Freeman.

Small, R. J. 1964. *The escarpment dry valleys of the Wiltshire Chalk.* Trans. Paps Inst. Br. Geogs Publn M34, 33–52.

Small, R. J. and J. Lewin 1965. The role of spring sapping in the formation of chalk escarpment valleys. *Southampton Res. Ser. Geog.* **1**, 3–29.

Stearns, R. G. 1967. Warping of the western Highland Rim Peneplain in Tennessee by groundwater sapping. Geol Soc. Am. Bull **78**, 1111–24.

Strahler, A. N. 1958. Dimensional analysis applied to fluvially eroded landforms. *Geol Soc. Am. Bull.* **69**, 279–300.

Tator, B. A. 1958. The aerial photograph and applied geomorphology. *Photogram. Engng* **24**(4), 549–61.

Terzahgi, K. 1931. Earth slopes and subsidence from underground erosion. *Engng News-Record,* 90–2.

Töth, J. 1980. Cross-formation gravity-flow of groundwater: a mechanism of the transport and accumulation of petroleum (the generalized hydraulic theory of petroleum migration). In *Problems of petroleum migration*, AAPG studies in geology No. 10. W. H. Roberts and R. J. Cordell (eds), 121–67.

4
Landforms and soils of the humid Tropics

Antonio V. Segovia and John E. Foss

Introduction

The relationship of landscapes and soils includes both dynamic and static components. Landscapes result from interaction of processes with materials that make up the uppermost part of the Earth's crust. Conversely, landscapes, in part determine the background that controls rate (intensity of action) of pedogenic processes. Among factors comprising this determining background are: nature of underlying sediments or bedrock (parent material of the pedologists), topography, age (time of action of processes above the threshold of effective intensity), groundwater conditions, geologic structure and tectonism, elevation, exposure and climate. These factors are not necessarily independent.

The factors listed above are commonly grouped together by pedologists in the first four factors of the general soils equation (Jenny 1941): SOIL = *F*(Parent Material, Time, Topography or Slope, Climate, and Biota).

Processes that lead to differentiation of horizons in a soil profile are grouped by pedologists into four categories, i.e. the "four basic kinds of changes" noted by Simonson (1959): additions, removals, transfers or translocations, and transformations. These changes are concerned mostly with the last four terms of the general soils equation. The first one, parent material and its emplacement, falls mostly in the realm of geologic or geomorphologic studies.

Factors and processes involved in the genesis and evolution of landscapes and soils can affect each other, sometimes synergistically and at other times antagonistically. For example, low relief can block or retard groundwater flow, which in turn reduces or even stops the rate of solution of carbonates. In other cases, an excess of a component of the parent material, dissolved in the circulating soil water, may reduce or block the solubility of another component, thereby producing marked variations in soils across facies boundaries in the parent material. Similarly, interactions among background factors and climate-controlled processes cause striking variations in landscapes and soils as one crosses climatic zones.

Most, if not all, of the processes that act upon surface materials in the temperate zone are also active in the tropical zone, but their rates of action and effectiveness are usually different. There may exist, however, some chemical thresholds that are crossed only in a given environment or, to be more rigorous, for which the great majority of the crossings occur in a given environment. This may be the case, e.g. with thresholds that intervene in the formation of bauxite deposits in the humid Tropics (Bishopp 1954).

Because landscapes represent a visible synthesis of many of the same factors and processes that are responsible for the formation of soils, an interpretation of landscapes is a very useful tool in soil surveying and evaluation, and should precede soils studies. Conversely, an investigation of soils, present and past, is also extremely useful in understanding the development of landscapes. Recent advances in the technology of remote sensing, and general availability of remotely sensed information, in the form of images and tapes, make evaluation of landscapes relatively inexpensive. In addition, the "remote sensing revolution" also provides soil scientists and geomorphologists with accurate and reliable base maps, as well as indirect or direct data on mineral and chemical composition, vegetation, and soil moisture conditions.

Morphogenic and pedogenic differences in temperate and humid-tropical zones

Figure 4.1 is a schematic diagram of the relationships among climatic factors and soil types. This chapter is concerned with the landscapes and soils that fall

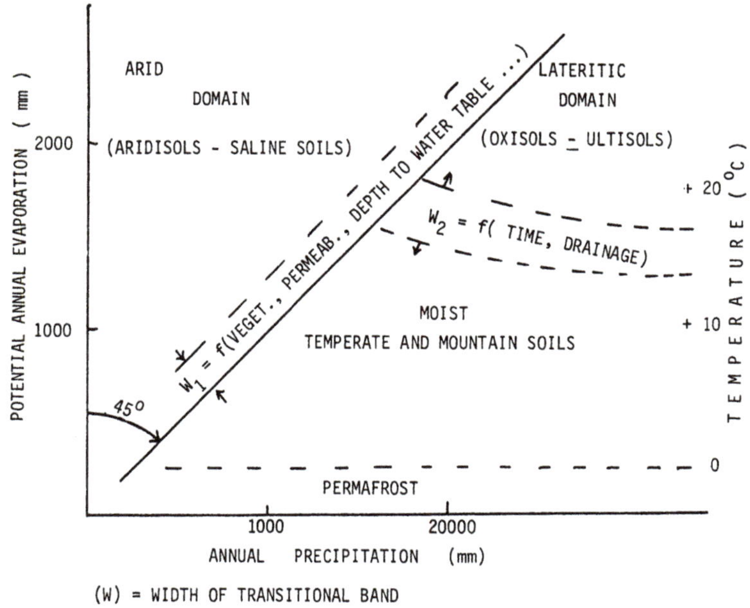

Figure 4.1 Climatic factors and general soil types.

within the "lateritic domain" of the graph. It may be helpful, however, to summarize in the following list the most important differences in morphogenic processes between temperate and humid-tropical environments:

(a) The effect of chemical processes of weathering is more visible and important in the tropical zone than in the temperate zone; higher mean annual tropical temperatures are responsible for higher velocities of chemical reactions. Deeper soil profiles, displaying very advanced leaching, develop in a shorter time in the Tropics than in the higher latitudes (Young 1976), hence, the common presence of oxisols and ultisols on landscapes of relatively young geologic age, some as young as 1–2 million years.
(b) The mechanical, abrasive action of gravels and very coarse sands in stream beds is usually missing or much less intense in the tropical zone than in temperate regions. Advanced chemical breakdown, deep soil profiles in areas of sediment provenance and lack of frost action account for a predominance of fine sediments in the bed load of tropical streams. As a result, adjustment of longitudinal stream profiles to changes in base level by abrasion of the bed takes longer than in the temperate zone. Rapids and waterfalls tend to persist on the tropical landscape for a long time (Tricart 1965).
(c) In littoral tropical areas, presence of coral and algal reefs, as well as beachrock, tends to retard beach erosion by wave action.
(d) Because temperatures are continuously above freezing, a micro- and macroscopic biota different from that of the temperate zone can subsist and prosper. Some of these organisms appear to play very active roles in modifying and stabilizing soils and landscapes.
(e) Some of the products of advanced pedogenesis, such as lateritic crusts, influence landscape development in the Tropics.
(f) Because of the greater distance of the areas affected by Pleistocene continental ice masses to the tropical zone than to the temperate zone, the effects of sea-level stands shown by paleo-littoral features between sea level and +200 m are generally free from distortions caused by Pleistocene and post-Pleistocene isostatic rebound, and thus are easier to identify and trace in tropical areas than in the higher latitudes.

Pedologic and morphologic investigations in the Tropics

Access to many tropical areas is costly and time consuming. Aerial photographs have become available for some tropical regions in the past 40 years, but access to them has often been restricted. In the past 10 years, images obtained by Side-Looking Radar (SLAR) and satellites (notably LANDSAT) have also become available. These tools provide usable images at scales of up to 1 : 25 000 for SLAR and 1 : 60 000 for commercially available satellite scenes. Thus, it is now possible to get a much better look at many tropical areas than it was 20 years ago. Our understanding of tropical soils and landscapes is improving, and some surprises are beginning to emerge. There will probably be many more in the future. The bibliography presented at the end of this chapter is a limited one but includes papers not cited in the text. The reader is referred to the references included in these papers for a more comprehensive view of the topic and of the extensive research on tropical pedology and geomorphology carried out by many investigators from European, Asian, Australian and Latin American institutions.

One of the promising areas of research for geomorphologists is that of Pleistocene sea-level changes. Satellite and radar images reveal bands of sand accumulation, at various levels from sea level to about 160 m, ringing the Orinoco and Plata river basins. The bands appear to represent dune systems and beach deposits associated with shorelines that range in age from late Tertiary to late Pleistocene. Accumulations of sands of probably similar age and genesis have also been reported from the Amazon Basin. In addition, successions of alluvial fans and delta fans have also been observed at various levels, formed in the past by ancestors of present-day tributaries of the Paraguay, Parana and Orinoco rivers. Because these areas were far enough from the centers of continental ice accumulation, the effect of crustal rebound after ice retreat should be minimal or non-existent in the tropical zone, removing one source of variation from the problem. The resulting conclusions should be of interest to investigators of coastal-plain development in temperate zones as well.

Tropical areas that lie at an elevation above 200 m are usually areas of erosion; whereas, those below that level are generally areas of aggradation, subjected to flooding, or have been so in the last 2 million years. A more rigorous separation can be made on the basis of drainage slope and distance from the point where the stream reaches base level. The percentage of land areas in both categories is very dissimilar for South America and Africa. A much greater percentage of the area lies below 200 m in tropical South America than in Africa. The older, strongly leached soils are present, in general, above 100 m in elevation. This difference suggests that soil resources and management – from the agronomist's point of view – must be very different for both continents.

Another promising area in tropical studies is that of reevaluation of soil resources, to eliminate the bias introduced into studies carried out in previous generations by factors such as inaccessibility, environmental health problems, and the preferential location of population centers in either coastal areas or in high and healthier terranes. Knowledge of low-lying tropical hinterlands is generally very scanty. This has resulted in serious errors in evaluating potential agricultural resources of some regions, and in addressing ecological problems caused by development in others. Until recently, e.g., it was thought that soils amenable for agricultural development represented only about 2% of the area of Venezuela. This was true for the northern part of the country, covered by existing soil surveys; but it neglected to take into account the large region between the Apure and Meta rivers, which was, and still is, largely inaccessible. Recent studies of satellite images reveal that the soils in that area previously excluded from pedologic investigations may represent a greater agricultural resource than all of the soils previously studied.

A popular misconception is that the soils of the Amazon Basin, if cleared from vegetation and put into agricultural use, would soon turn into a brick-like material that would render them unusable for agriculture, and the region would turn into a desert. In reality, it is very probable that a considerable part of the soils in that region are not lateritic and if properly managed, many millions of acres could be put into efficient and ecologically sound agricultural use (Sanchez & Buol 1975). On the other hand, the rates of soil erosion observed in some areas of the developing world are alarming (Soler 1981).

Joint landscape–soils studies

A joint or interdisciplinary geomorphology–soils approach to the study of landscapes has been very rewarding in the past, notably by Ruhe (1956, 1975), Ollier (1959), Ruhe *et al.* (1967), Lepsch *et al.* (1977), Sombroek (1966) and others. A general classification of landforms prepared for a multidisciplinary study normally includes three categories; (a) landforms of aggradation, in which the parent materials of soils are relatively recent sediments: floodplains, deltas, alluvial fans, bolsons, and dune fields, (b) erosional landforms, usually characterized by residual soils: pediplains, pediments, inselbergs, monadnocks, spurs, valleys, hogbacks, and areas of wind deflation, and (c) mixed or composite landforms, normally associated with polygenetic or polymorphic soils: areas in which the overall morphology indicates aggradation or degradation, but in which the topmost soil horizon shows predominance of processes of the opposite sign. This apparent contradiction stems from the fact that some short-lived (from a geologist's point of view) phenomena, such as climatic fluctuations, volcanic eruptions, etc., may have important effects on soils. In this category are included, e.g., erosional landscapes blanketed by volcanic ash (Soler 1981), or by layers of loess or wind-blown sand, scalped alluvial fans, and some bajadas or colluvial aprons. The composition of the underlying bedrock is not usually an important factor in understanding soil properties in areas of aggradation, whereas it may be a key factor in areas of erosion. On the other hand, the fertility of soils found in a landscape that presents strong overall erosional features may be surprisingly good if these features were blanketed by volcanic ash.

Erosion of tropical soils

In areas of patchy vegetation and in cultivated areas, tropical rains are more erosive than rains of the temperate zone, because tropical rains are of higher intensity (Lal 1977). In the undisturbed areas of the humid Tropics, thick vegetation buffers the mechanical action of raindrops. However, in the seasonally dry Tropics, at the beginning of the wet season, and where vegetation has been removed by man in the humid Tropics, the erodibility of soils is very great.

From the point of view of erosion, soils in the Tropics can be grouped into the following three classes (Sanchez 1976): (a) highly aggregated oxisols, andepts and oxidic families of other orders characterized by good internal drainage, and aggregates coated with oxides and organic matter, or in the case of volcanic debris, aggregates of small particles fused together. Soils in this class are generally resistant to erosion by runoff but are easily leached. (b) ultisols and alfisols with sandy topsoil textures. They can become severely eroded if uncovered, especially in slopes. (c) loamy to clayey soils (alfisols, ultisols, inceptisols and vertisols). They include soils that have great shrink–swell potential, are easily washed away in slopes, are prone to slides and mass movement in humid areas, and tend to develop salinity problems in semi-arid and seasonally arid regions.

In the semi-arid Tropics where potential evaporation is greater than precipitation, and in some karstic terranes where precipitation is greater than evapotranspiration but where most of the water is quickly lost to underground fissures, there is little or no through-flowing surface drainage, and there is no removal of materials from the basin. Salinity tends to increase in these soils. Salinity, in turn, affects vegetation. These areas are eventually characterized by a very specific type of vegetation adapted to high salinity, e.g., the "tintal" association in the "bajos" (lowlands) of northern Peten, Guatemala, and the "quebracho-palo santo–algarrobo" association of the Chaco plains of southeastern Bolivia and northwestern Paraguay.

Effect of climatic fluctuations on tropical landscapes

Climatic fluctuations could explain the presence of inselbergs and tors separated by wide areas of rotted rock described by Ollier (1959) from eastern and northern Uganda. Ollier interpreted that landscape as the result of weathering below the Gondwana erosional surface of King (1967) and later erosional carving of the African surface at about the end of the Tertiary, with erosion stopping in most areas before it reached the base of the weathered zone under the Gondwana surface. King (1967) suggested a similar origin for the sub-skyline and skyline tors in his interpretation of cyclic land surfaces in Europe and Asia. The establishment of the African surface would have been followed by a new period of weathering and development of the present soils. Similar cases of erosional remnants of fresh rock surrounded by plains of rotted rocks can be observed in the Guayana and Brazilian shields of South America.

Development and evolution of old landscapes (100–200 million years old?)

in cratonic regions of the Tropics merit further study as well. The puzzle of stepped topography and multiple departures from a smooth hydraulic curve for the longitudinal profiles of streams in such regions may, in some cases, be explained by lateritic steps (Tricart 1965), and perhaps in others by tectonic activity. However, in certain areas, this type of topography may have resulted from climatic fluctuations on tropical landscapes, and in still others from sea-level drops during the Pleistocene.

Past climatic changes towards aridity appear to have left their imprint on the landscape in regions that today are part of the humid Tropics. During those arid periods, there was probably extensive reduction of the vegetal cover and erosion of the soil on landscapes with clayey and sandy soils. Soil profiles in the humid Tropics can reach up to tens of meters in thickness (Young 1976). Because the stable profile for arid conditions is different from the profile for humid conditions, a succession of wet–arid climatic fluctuations would tend to cause multi-level surfaces. The lateral extent and width of the various levels would depend on the length of the arid periods. A series of episodes of erosion and weathering induced by climatic fluctuations could have generated the multiple surfaces observed in the Guayana Shield of southeastern Venezuela, as well as some of the higher-level waterfalls and rapids found there and in the western periphery of the Brazilian shield.

During arid or semi-arid episodes in the Tropics, vegetational cover would be greatly reduced and perhaps disappear entirely. Much of the soil cover and underlying weathered rock mantle would then be stripped away. The more intensely fractured areas, and those underlain by rocks more susceptible to chemical weathering, would have been decomposed to greater depth than the surrounding areas. These weaker zones would be carved to a lower elevation than the surrounding countryside. Streams would tend to shift their channels to these topographic lows. The slopes at the boundaries between chemically resistant and weak (or highly fractured) rocks would become steeper, and scarps or stepped landscapes would develop. Thus, the landscapes of the humid Tropics could now have an overall aspect closer to that associated with multiple pedimentation than to peneplanation. Such seems to be the case in many tropical areas, especially between elevations of ~100 m and 2000 m above sea level in the cratonic or shield regions. The extensive, shallow and swampy profiles of the Amazon Territory of Venezuela, resting on flat basement surfaces of low relief, can be explained by the recent establishment of a humid climate on a landscape characterized by a profile developed by pedimentation during a previous dry period.

During arid periods and/or transitions from arid to wet parts of climatic cycles, in areas where thick, hard laterites had developed, the cuirasses of the lateritic profiles would become exposed by erosion of the overlying part of the soil profile and become the equivalent of resistant cap rocks. Cuirass-capped landforms can be observed in the high Llanos of Venezuela and Colombia, and along tributaries of the Niger River in the Anwai area of Bendel State, Nigeria (Fig. 4.2).

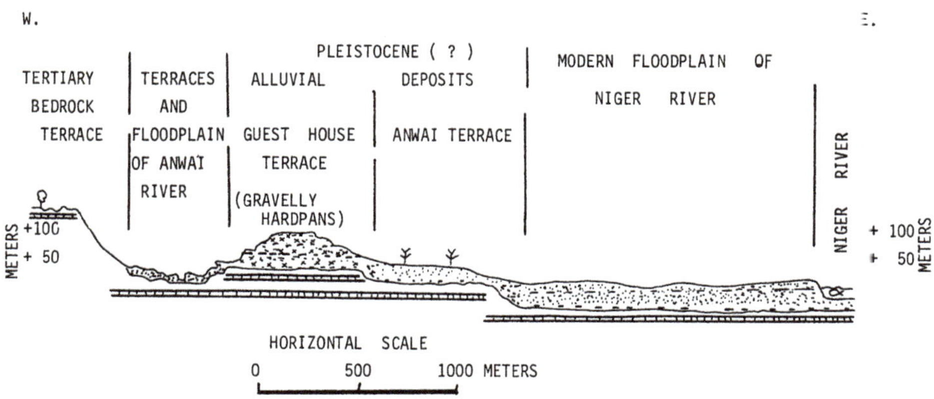

Figure 4.2 Schematic cross section at Anwai–Asaba School of Agriculture. Bendel State, Nigeria.

Climatic fluctuations, through their effects on soil development and erosion, can explain the presence of scarps, waterfalls, and stepped topography as superposed features on some of the old erosional landscapes of the humid and semi-humid Tropics. Evidence to prove that such fluctuations existed in the geologic past may, perhaps, be found by a careful evaluation of cores from depositional sequences such as that of the Chaco Basin and its northward extension, the depositional trough that follows the eastern side of the Andes Mountains as far north as Venezuela. There may exist recognizable paleosols within the parts of the section deposited under continental conditions. It may be more difficult to evaluate the influence of climatic fluctuations on the marine sediments, but perhaps not impossible. Fossil phytoliths, where recognizable, should be helpful. They have been reported from marine sediments dating as far back as the Paleocene by Baker (1960) and the Oligocene by Deflandre (1963). As these siliceous bodies can be ascribed to specific plant varieties, botanical assemblages associated with arid and wet chronologic events can probably be differentiated.

Depositional landforms

Stephens (1965) suggested that, tied to fluctuations in the extent of the area covered by glaciers and icecaps, there must have occurred during the Quaternary: (a) cycles of climatic change throughout the world, (b) development of soils in areas freshly exposed by oscillations of sea level, and (c) formation of suites of terraces by streams discharging into the sea, due to changes in base levels. The existence of climatically displaced red soils, that exhibit "lateritic, podzolic, solodic and red earth morphology" (Stephens 1965), in and below the southern Sahara and in many presently arid and semi-arid parts of Australia, is evidence supportive of severe climatic variations in the past, extending at least as far back as the Tertiary. The study of sedimentary cycles in Australia and New Zealand suggested to Loutit and Kennett (1981) that the inferences of Vail (1981) about global, eustatic sea-level changes during the Tertiary are valid. The changes indicated by these authors include several sea-level stands at about the +200 m level for the Oligocene–Paleocene interval. Several large-scale oscillations in sea level during the Cretaceous have also been interpreted on the basis of stratigraphic evidence on a world-wide scale. These may well have been accompanied by climatic variations on the continents.

Several generations of large, alluvial fans with slopes of the order of 0.002 can be observed on satellite images in the Chaco region of southeastern Bolivia, western Paraguay, and northwestern Argentina. Distance from the Andean mountain front to the apexes of the fans was probably controlled by precipitation on the eastern flank of the Andes. The rivers originating in that watershed, such as the Pilcomayo and Parapiti, built up the Chaco plains. The fans, as most of the rivers that flowed through the Chaco in the past, are now inactive. The Pilcomayo still flows through, but the apex of its present fan has migrated westward some 250 km in the past four or five decades, suggesting aridification. Unfortunately, enough human tampering with the stream has also been observed on satellite images of the period 1972–81, and in the field, to make it difficult to state without hesitation that the migration of the apex is a purely natural phenomenon. Multiple soil profiles (buried soils) found in the same Chaco region can be ascribed to either climatic fluctuations or lateral shifts in the position of the fans.

A similar series of multiple fans, with apexes determining lines parallel to the Andean trend, can be observed in northern South America in the western Llanos of Venezuela. Here, tectonism is still very active today and structural control of sedimentation in the Llanos may be difficult to differentiate from that of climatic control. In the same general region, but on the edges of the Guayana Shield south of the Orinoco River, large areas are covered by sandy, alluvial sediments and a landscape of low terraces, suggesting multiple episodes of alluviation.

Windblown silts derived from the arid regions of northern Africa are being added in quantities large enough to affect the color of topsoils at least as far south as Nigeria, and small amounts have also been recognized as far west as Barbados. Similar processes of addition to soil profiles by eolian transport probably occured in the past, the geographic setting dictated by climatic patterns.

The thickness of eolian deposits, barring the effect of local topographic irregularities, increases as one approaches the source of sediments. This is observed in the loessic accumulations southwest of the stabilized dune fields of southeastern Bolivia and northwestern Paraguay. Similar patterns can be expected on the lee side of the stabilized dune fields between the Papanaparo and Meta rivers, and the Guarico sand fields in southern Venezuela. The same effects can be observed on satellite images of an area west of the Kalahari

Desert in northern Namibia. There, the effect of wind erosion and deposition appears to be superposed on a pattern of deposition of sand ridges controlled originally by location of past shorelines.

Not all terraces and topographic steps can be attributed to widespread climatic fluctuations and base-level changes. Some sedimentary and erosional episodes were associated with tectonic movement with or without associated local climatic variation. Tectonic control of terrace formation east of the Andean front can be observed in the Beni region of northeastern Bolivia in the watershed of the Mamore River, tributary to the Amazon, where tilted, gravel-capped terraces can be observed to surround whale-back anticlines with cores of Tertiary sandstones. Similar patterns are reflected in the vegetation around the Agua Caliente Dome, adjacent to the Ucayali River in eastern Peru.

In a similar setting east of the Colombian Andes Orientales, a chain of low hills, Sierra de las Palomas, is the topographic expression of a long anticlinal structure with a core of late Tertiary sandstones. In the valley between this chain of hills and the Andes, there are some 14 levels of terraces whereas east of Sierra de las Palomas only 3 levels were counted (Segovia 1967). This has been interpreted as the result of control of fluvial downcutting and deposition by repeated uplift at Sierra de las Palomas in very late Tertiary and Quaternary times. Each episode of uplift in the Sierra raised the local base level of the Humea, Guacavia and Guayuriba rivers. These rivers anteceded the uplift of Sierra de las Palomas, and the older terraces in the valley are tilted westward, towards the Andes.

Time

Time, also, is a major factor in soil profile development. Precipitation rates and temperature affect soil development processes, as do the related phenomena of erosive, depositional and biological environments, but an evaluation of their effects would be meaningless if the variable time were left out of the analysis.

Continental sediments derived from the same watershed and found at elevations below 100 m above sea level in the Tropics can be expected to have the same lithologic provenance. The time available since the emergence of these landscapes above sea level was not enough to allow erosion to uncover very dissimilar rocks in the source area. Thus, the textural and compositional variations in soils found in the column below such elevation and in the area of influence of a given stream are related to: (a) changes in the energy of sedimentary processes associated with a particular microenvironment (stream bed, point bar, levee, back of floodplain, floodplain depression, fan, colluvial fan or apron), (b) climatic change, and (c) pedologic or diagenetic processes. Of these, the effects of climatic change on rates of sedimentation as well as on soil-forming processes can be expected to have been general over the region, and thus are useful in area-wide correlation.

Caution should be exercised in using any one criterion alone, as texture and composition of sediments contributed by tributaries draining small watersheds had great influence on floodplain deposits near their entrance into the larger stream valley. At times, the tributaries built fans on the larger floodplain and thus were predominantly responsible for local deposition. This effect is documented in the analysis of the geomorphologic history of tributaries to the Paraguay River and the Orinoco River, where the tributaries originate in the Brazilian Shield and in the Guayana Shield. Such local changes in texture and composition can produce great variation in the development of soils. Local influx of sediments with high clay content can cause the appearance of greater age. Evaluation of these effects is important in stratigraphic correlations for archeologic or paleoenvironmental studies.

References

Alexander, L. T. and J. G. Cady 1962. *Genesis and hardening of laterite soils*. US Dept Ag. Soil Cons. Serv. Tech. Bull. 1282.

Baker, G. 1959. Opal-phytoliths in some Victorian soils and "Red Rain" residue. *Aust. J. Bot.* **7**, 64–87.

Baker, G. 1960. Fossil opal-phytoliths. *Micropaleontology* **6**(1), 79–85.

Beinroth, F. H. 1982. Some highly weathered soils of Puerto Rico: 1. morphology, formation and classification. *Geoderma* **27**, 1–73.

Bishopp, D. W. 1954. *The bauxite resources of British Guiana and their development*. Br. G. Geol. Surv. Bull. 26.

Bloom, A. L. 1969. *The surface of the Earth*. Englewood Cliffs, NJ: Prentice-Hall.

Brazil Ministerio da Agricultura 1958. *Levantamento e reconhecimento dos solos do Estado de Rio de Janeiro e Distrito Federal*. Bol. Serv. Nac. Pesq. Agron. 11.

Bryan, K. and C. C. Albritton, Jr. 1943. Soil phenomena as evidence of climatic changes. *Am. J. Sci.* **241**(8), 469–90.

Buol, S. W., F. D. Hole and R. J. McCracken 1980. *Soil genesis and classification*. Ames, Iowa: Iowa State University Press.

Buringh, P. 1979. *Introduction to the study of soils in the tropical and subtropical regions*. Wageningen, Holland: Centre Agricultural Publishing & Documentation.

Cline, M. G. 1977. Historical highlights in soil genesis, morphology, and classification. *Soil Sci. Soc. Am. J.* **41**, 250–4.

Coffey, G. N. 1912. *A study of the soils of the United States*. US Dept Ag. Div. Soils. Bull. 85.

Deflandre, G. 1963. Les phytolithaires (Ehrenberg): *Protoplasma* **57**(1-4), 233–59. Vienna: Springer.

Derbyshire, E. (ed.) 1973. *Climatic geomorphology*. London: Macmillan.

Fairbridge, R. W. 1960. The changing level of the sea. *Sci. Am.* **202**, 70–9.

Fairbridge, R. W. 1961. Eustatic changes in sea level. *Phys. Chem. Earth* **4**, 99–185.

Fournier, F. 1963. The soils of Africa. In *A review of the natural resources of the African continent*. Unesco, Int. Doc. Serv. New York: Columbia University Press.

Goodland, R. J. A. and H. S. Irwin 1975. *Amazon Jungle: green hell to red desert? An ecological discussion of the environmental impact of the highway construction program in the Amazon Basin*. New York: Elsevier.

Harradine, F. 1966. Comparative morphology of lateritic and podzolic soils in California. *J. Soil Sci.* **101**, 142–51.

Hunt, C. B. 1972. *Geology of soils, their evolution, classification, and uses*. San Francisco: W. H. Freeman.

Jackson, M. L. 1965. Clay transformations in soil genesis during the Quaternary. *J. Soil Sci.* **99**, 15–22.

Jenny, H. 1941. *Factors of soil formation*. New York: McGraw-Hill.

Jordan, C. F. and R. Herrera 1981. Tropical rain forests: are nutrients really critical? *Am. Nat.* **117**(2), 167.

Kellog, C. E. 1936. *Development and significance of the great soils groups of the United States*. US Dept Ag. Misc. Publn 229.

King, L. C. 1967 *The morphology of the Earth*, 2nd edn. Edinburgh: Oliver & Boyd.

Lal, R. 1977. Analysis of factors affecting rainfall erosivity and soil erodibility. In *Soil conservation and management in the Tropics*. D. J. Greenland and R. Lal (eds) 49–56. New York: Wiley.

Lepsch, I. F., S. W. Buol and R. B. Daniels 1977. Soil–landscape relations in the Occidental Plateau of Sao Paulo State, Brazil: *Soil Sci Soc. Am. J.* **41**, 104–15.

Loutit, T. S. and J. P. Kennett 1981. New Zealand and Australian Cenozic sedimentary cycles and global sea-level changes. *Am. Assoc. Petrolm Geols Bull.* **65**(9), 1586–601.

McCaleb, S. B. 1967. The genesis of the red-yellow podzolic soils. *Proc. Soil Sci. Soc. Am.* **23**(1959), 164–8. Reprinted in *Selected papers in soil formation and classification*, No. 1, Soil Sci. Soc. Am. Spec. Publn Ser., J. V. Drew (ed.), 1967, 61–71.

Millard, C. E., L. M. Turk and H. D. Foth 1958. *Fundamentals of soil science*. New York: Wiley.

Mohr, E. C. J., F. A. Van Baren and J. Van Schuylenborgh 1972. *Tropical soils: a comprehensive study of their genesis*. The Hague: Mouton.

Monroe, W. H. 1980. *Some tropical landforms of Puerto Rico*. US Geol Surv. Prof. Pap. 1159.

Mueller, G. 1968. Genetic histories of nitrate deposits from Antarctica and Chile. *Nature* **219**, 1131–4.

Munoz, N. G., A. V. Segovia, and J. E. Foss 1980. Geology, morphotectonic analysis and soils of central Haiti, based on Landsat image and aerial photographs. *Proc. XIV Int. Symp. Remote Sens. Environ.* San Jose, Costa Rica, 1529–35. Ann Arbor, Mich.: Environmental Research Institute of Michigan.

Ollier, C. D. 1959. A two-cycle theory of tropical pedology. *J. Soil Sci.* **10**, 137–48.

Ollier, C. 1967. *Weathering*. Edinburgh: Oliver & Boyd.

Ruhe, R. V. 1956. Geomorphic surfaces and the nature of soils. *J. Soil Sci.* **82**, 441–55. Reprinted in *Selected papers in soil formation and classification*, Soil Sci. Soc. Am. Spec. Publn Ser., J. V. Drew (ed.), 1967, 270–85.

Ruhe, R. V. 1975. *Geomorphology*. Boston: Houghton Mifflin.

Ruhe, R. V., R. B. Daniels and J. G. Cady 1967. *Landscape evolution and soil formation in southwestern Iowa*. US Dept Ag. Tech. Bull. 1349.

Sanchez, P. A. 1976. *Properties and management of soils in the Tropics*. New York: Wiley.

Sanchez, P. A. and S. W. Buol 1975. Soils of the Tropics and the World food crisis. *Science* **188**, 598–603.

Segovia, A. V. 1967. Geology of Plancha L-12, Colombia, South America: a reconnaissance. *Geol Soc. Am. Bull.* **78**, 1007–28.

Segovia, A. V., J. E. Foss and E. A. Soler 1980. Evaluation of soils and landscapes of the seasonal tropics by means of remote sensing: an interdisciplinary study. *Proc. XIV Int. Symp. Remote Sens. Environ.*, San Jose, Costa Rica, 915–27. Ann Arbor, Mich.: Environmental Research Institute of Michigan.

Simonson, R. W. 1959. Outline of a generalized theory of soil genesis. *Proc. Soil Sci. Soc. Am.* **23**, 152–56.

Simonson, R. W. 1980. Soil survey and soil classification in the United States. *Proc. 8th Nat. Congr. Soil Sci. Soc. S. Afr.* Pietermaritzburg, 1978, 10–21.

Soil Survey Staff 1960. *Soil classification: a comprehensive system, 7th approximation*. Washington, DC: US Department of Agriculture.

Soil Survey Staff 1975. *Soil taxonomy: a basic system of soil classification for making and interpreting soil surveys*. US Dept Ag. Handbook 436.

Soler, E. A. 1981. *Erosion reconnaissance survey of Costa Rica using remote sensing methods*. MS thesis. University of Maryland.

Sombroek, W. G. 1966. *Amazon soils*. Wageningen, Holland: Centre Agricultural Publishing & Documentation.

Stephens, C. G. 1965. Climate as a factor of soil formation through the Quaternary. *J. Soil Sci.* **99**, 9–14.

Thomas, M. F. 1974. *Tropical geomorphology*. New York: Wiley.

Tricart, J. 1965. *Le modelé des regions chaudes, forets, et savanes*. Paris: Soc. d'Edition d'Enseignement Superieur. (English translation: C. J. Kiewit de Jonge, 1972, New York: St. Martin's Press).

Vail, P. R. 1981. Cenozoic sea level changes: introduction to colloquium 3, Geology of continental margins. *Proc. 26th Int. Geol Congr.* 1587–92. Paris. Extracted in Loutit & Kennett (1981).

Villiers, J. M. de 1965. Present soil-forming factors and processes in tropical and subtropical regions. *J. Soil Sci.* **99**, 50–7.

Weyl, P. K. 1970. *Oceanography: an introduction to the marine environment*. New York: Wiley.

Young, A. 1976. *Tropical soils and soil survey*. Cambridge: Cambridge University Press.

5
Role of subterranean water in landform development in tropical and subtropical regions

C. Rowland Twidale

Introduction

Only about 0.614% of all the water on Earth occurs as subsurface water on the continents (Nace 1960). Of this, a small part (0.306%) is found in the shallow groundwater zone, arbitrarily defined as within about 800m of the land surface, and only 0.0018% is in the root zone. Yet proportionately minute as it is, this soil moisture zone forms a vital, if incredibly thin, life jacket. It is also the scene of much geomorphological activity and is of considerable geological significance. Hills (1954), in a brief review, pointed out that water is an important constituent of rocks and that it has a marked influence on their behavior under stress. He also indicated briefly, some of the geomorphic effects of subterranean waters.

Many of the geomorphological consequences of soil moisture are so obvious as to require no more than passing mention. Without soil moisture, there would be little or no vegetation. According to many authorities, this would cause a diminution in the amount of oxygen in the atmosphere, and this in turn would entail a decrease in oxidation. Certainly, both fluvial and aeolian erosion would be much more pronounced than they now are. The many extant accounts of accelerated soil erosion (see, e.g. Duce 1918, Ratcliffe 1936, 1937, Jacks & Whyte 1939, Sharpe 1941, Holmes 1946, Vogt 1953, Bennett 1960) only begin to convey the importance of plants in protecting the land surface. Unfortunately, the same point is made even more succinctly and poignantly in the landscape over extensive areas of several continents, particularly in the so-called Third World (Fig. 5.1). Russell (1958) has pointed out that without a soil cover, for the formation of which moisture and organisms that are moisture dependent are essential, the landscape would present a very different face, and that in particular, rounded, graded hillslopes would be less common than they now are. And so on – soil moisture has several indirect or general geomorphological effects.

Specific and direct geomorphological effects of subterranean moisture can be considered under several distinct headings. First, water is a vital component

Figure 5.1 This gullying in the Transkei, southern Africa, is due, in part, to the deliberate felling of trees for use as domestic fuel, partly to the regular and long-continued tramping of people and stock from the safer hill-top settlements to the water supply in the valley floor.

in many important types of rock weathering. For most purposes, weathering may conveniently be defined as the disintegration or alteration of rocks at and near the Earth's surface, and in the range of temperatures found there. In this context, the complex of processes assumes considerable importance as a necessary precursor to erosion and hence to deposition.

Even in temperate lands, however, it has a wider significance. Bloom (1978, p. 103) states, "Weathering does not make landforms, but only altered or broken rocks from which landforms are shaped." Undoubtedly, the implied destructive aspect of weathering is very important the world over, but many landforms directly reflect the exploitation by erosional agencies of contrasts in susceptibility caused by weathering. The lower or lateral limit of weathering is called the **weathering front** (Mabbutt 1961a), and the front exposed by the stripping of the regolith is an **etch surface**. Such etch forms and surfaces are particularly commonplace and extensive in tropical and subtropical regions, though they are widely represented in temperate lands also. This is a second way in which subsurface moisture influences landform development, and to this extent, exception must be taken to Bloom's assertion.

In addition, and particularly in the lower latitudes, weathering has a constructive as well as a destructive aspect, and soil moisture, again, is crucial to the activity of the various processes. Subterranean waters are a medium of transportation. Salts are translocated in solution and are transported both vertically within the weathering profile and laterally in the drainage basin before being preferentially deposited in distinct horizons within the regolith. The mineral concentrations become as hard as cement when dry. The various **duricrusts** so formed are geomorphologically as well as economically significant in many parts of tropical and subtropical lands. They are not unknown, however, in temperate regions, and it is worth recalling that the term calcrete was coined by Lamplugh (1902) to describe calcareous developments in glacial drift near Dublin. For all that, however, it is in the lower latitudes that duricrusted surfaces are widely developed and preserved.

Finally, subterranean waters form a vital, though variable, constituent of the regolith.

Thus, weathering has not one face but four. Several of the features discussed are not confined to the Tropics, but they have their greatest expression there.

Subterranean waters and rock weathering

Although there are details that are controversial, and have different emphases, it is generally agreed that water is the most important single agent in rock weathering. It is vital to such processes as gelifraction (but see also White 1973), may have a catalytic effect in some mechanical processes (Griggs 1936), and even in arid lands is more effective in rock weathering than are processes of physical disintegration (Barton 1916). All common rock-forming minerals are to a greater or lesser extent vulnerable to attack by water. Even in the semi-arid and arid regions of the Tropics and subtropics, water is the most significant weathering agent.

Water is a superb solvent. Solution is a particularly significant weathering process:

> Solution is essential to chemical weathering. Minerals break down primarily because some of the constituent atoms and ions are dissolved and effectively removed from the environment. The loss of these constituents renders the existing mineral structures unstable and new crystalline phases tend to form in their stead. (Loughnan 1969, p. 61).

Solution is thus an essential step in chemical attack. It leaves many minerals vulnerable to further weathering processes, prominent among which are hydration and particularly, perhaps, hydration shattering (White 1973 – the *Hydratationssprengung* of Wilhelmy 1958, p. 52), and hydrolysis, which affect many common rock-forming minerals and especially the feldspars and the micas. But even such comparatively stable minerals as quartz are attacked, particularly along microfractures that are probably due to tectonic stresses (see Moss 1973, Moss *et al.* 1973, Moss & Green 1975, also Turner & Verhoogen 1960, p. 476).

Thus, moisture attack is widespread and commonplace. Water is the most effective single weathering agent. No one can dispute the overwhelming importance of moisture in bringing about the decomposition of rocks. Coherent, strong rocks are transformed into weak, friable aggregates that are eroded with comparative ease. The subject is dealt with at length not only in general texts but also in various accounts of rock weathering (see, e.g. Merrill 1897, Reiche 1950, Ollier 1969, Birkeland 1974, and especially, Loughnan 1969).

Without ground moisture, weathering would be much slower than it now is. It would conceivably be outpaced by erosion so that bare rock surfaces would be a general feature. Without groundwater weathering, rocks like granite would remain massive and coherent; there would be none of the granite pediments, peneplains and ultiplains so well developed and preserved in various parts of the tropical and subtropical world.

Some particular aspects are, however, worthy of further discussion in the immediate context of subterranean waters and landform development. Solution has spectacular effects in rocks that are susceptible to dissolution. Calcareous materials in particular, as is well known, are readily attacked by water. There is no necessity to elaborate on karst processes and forms, for they are considered by several other contributors to this volume. There is, in any case, an impressive karst literature. This is not to suggest that all details of karst chemistry or morphology are fully understood, but the broad outlines involving the reaction of carbonates and water are well enough known.

Several writers have, however, over the years adverted to pseudokarst features, forms that though morphologically similar to those evolved on limestone are nonetheless developed on siliceous rocks: quartzite, sandstone or granite. Thus, more than 150 years ago Caldclough (1829) described silica stalactites from near Rio de Janeiro, and more recently, Renault (1953) reported similar forms in sandstone caves in the Sahara; Klaer (1957) described *Rinnen* and *Karren*, as well as pans and basins, from granitic rocks in Corsica; Tschang (1961, 1962) reported *pseudokarren* from Singapore and Malaysia; pans, flares, tafoni and *Rille* occur on sandstone outcrops in the Drakensberg of South Africa (Twidale 1980) and on arkose at Ayers Rock in central Australia (Twidale 1978b); and several writers, notably, White, Jefferson and Haman (1966), Urbani and his associates (Urbani 1977, Szczerban *et al.* 1977) and most recently, Pouyllau and Seurin (1981) have described caves and *Rille* from the Roraima quartzite region of southern Venezuela.

There are few difficulties explaining the detailed sculpture of granitic rocks by moisture, particularly, if it is appreciated that much of the initial etching takes place in the moist subsurface, at the weathering front (Logan 1849, 1851, Twidale 1962, 1971, 1976a, 1978a, 1980, Boyé & Fritsch 1973, Twidale & Bourne 1975, 1976a). What is fascinating is the formation, apparently by solutional processes, of caves, *Rille*, pans and pipes in highly siliceous rocks in areas as contrasted as humid Venezuela and arid Australia.

The South American sites have already been referenced. A few Australian examples may briefly be mentioned. On the Brachina pediment in the western piedmont of the Flinders Ranges, the mantle consists largely of cobbles and boulders of limestone and quartzite. The waters that wash over the pediment are alkaline, and many of the clasts carry a pellicle of lime. The limestone cobbles are, however, little modified by weathering since transport and deposition, whereas the quartz fragments are notably fretted and many have well developed flared margins (Fig. 5.2). Again, hemispherical basins have been formed in silcrete exposed near the southern extremity of Lake Torrens, a salina with a crust that consists mainly of gypsum but with some halite. In the Ashburton Valley and in the Hamersley Ranges of the northwest of Western Australia, the opaline cappings of mesas and plateaus display intricate piping, and on quartzite and silcrete cappings located in the piedmont of the MacDonnell Ranges and on the adjacent Emily Plain (southeast of Alice

Figure 5.2 This small quartzite block sitting on the Branchina pediment, in the western piedmont of the Flinders Ranges, South Australia, was until recently partly buried by soil. The alkaline soil moisture (many cobbles and other fragments carry a coating of lime) caused the flanks to be weathered. There has been soil erosion on the order of 10 cm.

Springs) pipes (Fig. 5.3), basins and gutters are well developed. Geomorphological evidence suggests that the forms developed in humid warm conditions during the late Mesozoic and early Tertiary, and thus that they are relic or inherited features, but the chemistry of the processes responsible for their formation is nevertheless difficult. Here is a fertile field for further research.

Subterranean moisture is unevenly distributed in the near-surface zone so that its effects are greater in some locations than in others. Thus, it has been noted that in arid and semi-arid lands, in particular, the rocks in the piedmont zone are more deeply and intensely weathered than are similar materials exposed either in the adjacent hills or a few score meters away on the plains.

In the Brachina area of the western piedmont of the Flinders Ranges in South Australia, for instance, the siltstone exposed in the scarp-foot zone is

Figure 5.3 Clay-filled solution pipe and associated horizontal tube (×−×−×) developed in chalcedonic capping of mesa standing above the Todd Plain, some 50 km ESE of Alice Springs, central Australia.

bleached and kaolinized to a depth of 10–15 m, whereas the limestone and quartzite of the backing scarps are fresh, as is the siltstone underlying the pediments and exposed only 100–150 m from the hill base.

Again, Ucontitchie Hill may be taken as typical of many granitic inselbergs in that rock of the residual is fresh and that exposed in the pediment a couple of hundred meters downslope from the scarp foot is rather crumbly and slightly altered. It nevertheless stands in stark contrast with the granite of the scarp foot, which has been altered to a puggy clay that is gritty on account of the contained quartz fragments.

Such pronounced scarp-foot weathering has been attributed to the concentration of water that either runs down the backing scarps or seeps out at the hill base and percolates into the subsurface. Scarp-foot weathering gives rise to such specific forms as flared slopes (Figs 5.4 & 5) and associated platforms (Twidale 1962, 1968, 1978a). It has been advanced as a common cause of the basal steepening of hillslopes, the development both of the piedmont angle or nick (Twidale 1960, 1967) and of a tendency to scarp recession (Twidale 1960, see also Fisher 1866, 1872). Some cliff-foot caves or tafoni may be initiated in the scarp foot by soil moisture, for flares and such tafoni merge one with

Figure 5.4 Flared slope, 7–8 m high and associated fringing platform on the western side of Ucontitchie Hill, northwestern Eyre Peninsula, South Australia.

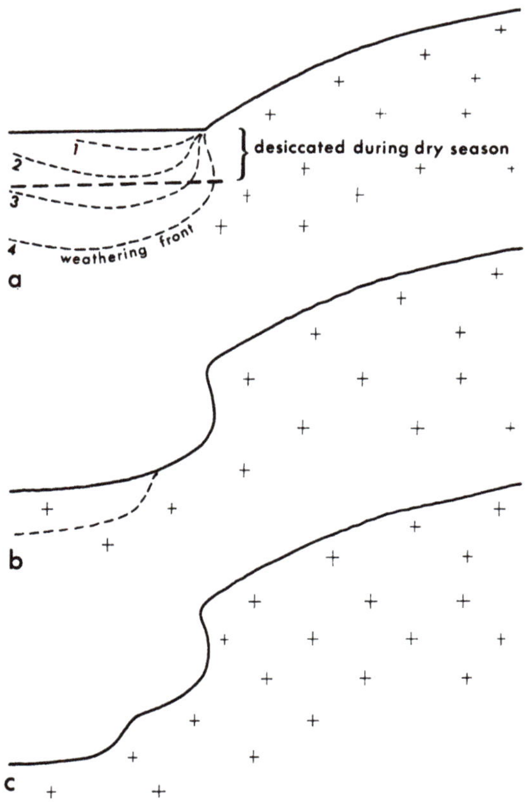

Figure 5.5 Development of flared slopes.

the other (Fig. 5.6) at Ayers Rock (Northwest Territories) Kokerbin Hill (Western Australia) and elsewhere (Twidale 1978b, see also Jennings 1976, Twidale & Bourne 1976a), though at some sites (Fig. 5.7) they appear to be caused by scarp-foot or subsurface seepage (Peel 1941, Twidale 1964, Jennings 1969, Smith 1978).

In addition to such specific features, many boulders and blocks display

---

Figure 5.6 (a) Cliff-foot cave associated with flared basal slope on the western side of Ayers Rock, a large arkosic inselberg that stands above the desert plains of central Australia. (b) Flared slopes and tafoni developed at the base of Kokerbin Hill, in the southwest of Western Australia.

(a)

(b)

Figure 5.7 Shelter, or elongate cliff-foot cave, about 2 m high and developed at the base of a sandstone cliff in the southern Drakensberg, South Africa.

irregular basal fretting, which is due to similar soil moisture attack (Figs 5.7 & 8) and eventually leads to the formation of mushroom rocks. It may be mentioned in passing that some, at least, of the zeugen attributed to aeolian sand blasting are more likely to result from soil-moisture attack. Some of the forms illustrated by Peel (1966, Pl. IV & XI) for example, and the hourglass form on basalt featured in Death Valley Natural Monument (Twidale 1976a, p. 292) are two examples that come readily to mind.

The weathered material of the scarp foot has in places been eroded to give scarp-foot valleys or depressions (Fig. 5.9). At some sites, they form annular depressions surrounding isolated hills. Known as linear depressions or *Bergfussniederungen*, they have been described from West Africa (Clayton 1956, Pugh 1956), the Egyptian desert (Dumanowski 1960), the Sudan (where they are known as "fules," see Ruxton & Berry 1961), central Australia (Mabbutt 1965) and Eyre Peninsula (Twidale *et al.* 1974). They are also developed in central Namibia, Angola and the Mojave Desert of southern California. The topographic lows are due to scarp-foot weathering, but whether stream erosion, subsurface flushing and evacuation of fines (Ruxton 1958), or decrease of volume consequent on weathering followed by settling, compaction and subsidence is responsible for the actual lowering of the land surface is debatable, though the causation may vary from site to site.

(a)

(b)

Figure 5.8 (a) Subsurface moisture attack has caused this block of coarse porphyritic granite located on eastern Eyre Peninsula between Whyalla and Cowell, to be peripherally weathered so that after the lowering of the adjacent plains its margins appear to be fretted. (b) The Cumberland Stone, the vantage point used by the Duke of Cumberland during the Battle of Culloden Moor in April 1746, has fretted lower margins (British Tourist Authority).

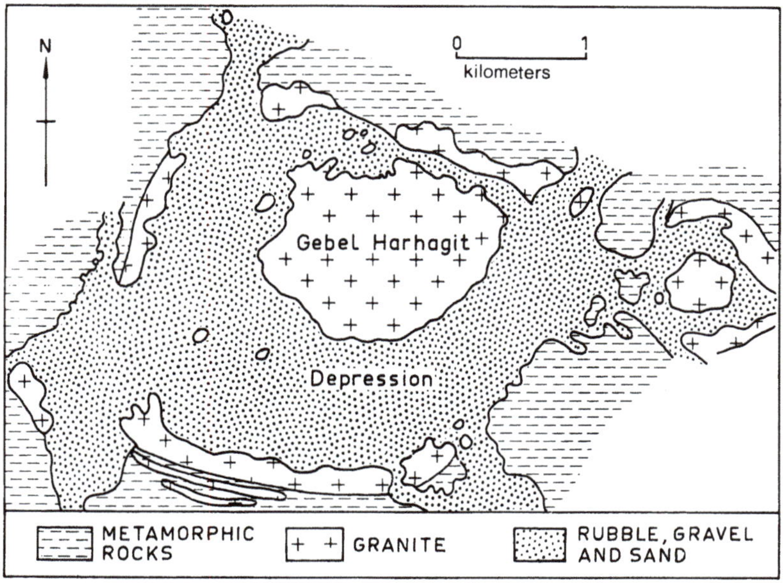

Figure 5.9 Annular scarp-foot depression at Gebel Harhagit, Egypt (after Dumanowski 1960).

Translocation of salts

Moisture not only contributes substantially to rock disintegration and alteration but is also the medium through which salts and minerals are translocated either within the regolith or within the drainage basin. Immense volumes of salts and minerals are in motion in solution at any given moment. According to Livingstone (1963), the average river carries 120 ppm dissolved salts, with carbonates (58.4 ppm) predominant, but with considerable amounts of silica (13.1), sulphate (11.2) and chlorine (7.8); these figures, of course, take no account of the vast quantities of salts in the regolith either in transit or more-or-less stored there.

Carbonates, silica, iron, alumina and gypsum, in particular, are precipitated out preferentially to form distinct concentrations at the land surface and in weathering profiles.

Some such carapaces are of great antiquity. Leith (1925), for example, has described silicified erosion surfaces of Precambrian age, preserved largely in unconformity in the Lake Superior region, and others of Paleozoic age, preserved in the Mississippi Valley. They are especially well developed in calcareous environments and are considered by Leith to be partly due to solution and partly to preferential erosion.

Water rises to the surface or near-surface zones from great depths bringing with it salts, which are precipitated around the surface springs. In arid climates such as obtain in central Australia, the salts are not redissolved by meteoric waters at least for some time and large circular or oval mounds up to 10 m high and called mound springs come to be formed.

The precipitation of salts at preferred sites has given rise to a distinctive suite of minor forms. Box patterns, as developed for example on the tesselated pavement at Eagle Hawk Neck, southeastern Tasmania, result from the deposition of limonite along joints. Again, limonite deposition in fine-grained sandstones at coastal sites in New Zealand, Victoria and South Australia has given rise to "cannonballs" and cauldrons with raised rims (Figs 5.10 & 11).

They are commonly known as **case hardening** because they are frequently associated with visors and the other minor features indicative of relative resistance. Despite Smith's (1978, p. 35) claim that there is a generally accepted origin for the feature, there is in reality considerable controversy (see, e.g. Merrill 1898, Blackwelder 1954, Engel & Sharp 1958, Scheffer *et al.* 1963, Höllermann 1963), but several of the suggested explanations involve salts

Figure 5.10 Iron-indurated "cannonball" developed in fine-grained sandstone at Shag Point, Palmerstone South, South Island, New Zealand.

Figure 5.11 Large doughnut or cauldron with raised rim, formed in iron-indurated sandstone, Shag Point, New Zealand.

translocated by subterranean waters. Merrill (1898, p. 380), for instance, suggested that desert varnish is an induration "brought about through the deposition of matter held in solution by ground water which is brought to the surface by capillarity and there evaporated." Several other writers, particularly those experienced in desert geomorphology, concurred (e.g. Anderson 1931, Blackwelder 1954). Others consider this an oversimplified interpretation; White (1944) for instance, suggests that iron derived from the breakdown of biotite hornblende and other ferromagnesian minerals is taken into solution in ferrous form and is later oxidized to insoluble hematite (and other similar materials).

Thin layers and horizons rich in iron oxides are commonly developed at the weathering front in granitic rocks (Fig. 5.12). It is derived from the breakdown of biotite and other ferromagnesians and has been illuviated to the impermeable fresh rock. The weathered rock is readily eroded, and it may be that its removal leaves exposed the iron-rich layer as the rind or hardened skin.

Where such salt precipitation (or clay illuviation) has occurred in specific horizons **hardpans** (using the term to imply a hard, impenetrable layer, as distinct from using the term as synonymous with duricrust – see below) are formed. They have the effect of impeding the subsurface infiltration of water and so increasing the likelihood of flooding during and following heavy rainstorms. Such effects are not confined to the low latitudes. For example, the local but disastrous floods that affected Lynmouth in North Devon in 1952

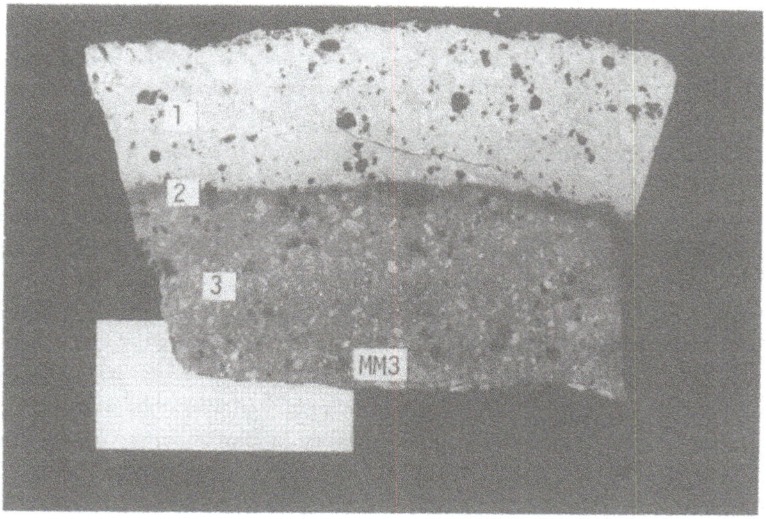

Figure 5.12 Iron-rich horizon accumulated at the weathering front (2) in porphyritic granite (from Mount Monster, South East district, South Australia). Zone 1, kaolinized rock; Zone 3, fresh rock. Cm/mm scale (A. R. Milnes).

were due to a heavy rain following an extended wet period (Kidson 1953), but the impact of the rains was enhanced by the common development of a ferruginous hardpan in the soils of the poor plateau.

Duricrusts and their significance

On desiccation (as for instance following uplift and stream dissection), these mineral concentrates cement the weathered rock and form tough, resistant horizons or carapaces, which when exposed are called *duricrusts*. Several types of duricrusts have been described in the literature, but the most important, because they are extensively developed, are laterite, bauxite, silcrete, calcrete and gypcrete.

Descriptions

Laterite is a highly weathered material rich in secondary oxides of iron and alumina or both (see e.g. Prescott & Pendleton 1952, Alexander & Cady 1962, Sivarajasingham *et al.* 1962 , Maignien 1966). It is virtually devoid of bases and primary silicates, although in some samples there are large amounts of quartz and kaolinite. Laterite is soft when wet, but hardens on exposure and drying. This is indeed the origin of the term laterite, for it is used as a building material in southern India. This was observed by the early Europeans (see, e.g. Buchanan 1807, Babington 1821) who, calling upon their knowledge of the Classic languages, coined the term laterite (from the Latin *later*, a brick).

A typical laterite profile consists of a sandy A horizon, 6–7 m thick, but more frequently missing as a result of erosion of the loose friable material. This is underlain by a ferruginous B horizon up to 5 m thick. It is this ferruginous zone that commonly is exposed in the landscape. It is commonly pisolitic and vesicular and in many places displays a rudimentary stratification. It has a shiny-brown or brownish purple exterior, but inside the pisoliths are white, brown, yellow and red, and most commonly display several of these colors. Beneath the iron-rich layer is a zone, up to 30 m thick, consisting essentially of kaolinized clay. White, with splashes of yellow, red and brown, the clay contains irregular, usually small, masses of chalcedony. Many workers subdivide the kaolinized zone into an upper and lower pallid zone, but the two are frequently indistinguishable, patchily distributed or occur in reverse order, with the mottled material below the pallid. Below the kaolinized material is the weathered bedrock.

Bauxite is similar to laterite. It is rich in alumina and is the result of intense desilication. **Ferricrete** is ferruginous encrustation lacking the profile development of true laterites. There are obvious terminological difficulties, for Goudie (1973) has suggested, logically but with utter disregard for historical precedent, that the various duricrusts be named according to their dominant chemistry and with the suffix *crete* to denote hardness. Thus, laterite becomes ferricrete, bauxite, alcrete, and so on. In these two particular instances, the present writer prefers to recall that there is no one logic but many, and to be reminded of bricks being made from laterite, and of bauxite occurring at Les Baux. Moreover, there is the practical advantage of then retaining the term ferricrete to denote ferruginous carapaces that are not underlain by mottled and pallid zones and that differ genetically from laterite.

Many hypotheses have been offered in explanation of laterite (see Maignien 1966). Some early workers, observing their vesicular structure, thought they were of volcanic origin. Now, however, though there are controversies as to detail, it is agreed that laterites are weathering profiles. Both laterite and bauxite are presently forming in the humid Tropics and particularly in such monsoonal areas as southern India, Malaysia and Indonesia. They are formed in regions where high temperatures and rainfall and abundant products of vegetational decay induce intense desilication. The mechanism of iron-oxide precipitation is a matter of debate, but it is generally considered to be optimized by a fluctuating water table, though as is so often the case, the feature may evolve in different ways at different sites.

Silcrete is a silica-rich rock consisting of 95% or more quartz (Frankel & Kent 1937, Hutton *et al.* 1978). It is characterized by the presence of porphyroclasts of quartz set in crystalline quartzose matrix. Silcrete is grey or yellow brown with patches of red and dark-brown material. It has a vitreous sheen and a good conchoidal fracture. Silcrete characteristically occurs in thick sheets of considerable extent. Pebbles, cobbles and even boulders of exotic material are common inclusions, though rounded silcrete fragments also occur.

The siliceous materials are frequently underlain by bleached kaolinized bedrock, but there is no necessary genetic connection between the two. Their total chemistries are in some instances such that the silcrete cannot have been derived from the substrate. Thus to take an obvious example, in the valley of Capowie Creek south of Quorn in the southern Flinders Ranges, South Australia, silcrete rests on limestone.

The extensive developments of silcrete are probably of alluvial origin and involve the long-distance translocation of silica (see Stephens 1964, 1970). But other occurrences, minor in volume and less common than the sheet materials, occur as skins developed on blocks and boulders, usually of quartzite, and are most likely residual features left behind by the long-term solution and evacuation of silica, and the resultant concentration of such resistates as titanium, niobium and zirconium. These skin silcretes are devoid of aluminium, potassium and iron, but may contain up to 25% by volume of titanium oxide (Hutton *et al.* 1972, Hutton *et al.* 1978).

Because the major occurrences of silcrete take the form of extensive plains and plateaus, it has been assumed that they are remnants of once contiguous sheets, but this is not necessarily so. The remnants are in all probability former river floodplains with silicified alluvium as a veneer. The situation is seen in miniature in the Beda Valley, southern Arcoona Plateau, South Australia (Twidale *et al.* 1970), where the sheet silcrete with pebbles is restricted to the valley floor (Fig. 5.13). Clearly, it and other, more extensive sheets represent old valley floors, and their present topographic position is the result of relief

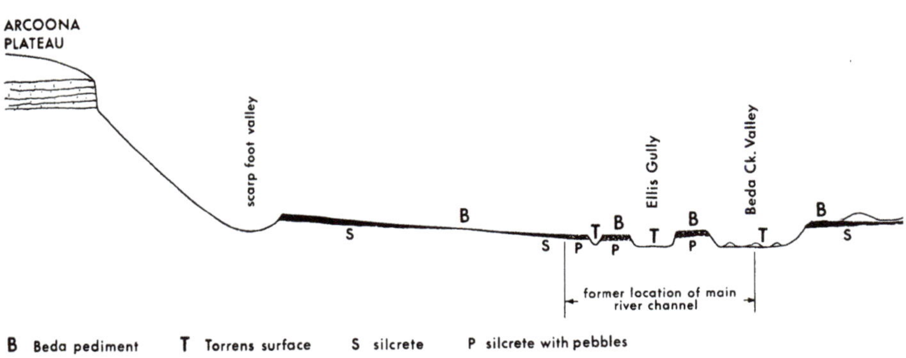

Figure 5.13 Section through the Beda pediment, southern Arcoona plateau, South Australia (after Twidale, Shepherd & Thomson 1970).

inversion on a regional scale (Fig. 5.14). (Similar but local relief inversion involving travertine has been described from Arabia – see Miller 1937). Silcrete does not appear to be forming at present anywhere in the world; it is a truly relic accumulation.

Several theories, not mutually exclusive, have been proposed in explanation of silcrete. Some authors have suggested short-distance translocation followed by concentration and precipitation of silica (Jack 1915). Others consider that alkaline conditions are conducive to the coagulation of colloidal solutions of silica involved a short-distance transport (Frankel & Kent 1937). Long-distance translocation has been used by Stephens (1964, 1970), who with respect to the Lake Eyre Basin has suggested that evaporation in the arid interior caused silica precipitation. Precipitation from silica-rich gels in

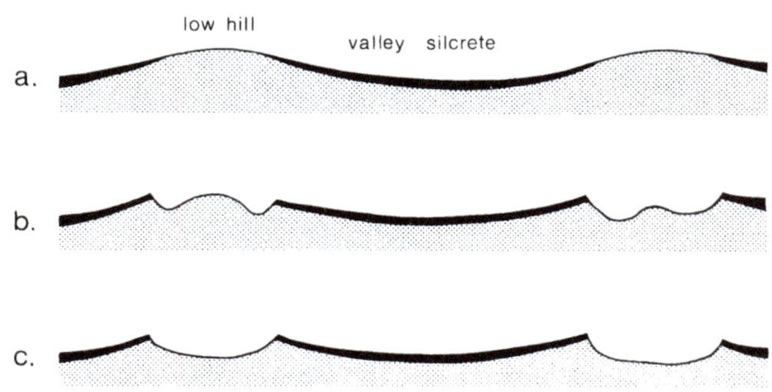

Figure 5.14 Development of inverted relief in area subjected to silcrete formation and then dissected.

lacustrine conditions (Öpik 1954) and concentration by primitive plants (Lovering 1959) are other mechanisms that have been proposed. Gradual crystallization of silica on stable surfaces of low relief over very long periods of geological time, probably aided by plants, is yet another suggestion (Hutton *et al.* 1978).

Calcrete, also known as kankar, kunkar and caliche, is a pedogenic accumulation rich in calcium carbonate. It commonly contains 80% or more calcium carbonate with minor amounts of quartz, dolomite and clays. Calcrete takes various forms, which can be viewed as an evolutionary sequence (Netterberg 1971). Some calcrete consists of isolated nodules of concentric structure, and these obviously grow and coalesce so that the rock consists of a mass of lime nodules cemented together. Continuation of the process leads to the formation of a solid sheet or hardpan of lime. Some calcrete has a honeycomb structure, and in many deposits, old roots and other structures are preserved.

Calcrete is distinguished from travertine, which is lime precipitated from springs and stream waters. It also differs genetically from the so-called valley calcretes of Western Australia, which are of commercial interest because of their associated uraniferous (carnotite) deposits. According to Butt *et al.* (1977) and Mann and Horwitz (1979), these deposits have accumulated from below, from groundwaters, but along stream lines.

Calcrete is of arid as well as semi-arid provenance. The sheet (nodular, honeycomb, hardpan) forms either through precipitation of carbonate because of variations in pore pressure, through capillary rise and evaporation, through water-table oscillations, or because of organic activity.

In addition to these major occurrences, **gypcrete** (Wopfner & Twidale 1967) has accumulated on a Pleistocene land surface over considerable areas to the west and southwest of Lake Eyre. It consists of 2–3 m of crystalline (butterfly twinned) gypsum underlain by gypsiferous silts. Similar Quaternary gypseous duricrusts have been described from several other areas, as for instance northern Iraq (Tucker 1978) and Kansas, where the Gypsum Hills are capped by gypsum formed in Permian lakes and marine embayments (Grimsley 1897).

Duricrusts as caprocks

Whatever their condition when wet, duricrusts become hard when dry. In laterite, the ferruginous materials develop a crystalline continuity (Alexander & Cady 1962) that is irreversible even if the rock is again wetted. Thus, once dissected and left above the local or regional water table, these materials come to form resistant cappings on plateaus, mesas and buttes. Gypcrete and calcrete are more ephemeral than laterite or silcrete because of their physical softness and chemical reactivity. They persist only in hot, arid conditions.

Laterite and silcrete, however, survive longer (indeed for very long periods in some instances) and in a wide range of climates, though their preservation is again aided by heat and aridity (Fig. 5.15). The sandy A horizon of laterites is usually stripped away to expose the pisolitic ferruginous material, though in some areas, such as the Gilberton Plateau of northwest Queensland (Twidale 1966), it is preserved beneath the eucalypt woodland. Most laterites and silcretes are essentially flat lying, so that plateau forms predominate (Fig. 5.16). Elsewhere, they are folded, and in still other areas, gently warped though it is in some instances difficult to distinguish between such distortions and the original relief. But where they are folded, they act as resistant strata

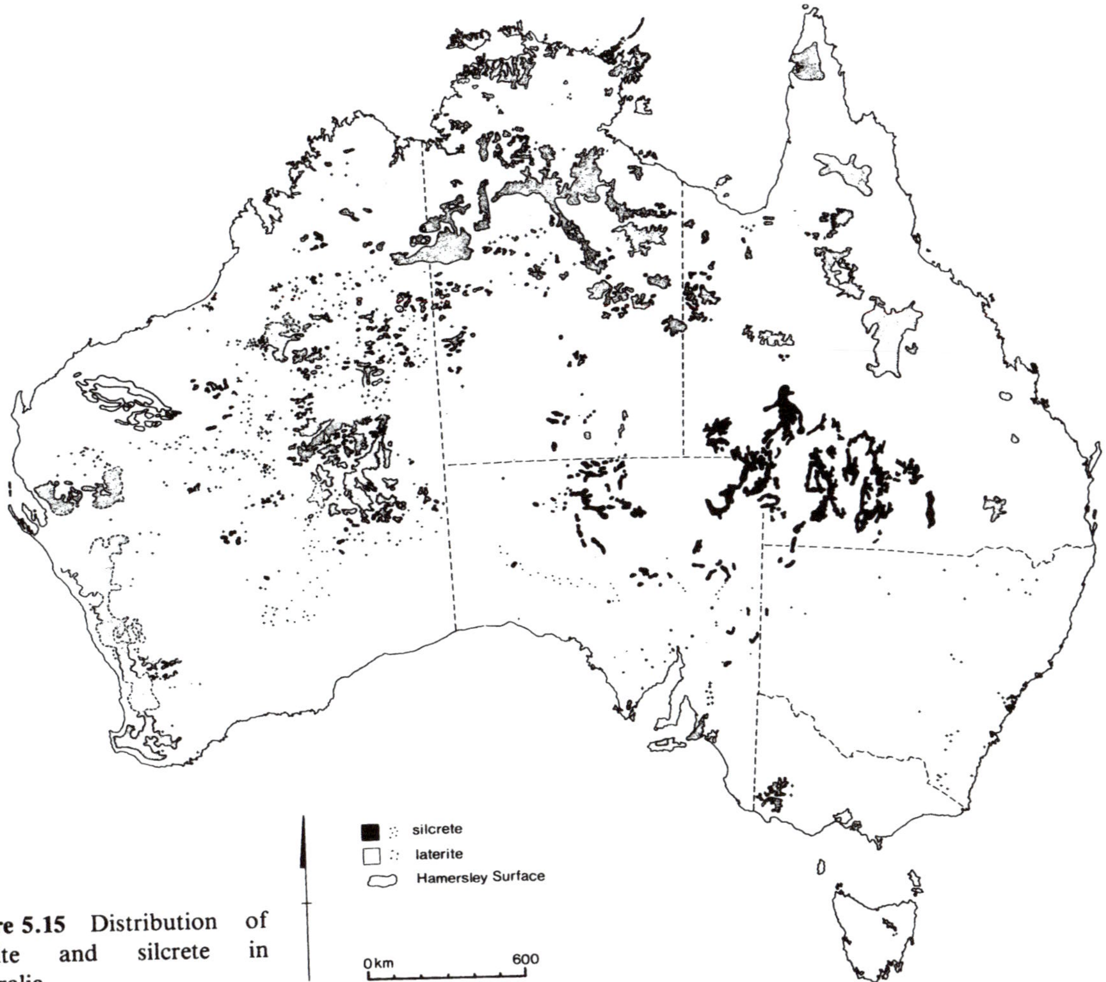

Figure 5.15 Distribution of laterite and silcrete in Australia.

(a)

(b)

Figure 5.16 (a) Butte and mesas capped by laterite developed on granite, south of Cloncurry, northwest Queensland (CSIRO). (b) Silcrete-capped butte and mesa near Rumbalara, central Australia (CSIRO).

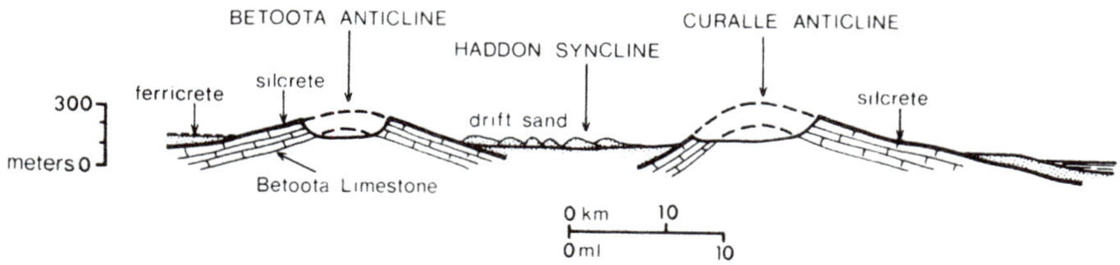

Figure 5.17 Section through Curalle and Betoota anticlines showing various ridge forms capped by silcrete that has been folded and dissected (after Wopfner 1960).

(a)

(b)

Figure 5.18 (a) Calcrete-capped mesa, Ashburton Valley, northwest of Western Australia (R. C. Horwitz). (b) Calcrete-capped Cretaceous sediments located south of Lake Eyre and in the southwestern extremity of the Great Artesian Basin.

Figure 5.19 The western edge of Lake Eyre is formed by gypcrete-capped cliffs some 7–8 m high. The bed of the salina is encrusted with salt that normally gleams white, but at the time this photograph was taken, in the early 1960s, it was coated with red dust.

and form ridges (Fig. 5.17), the precise morphology of which depends on the disposition of the duricrust.

Where dissected, as in the Ashburton Valley of Western Australia (Fig. 5.18a), and in the lower parts of the tectonically subsiding Lake Eyre Basin (Fig. 5.18b), calcrete also gives rise to plateaus, mesas and buttes, but elsewhere, as in the western Murray Basin, it has a stabilizing effect, preserving against weathering and erosion a high plain cut in Pliocene calcarenite and of Pliocene–early Pleistocene age (Twidale *et al.* 1978). Again, on Yorke and Eyre peninsulas, the later Pleistocene planation surface is effectively protected by a veneer of calcrete (Twidale *et al.* 1974, 1976).

To refer to gypcrete as a caprock may seem absurd in view of the physical softness of the constituent mineral (gypsum has a physical hardness of 2 on Moh's scale) and the solubility of hydrous calcium sulphate. But it must be remembered that though powdered (flour) gypsum is fairly insoluble, the crystalline form is not readily dissolved so that in an arid climate, such as that which obtains in central Australia, the material is more resistant than would at first seem the case. The massive gypcrete stratum exposed in the cliffs on the western side of Lake Eyre (Fig. 5.19) is certainly more resistant to weathering and erosion than the gypsiferous silts that underlie it, and it here forms a caprock and holds up precipitous cliffs.

Duricrusts as morphostratigraphic markers

In addition to their direct bearing on landform development, duricrusts are useful as morphostratigraphic markers that allow landforms and landscapes to be dated. This is not to suggest that all laterites, e.g., can be correlated, but

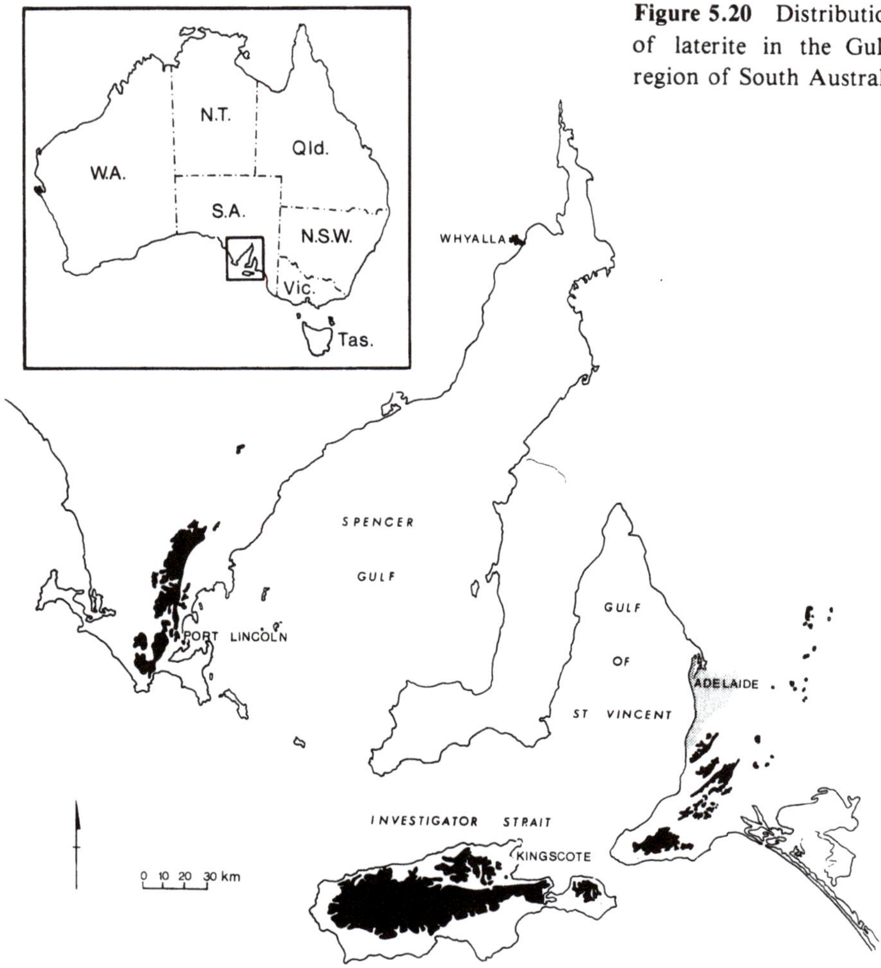

Figure 5.20 Distribution of laterite in the Gulfs region of South Australia.

at the regional scale, and provided that the duricrusts have been dated stratigraphically, they form valuable marker horizons.

A distinction must be made between the laterite (and bauxite) of southern Australia and that of northern regions. Laterite forms a capping to uplands in the "Gulfs region" of South Australia (Fig. 5.20). The gross morphology is determined by faulting, and it is clear from the stratigraphic evidence (Glaessner 1953, Glaessner & Wade 1958) that the most recent phase of major faulting, and the faulting responsible for the uplift of the present Mount Lofty Ranges, e.g., occurred in the latest Cretaceous or earliest Eocene. As the lateritized summit or high plain surface was uplifted by this faulting, the laterite obviously developed during the later Mesozoic.

Some of it, at least, is of much greater antiquity. The laterite capping the plateau or high plain that occupies the major part of Kangaroo Island is developed on strata of Cambrian and Permian age. The latter are of glacigene origin and consist of sands and clays that are readily eroded. West of Kingscote, basaltic lavas were extruded on a plain cut in mottled zone materials developed on these glacigene materials (Fig. 5.21), though to the

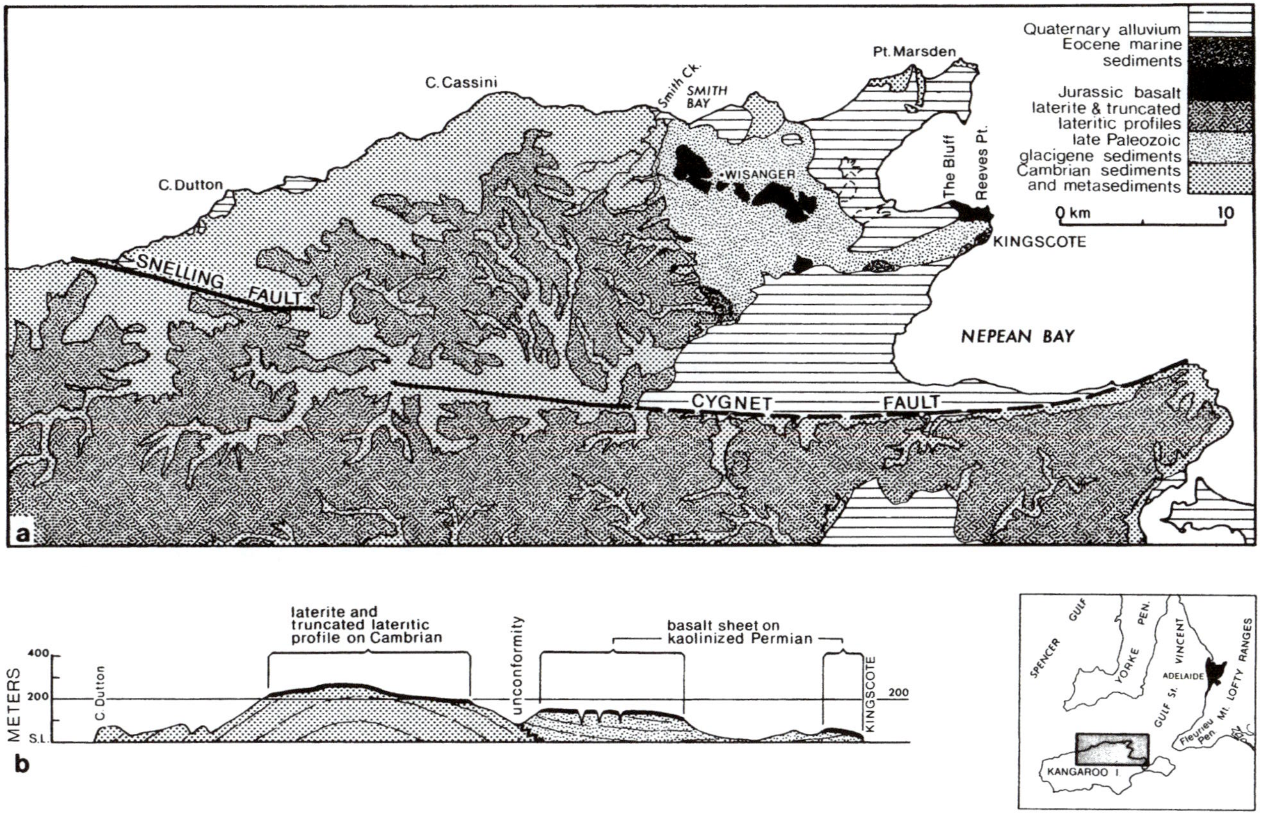

Figure 5.21 Geological map of part of Kangaroo Island, South Australia, showing distribution of laterite and (section) relationship with Jurassic basalt extrusion (after Daily *et al.* 1974).

east, near Penneshaw, the basalt overlies the ferruginous zone of the laterite. The laterite is thus clearly older than the basalt, but younger than the Permian strata on which it is developed. Radiometric datings show that the basalt is of Middle Jurassic age (Wellman 1971), and the "best" age for the laterite, taking into account both geological and paleoclimatic data, is Triassic (Daily *et al.* 1974). The laterite that forms a capping to the high plains of southern Eyre Peninsula and the Mount Lofty Ranges occupies a similar topographic situation and is presumably of a similar age (Fig. 5.20).

In Western Australia silicification commenced in the Oligocene (Fairbridge & Finkl 1978). The date of its termination is not known with precision, but it predates the Quaternary. Near Mount Magnet and elsewhere, silcrete forms in-siderations, both geological and geomorphological, have caused that conclusion to be questioned. More recent stratigraphic and tectonic analyses suggest that lateritization in the area extended through the later Mesozoic and the Eocene (Fairbridge & Finkl 1978). Thus in southern Australia, the lateritized land surfaces are of considerable antiquity.

Lateritic and bauxitic duricrusts of similar though variable Mesozoic ages have been reported from West Africa, central Africa and southern India (Wayland 1934, Demangeot 1978, Fritsch 1978, Michel 1978), one such surface in Senegal being assigned to the Jurassic (Michel 1978).

In northern Australia, the laterite has a considerable age range, but its formation appears to have begun during the Cretaceous (Hays 1967) and lasted until the Miocene (see Twidale 1956, 1966). Certainly, Cretaceous and early Tertiary strata have been lateritized, but rocks of later Cenozoic, certainly Pleistocene and possibly Pliocene, age have not.

Silcrete has a similar age range. Though Jurassic silcretes have been reported (Wopfner 1978) and strong silicification was taking place in central Australia during the Cretaceous (Wopfner *et al.* 1974), the main period of silcrete development was again the early–middle Tertiary (Wopfner & Twidale 1967). The process had ceased by the Pliocene, for the Etadunna Formation of the Miocene or later age is not affected.

In Western Australia silicification commenced in the Oligocene (Fairbridge & Finkl 1978). The date of its termination is not known with precision, but it predates the Quaternary. Near Mount Magnet and elsewhere, silcrete forms infillings in joints in the laterite.

The northern laterite and the silcrete have the same age range, and both developed under humid tropical conditions (Ludbrook 1969, Twidale & Harris 1977, Kemp 1978), but whereas the laterite mainly occupies regions of exoreic drainage, the silcrete occurs in basins of interior drainage (Fig. 5.14). In the south, the silcrete is younger than the laterite and consistently occupies lower topographic positions. Again however, major elements of the present landscape are of considerable antiquity.

Ferricrete, gypcrete and calcrete are much younger than the laterite and silcrete. Ferricrete is developed in valley floors and basins in the Mount Lofty Ranges and on plains eroded in folded silcrete in southwest Queensland (Fig.

5.16) and on northeastern Eyre Peninsula (Twidale *et al.* 1976). Thus, it clearly post-dates the Miocene, and stratigraphic evidence from Yorke Peninsula (Horwitz & Daily 1958) suggests that the iron encrustations, both buried and exposed, of that and other areas are of Plio–Pleistocene age. Certainly, this is consistent with all the evidence so far adduced (see also Horwitz 1960).

The gypcrete developed on a Pleistocene land surface is associated with Pleistocene fossiliferous sediments (Wopfner & Twidale 1967) and predates the later Pleistocene, but no greater precision is at present possible. Calcrete is still forming and has been developing for several tens of thousands of years and possibly longer. Again, it is not possible to be more precise at the present time.

Etch forms and surfaces

The weathering front

In some lithological environments, the transition between weathered and fresh rock is gradational, but in others, is quite sharp. The junction between weathered and fresh rock is, as mentioned earlier, the weathering front (Mabbutt 1961a). In crystalline rocks such as granite, the transition zone is commonly very narrow because granite, like many other rocks, is susceptible to attack by moisture, but is stable when dry. Fresh granite is of exceedingly low porosity and permeability, but once water has penetrated into the rock by way of mineral cleavages and other microfissures, alteration and disintegration are rapid. As a result, porosity and permeability increase dramatically (Kessler *et al.* 1940) and further weathering proceeds apace. Thus, granite tends to be either fresh or quite weathered. As the weathering front, which is, in effect, the moisture front, advances, the granite is altered, and there are pronounced changes in permeability so that further weathering rapidly develops. Thus, the weathering front comes to be a sharply defined line or narrow zone rather than a broad zone of gradation.

Etch surfaces (sensu lato)

Most regoliths consist of loose, friable, weak aggregates that are markedly less cohesive and resistant to erosion than the fresh rock. Hence where baselevel has been lowered, or where stream gradients have adjusted to the finer caliber of material produced by weathering, the regolith has been stripped, exposing the weathering front in fresh rock. Such exposed weathering fronts are called etch surfaces.

Etch surfaces and forms occur in temperate lands and even at high latitudes, the tors or castellated inselbergs of Dartmoor and the Bohemian massif being examples that come to mind as having developed by subsurface weathering under humid tropical conditions in Tertiary times (Linton 1955, Demek 1964). Again, Boyé (1950) suggested that ice sheets act as bulldozers and essentially remove the preexisting regolith to reveal the weathering front, so that the high plain of Labrador Peninsula, for instance (Tanner 1944), may be in reality an extensive etch plain, though there is some suggestion that, in parts at any rate, it is exhumed (see Ambrose 1964, Cowie 1961).

But for all that, etch forms and surfaces are more widely recognized in the lower latitudes, but probably only because the evidence necessary for their interpretation is better preserved there.

Examples of etch forms

Granite is exposed over about 15% of the continental areas. One of the most common residual forms associated with granite exposures is the boulder (or tor – see Twidale 1971, p. 14–17). As has been appreciated for almost two centuries (Hassenfratz 1791, see also Twidale 1978c), many, perhaps most, boulders develop in two stages. The first involves the infiltration of water

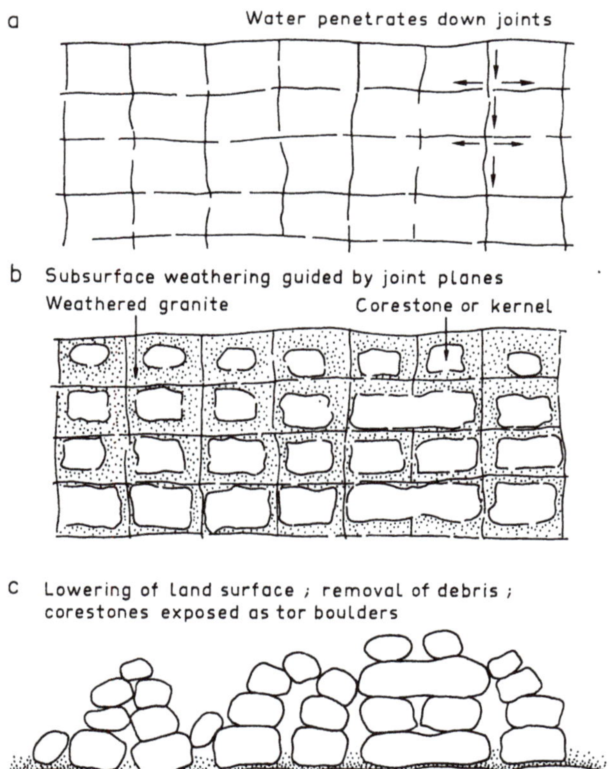

Figure 5.22 Two-stage development of boulders.

down the orthogonal fracture sets that are characteristic of granite and the weathering of the rock with which it comes into contact to produce granite sand and grus (Fig 5.22). As both MacCulloch (1814) and Logan (1849) appreciated, weathering is greater at corners and edges than on plane faces, so the erstwhile cubic or quadrangular blocks are converted into spherical or ellipsoidal masses set in a matrix of friable grus (Fig. 5.23). The grus is removed by running water or by mass movements leaving the corestones exposed as boulders, the surfaces of which are commonly pitted (Twidale & Bourne 1976b) as a result of preferential weathering of feldspar and mica leaving the quartz crystals in microrelief (Fig. 5.24). These pitted surfaces are etch surfaces, for they are exposed weathering fronts.

Figure 5.23 Granite corestone and grus (weathered granite) exposed in a road cutting near Lake Tahoe, eastern Sierra Nevada, California.

(a)

(b)

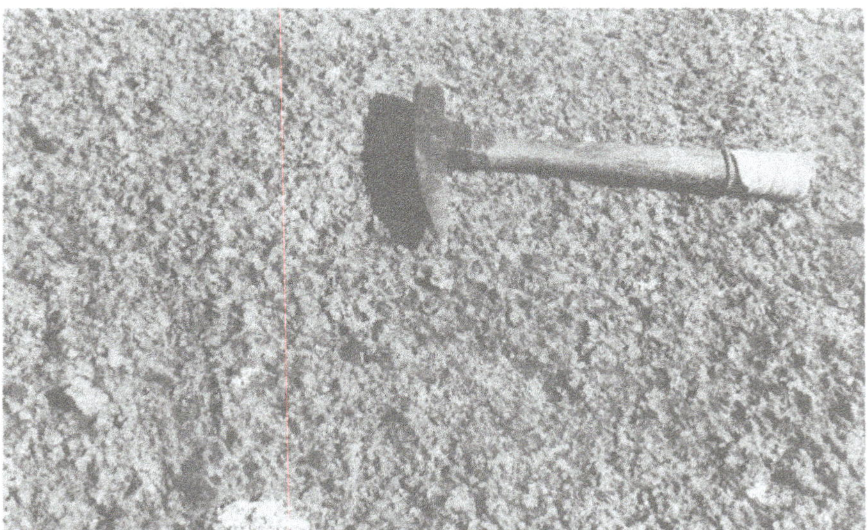

Figure 5.24 (a) Part of the northern slop of Pildappa Rock, a low whaleback on western Eyre Peninsula. The region is arid, and the bare rocks offer possibilities for water conservation. The soil has been cleared from lower slopes and low walls built partly to prevent run-off being dissipated on the adjacent plains, partly to guide the water into storages. The clearing of the soil exposed pitted granite surfaces that are due to differential moisture attack at the crystal scale. (b) Detail of pitted surfaces in Figure 5.24a.

(a)

(b)

Figure 5.25 Domed inselbergs or bornhardts (a) near Nanutarra, in the northwest of Western Australia, where the residuals and the plains are eroded in strongly porphyritic Archaean gneiss (b) Paarlberg, in the western Cape Province, South Africa.

On a larger scale, some bornhardts have evolved in a similar manner, subsurface weather exploiting weaknesses in the country rock to leave the resistant masses as towers protruding upwards into the deep regolith. These projections, when exposed by the erosional stripping of the regolith, become inselbergs, and most commonly, domed inselbergs or bornhardts (Fig. 5.25). Again, this mode of development has been appreciated for many years. Thus, Falconer (1911, p. 246) wrote:

A plane surface of granite and gneiss subjected to long-continued weathering at base level would be decomposed to unequal depths, mainly according to the composition and texture of the various rocks. When elevation and erosion ensured, the weathered crust would be removed, and an irregular surface would be produced from which the more resistant rocks would project. Those rocks which had offered greatest resistance to chemical weathering beneath the surface would upon exposure naturally assume that configuration of surface which afforded the least scope for the activity of the agents of denudation. In this way would arise the characteristic domes and turtlebacks which suffer further denudation only through insolation and exfoliation. Their general elliptical outlines, which Merrill would ascribe very largely to the influence of crustal stress and strain, are probably in great part due simply to the modifications by weathering of original phacolithic intrusions.

In addition, many of the minor forms typical of bornhardts are demonstrably initiated at the weathering front as a result of moisture attack. Pitting has been mentioned, as have flared slopes and platforms, but *Rille* (Fig. 5.26) also have been found on newly exposed etch surfaces (Twidale 1971 p. 90, 1976b, p. 223–4, Boyé & Fritsch 1973, Twidale & Bourne 1975, see also Logan, 1849, 1851). Contrary to the assertion of Hedges (1969), gnammas or weather pits are also initiated beneath the regolith (Twidale & Bourne 1975). Some tafoni also appear to be initiated at the weathering front (Twidale 1978b, Twidale & Bourne 1975, 1978).

Similar forms in sandstone environments (Fig. 5.27) also apparently form beneath the soil cover, examples having been described from South Australia, central Australia (Ayers Rock) and South Africa (Twidale 1978b, 1980, see also Verrall 1975).

Waves no less than rivers exploit contrasts in rock resistance, and some shore platforms appear to be etch character. Thus, the very broad platforms (they extend 200 m from the base of the cliff – Fig. 5.28) cut in granite at Smooth Pool near Streaky Bay, on the west coast of Eyre Peninsula, and which by their very extent and regularity form exceptions to Jutson's (1940) and Hills' (1949, 1971) assertions that platforms are not well developed in fresh granite, are exposed weathering fronts that happen to fall within the pre-

Figure 5.26 The excavation of a reservoir at Dumonte Rock, near Wudinna, northwestern Eyre Peninsula, revealed that the weathering front is scored by channels that are continuations of those on the naturally exposed rock surface.

Figure 5.27 Shallow channels on a recently exposed sandstone platform in northern Lesotho, southern Africa.

Figure 5.28 Shore platform eroded in granite, and effectively exposing the former weathering front, Smooth Pool, near Streaky Bay, western Eyre Peninsula.

5.26

5.27

5.28

sent tidal range and thus have been exposed (Twidale *et al.* 1977). Again, on parts of the central California coast, the platforms cut in gently dipping sediments appear to be of etch character (see Bradley & Griggs 1976, at Fig. 4), and Bartrum (1926) has made the general point that waves exploit the zone of fluctuating water table where it merges at the coast to form what he called "abnormal" platforms.

Etch surfaces of low relief

Discussing the erosional surfaces of Uganda in 1934, Wayland suggested that not all of the planate surfaces of that central African country were true peneplains, in the sense that they were not due to the erosion of a stable landmass. He argued that the deep saprolite (or regolith) produced by the chemical alteration of the bedrock would be "largely removed by denudation if and when land elevation supervenes..." (Wayland 1934, p. 79) and went on to sug-

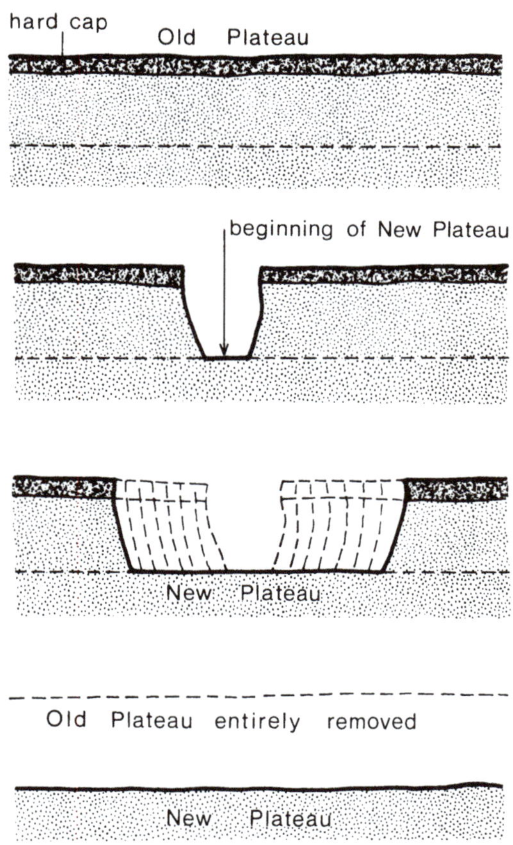

Figure 5.29 Development of New Plateau by the stripping of lateritic regolith formed beneath the Old surface (after Jutson 1914).

gest that the peneplains other than the highest and oldest, what was known as Peneplain I, are not true peneplains but rather "... they are etch plains...indicative not of tectonic stability and quiescence but of instability and upward movement" (p. 79). The erosion levels below Peneplain I are "etch platforms formed during periods of slow elevation."

Where duricrusts and other deep weathering profiles are well and widely developed, the notion of the stripping of the regolith has found increasing favor as an explanation of various forms (Hills 1954). Jutson (1914) had the idea in his grasp and indicated that the New Plateau of the southwest of Western Australia is merely an etch plain from which the laterite, remnants of which form the Old Plateau, had been stripped (Jutson 1914, p. 96–7, particularly Figs. 47–50) (Fig. 5.29). He was not explicit on the point, and it is possible to read much into his work. He seems to have been thinking in terms of the erosional exploitation of lithological contrasts induced by weathering.

In any event, Mabbutt (1961b) clearly demonstrated that the weathering front preserved at the base of many Old Plateau mesa residuals capped by laterite is coincident with the level of the high plain that is the New Plateau and which is underlain by intrinsically fresh rock. Similarly, it has been shown that the high plain surface of the Mount Lofty Ranges in South Australia (Fig. 5.30) is in large measure an etch surface resulting from the stripping of the lateritic regolith over wide areas (Twidale & Bourne 1975, Twidale 1976b).

The widespread development of duricrusts in central and northern Australia suggests that etch surfaces may be common there too. Indeed, Wright (1963) has described planation surfaces from the Daly River basin of the Northern Territory that correspond not with the weathering front but rather with (silicified) horizons within the lateritic regolith. More commonly, it is the stripping of the entire zone of weathering that has taken place. In argillaceous bedrock, it is more difficult to identify etch surfaces with certainty because the base of the regolith is diffuse, but it is clear that in northwest Queensland, for instance the Julia Plains (Twidale 1956, 1966), cut in Cretaceous strata and with plateaus capped by laterite standing above them, may be of etch character. The plains extend to the base of the laterite-capped mesas, but there is a gradation between the pallid material and the bedrock.

Variations in rock volume and characteristics

Apart from the development of depressions consequent on the withdrawal of groundwaters by man in such areas as Mexico City (Fox 1965), the Central Valley of California (Lofgren 1965, Meade 1968, Lofgren & Klausing 1969, Riley 1970) and Shanghai, Trendall (1962) has attributed the formation of deeply weathered (lateritized) surfaces, which he calls "apparent peneplains," to the evacuation in solution of salts released by weathering, volume decrease and settling and compaction. The notion has been adversely criticized by Ollier (1969, p. 181–2), but the idea has, nevertheless, some merit.

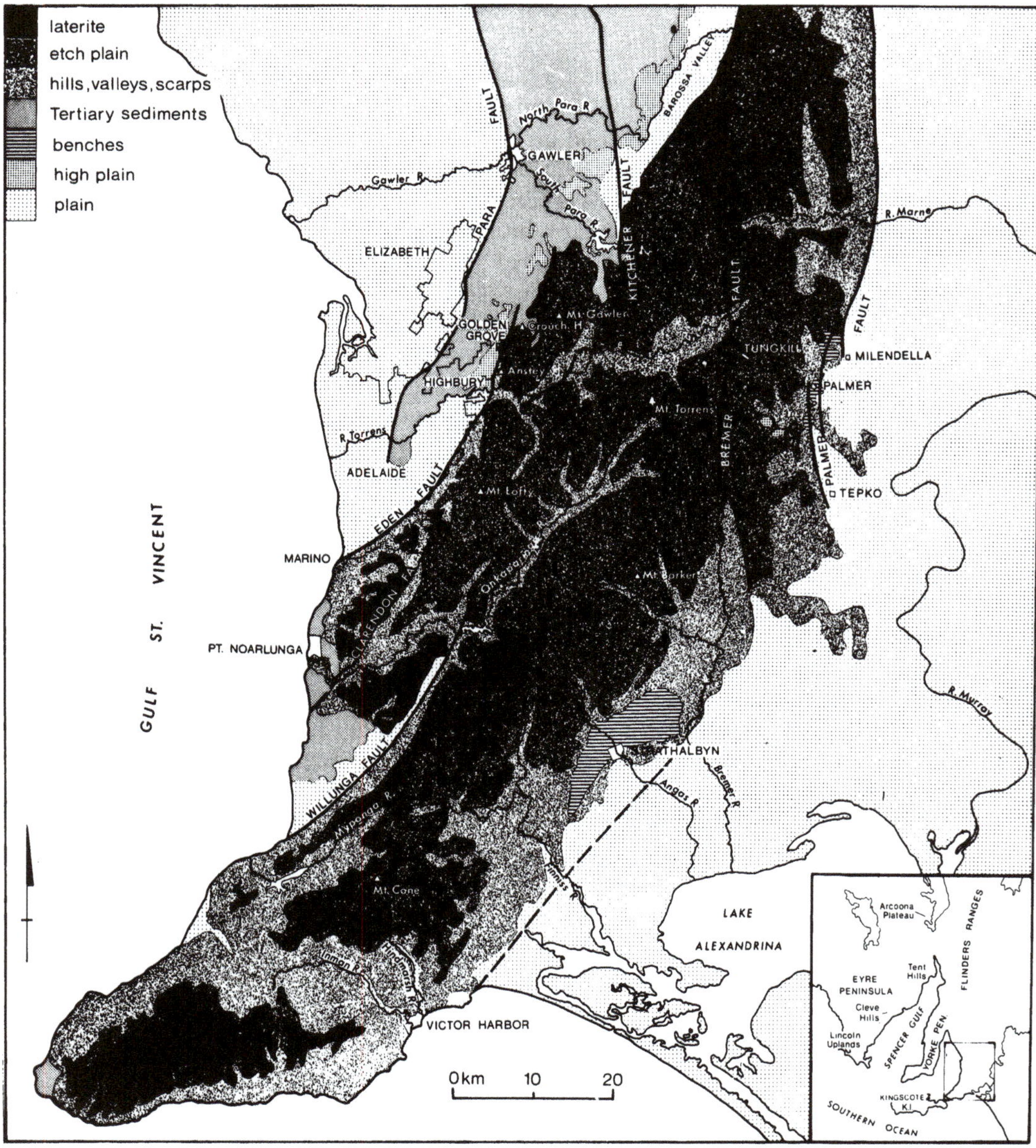

Figure 5.30 Distribution of laterite and associated etch plain in the Mount Lofty Ranges, South Australia (after Twidale 1976b).

Subsurface solution in carbonate rocks is, of course, fully accepted, but its indirect effects are less well known, though Thomas (1954, 1963), e.g., has documented cases of the regional subsidence of strata caused by solution and collapse of underlying limestone (Fig. 5.31), and similar forms have been observed in quartzite underlain by dolomite in the southern piedmont of the MacDonnell Ranges in central Australia.

Water can and does introduce salts to horizons as well as evacuate them, and

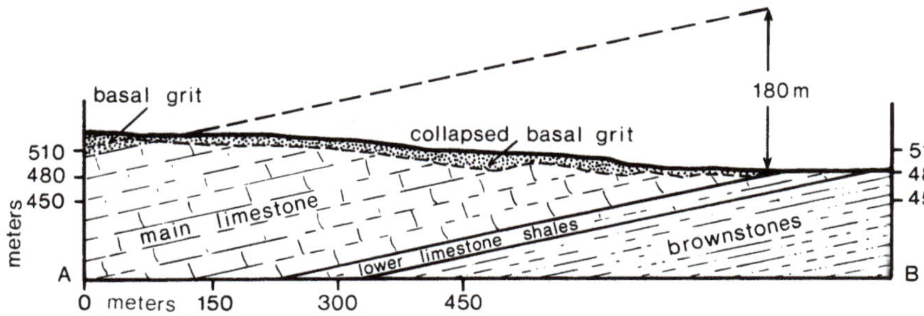

Figure 5.31 Redistribution of basal grit as result of subsidence consequent on solution of underlying limestone in South Wales (after Thomas 1963).

minor anticlinal forms due to precipitation of carbonates and consequent swelling and arching of the materials have been described from Texas (Price 1925), the Fitzroy Basin of northwestern Australia (Jennings & Sweeting 1963) and the interior of Western Australia (Mann & Horwitz 1979). Such pseudo-anticlines are geometrically anomalous with respect to the structure of the underlying bedrock and form distinct, although minor, landforms.

Underground water running along impervious interfaces is responsible for the flushing of fine solids, as well as solutes, causes volume decrease (Ruxton, 1958), and in particular, produces camber effects (Hollingworth *et al.* 1944). It also renders clays plastic and mobile. Thus in the Northamptonshire ironstone field, the rivers have eroded deep valleys into the plateaus capped by limestone, ironstone and sandstone, and underlain by soft clays. The weight of the overlying competent strata has caused the weak plastic clays to flow outwards from beneath the plateaus into the valleys where they form distinct bulges, some of them bounded by faults (Figs 5.32a & b).

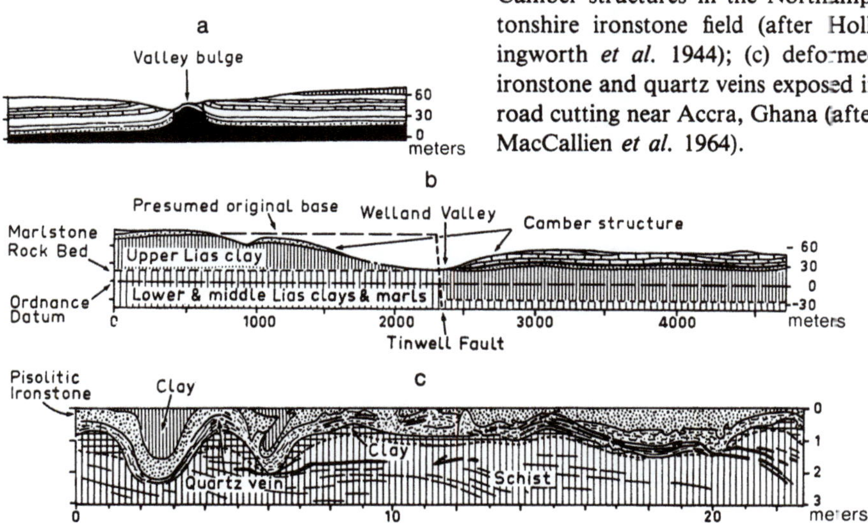

Figure 5.32 (a) Valley bulge. (b) Camber structures in the Northamptonshire ironstone field (after Hollingworth *et al.* 1944); (c) deformed ironstone and quartz veins exposed in road cutting near Accra, Ghana (after MacCallien *et al.* 1964).

Near Accra in Ghana, a ferruginous pisolite has formed on a weathered gneiss. The weathering front is uneven, and the weight of the pisolite has caused the clays of the decomposed gneiss to be squeezed from the small depressions or basins in the weathering front. This, in turn, has caused intense distortions (Fig. 5.32c) in the regolithic material, including the formation of faults and globular, mushroom-shaped and tongue-like structures (MacCallien *et al.* 1964).

These distortions are not of tectonic origin, for they are related not to deep-seated pressure but to superficial pressures caused by the interaction of competent strata and strata rendered incompetent by subterranean water.

Alternations of wetting and drying, of expansion and contraction lead to the development of patterned ground, known in Australia as gilgai, in its many forms (see, e.g. Hallsworth *et al.* 1955, Springer 1958, also Washburn 1956). It is also instrumental in the development of superficial instability, which in some sites is sufficient to cause landslides and other mass movements of debris. The soles of many such slides are coincident not with any primary geological boundary or discontinuity but rather with the weathering front (Fig. 5.33). The mobile material is wholly regolithic (Twidale 1976c).

Figure 5.33 Landslide near Inglewood, Mount Lofty Ranges, South Australia. Only superficial weathered debris is involved (J. Selby, S. Aust. Dept Mines & Energy).

Conclusion

If nature were like a connoisseur of Scotch and had no water, the world would be a very different place from the one we have come, albeit imperfectly, to know. If the Earth's surface were utterly impermeable and impervious and there were no subterranean water, the geomorphological world would scarcely be recognizable. There would be less vegetation, less regolith and less soil. Most of the forms due directly or indirectly to rock weathering would be unknown. There would be no etch forms, no underground karst, none of the morphological varieties developed on granite. There would be less erosion by rivers and streams, but because of the lack of soil moisture and of vegetation, aeolian processes and forms would be more widely developed.

Subterranean water contributes enormously to the development of landforms. It is an agent of alteration, a medium of translocation of salts, a lubricant, and an important constituent of rocks and of the regolith. Of course, this is no new viewpoint. Shakespeare has the sexton in *Hamlet* (Act V, Scene i) point out that "your water is a sore decayer," and the same sort of rather gruesome image was invoked by MacCulloch (1814, p. 72) when describing the weathered granite of Cornwall as "gangrenous." But these descriptions indicate only one of the several important facets played by subterranean water in landscape and landform development. Subterranean water not only plays an all-important part in the destructive aspects of rock weathering, it has a significant role as a medium of transportation, in the constructive role of weathering, and is a significant if varied and valuable component of both rocks and the regolith. It is not too much to say that we should not recognize a world devoid of subterranean water.

References

Alexander, L. T. and J. G. Cady 1962. *Genesis and hardening of laterite in soils.* US Dept Ag. Tech. Bull. 1282.

Ambrose, J. W. 1964. Exhumed palaeoplains of the Precambrian Shield of North America. *Am. J. Sci.* **262**, 817–57.

Anderson, A. L. 1931. *Geology and mineral resources of eastern Cassia County, Idaho.* Idaho Bur. Mines Geol. Bull. 14.

Babington, B. G. 1821. Remarks on the geology of the country between Tellicherry and Madras [1819]. *Trans Geol Soc.* **5**, 328–39.

Barton, D. C. 1916. Notes on the disintegration of granite in Egypt. *J. Geol.* **24**, 382–93.

Bartrum, J. A. 1926. "Abnormal" shore platforms. *J. Geol.* **34**, 798–806.

Bennett, H. H. 1960. Soil erosion in Spain. *Geogrl Rev.* **50**, 59–72.

Birkeland, P. W. 1974. *Pedology, weathering and geomorphological research.* New York: Oxford University Press.

Blackwelder, E. 1954. Geomorphic processes in the desert. In *Geomorphology in geology of southern California*, R. H. Jahns (ed.), 11–20. Calif. Div. Mines Bull. 170.

Bloom, A. L. 1978. *Geomorphology: a systematic analysis of Late Cenozoic landforms.* Englewood Cliffs, N J: Prentice-Hall.

Boyé, M. 1950. *Glaciaire et périglaciaire de l'Ata Sund, Nord Oriental Groenland.* Paris: Hermann.

Boyé, M and P. Fritsch 1973. Dégagement artificiel d'un dôme crystallin au Sud-Cameroun. *Trav. Doc. Géogr. Trop.* **8**, 33–63.
Bradley, W. C. and G. B. Griggs 1976. Form, genesis, and deformation of central California wave-cut platforms. *Geol Soc. Am. Bull.* **87**, 433–49.
Buchanan, F. 1807. *A journey from Malabar through the countries of Mysore, Canara and Malabar*, Vol. 2. London: East India.
Butt, C. R. M., R. C. Horwitz and A. W. Mann 1977. *Uranium occurrences in calcrete and associated sediments*. CSIRO Div. Mineral Rep. FP 16.
Caldclough, A. 1829. On the geology of Rio de Janeiro. *Trans Geol Soc. Lond.* **2**, 69–72.
Clayton, R. W. 1956. Linear depressions (*Bergfussneiderungen*) in savannah landscapes. *Geogrl Stud.* **3**, 102–26.
Cowie, J. W. 1961. Contributions to the geology of North Greenland. *Med. Grøn.* **164**(3).
Daily, B., C. R. Twidale and A. R. Milnes 1974. The age of the lateritised summit surface on Kangaroo Island and adjacent areas of South Australia. *J. Geol Soc. Aust.* **21**, 387–92.
Demangeot, J. 1978. Les reliefs cuirasses de L'Inde de Sud. *Trav. Doc. Géogr. Trop.* **33**, 97–111.
Demek, J. 1964. Castle koppies and tors in the Bohemian Highland (Czeckoslovakia). *Biul. Peryglac.* **14**, 195–216.
Duce, J. T. 1918. The effect of cattle on the erosion of cañon bottoms. *Science* **47**, 450–2.
Dumanowski, B. 1960. Comments on the origin of depressions surrounding granite massifs in the westen desert in Egypt. *Bull. Acad. Pol. Sci.* **8**, 305–12.
Engel, C. G. and R. P. Sharp 1958. Chemical data on desert varnish. *Geol Soc. Am. Bull.* **69**, 487–518.
Fairbridge, R. W. and C. W. Finkl 1978. Geomorphic analysis of the rifted cratonic margins of Western Australia. *Z. Geomorph.* **22**, 369–89.
Falconer, J. D. 1911. *Geology and geography of northern Nigeria.* London: Macmillan.
Fisher, O. 1866. On the disintegration of a chalk cliff. *Geol Mag.* **3**, 354–6.
Fisher, O. 1872. On cirques and taluses. *Geol Mag.* **8**, 10–12.
Fox, D. J. 1965. Man–water relationships in metropolitan Mexico. *Geogrl Rev.* **55**, 523–45.
Frankel, J. J. and L. E. Kent 1937. Grahamstown surface quartzites (silcretes). *Trans Geol Soc. S. Afr.* **40**, 1–42.
Fritsch, P. 1978. Chronologie relative des formations cuirassées et analyse géographique des facteurs de cuirassement au Cameroun. *Trav. Doc. Géogr. Trop.* **33**, 113–32.
Glaessner, M. F. 1953. Some problems of Tertiary geology in southern Australia. *J. Proc. R. Soc. NSW* **87**, 31–45.
Glaessner, M. F. and M. Wade 1958. The St. Vincent Basin. In *The geology of South Australia*, M. F. Glaessner and L. W. Parkin (eds), p. 115–26. Melbourne: Melbourne University Press.
Goudie, A. 1973. *Duricrusts in tropical and subtropical landscapes.* Oxford: Clarendon Press.
Griggs, D. T. 1936. The factor of fatigue in rock exfoliation. *J. Geol.* **44**, 783–96.
Grimsley, G. P. 1897. Gypsum deposits in Kansas. *Geol Soc. Am. Bull.* **8**, 227–40.
Hallsworth, E. G., G. K. Robertson and F. R. Gibbons 1955. Studies in pedogenesis in New South Wales VII. The gilgai soils. *J. Soil Sci.* **6**, 1–31.
Hassenfratz, J-H. 1791. Sur l'arrangement de plusieurs gros blocs de différentes pierres que l'on observe dans les montagnes. *Annls Chim.* **11**, 95–107.
Hays, J. 1967. Land surfaces and laterites in the north of the Northern Territory. In *Landform studies from Australia and New Guinea*, J. N. Jennings and J. A. Mabbutt (eds), 182–210. Canberra: ANU Press.
Hedges, J. 1969. Opferkessel *Z. Geomorph.* **13**, 22–55.
Hills, E. S. 1949. Shore platforms. *Geol Mag.* **86**, 137–52.
Hills, E. S. 1954. Underground water as a factor in geological processes. In *Proceedings of the Ankara symposium on arid zone hydrology, Pt. V.: relationship between the hydrology of underground water and other solutes*, Vols 2 & 3 262–4. Paris: Unesco.
Hills, E. S. 1971. A study of cliffy coastal profiles based on examples in Victoria, Australia. *Z. Geomorph.* **15**, 137–80.
Höllermann, P. 1963. "Verwitterungsrinden" in den Alpen. *Z. Geomorph.* **7**, 172–7.

Hollingworth, S. E., J. H. Taylor and G. A. Kellaway 1944. Large-scale superficial structures in the Northamptonshire ironstone field. *Q. J. Geol Soc. Lond.* **100**, 1–35.
Holmes, J. McD. 1946. *Soil erosion in Australia and New Zealand*. Sydney: Angus & Robertson.
Horwitz, R. C. 1960. Géologie de la région de Mt. Compass (Fueille Milang), Australia méridonale. *Ecol. Géol Helv.* **53**, 211–63.
Horwitz, R. C. and B. Daily 1958. Yorke Peninsula. In *The geology of South Australia*, M. F. Glaessner and L. W. Parkin (eds), 46–60. Melbourne: Melbourne University Press.
Hutton, J. T., C. R. Twidale and A. R. Milnes 1978. Characteristics and origin of some Australian silcretes. In *Silcrete in Australia*, T. Langford-Smith (ed.), 18–39. Armidale: University New England Press.
Hutton, J. T., C. R. Twidale, A. R. Milnes and H. Rosser 1972. Composition and genesis of silcretes and silcrete skins from the Beda Valley, southern Arcoona Plateau, South Australia. *J. Geol Soc. Aust.* **19**, 31–9.
Jack, R. L. 1915. *Geology and prospects of the region to the south of the Musgrave Ranges and the geology of the western part of the Great Artesian Basin*. Mines Dept S. Aust. Bull. 5.
Jacks, G. V. and R. O. Whyte 1939. *The rape of the Earth*. London: Faber & Faber.
Jennings, J. N. 1969. Karst of the seasonally humid tropics in Australia. In *Problems of the karst denudation*, O. Stelcl (ed.), 149–58. Studia Geogr. Clské Akad. Ved. Geograficky Ustar, Brno.
Jennings, J. N. 1976. A test of the importance of cliff-foot caves in tower karst development. *Z. Geomorph. Suppl.* **26**, 92–7.
Jennings, J. N. and M. M. Sweeting 1963. Caliche pseudoanticlines in the Fitzroy Basin, Western Australia. *Am. J. Sci.* **259**, 635–9.
Jutson, J. T. 1914. *An outline of the physiographical geology (physiography) of Western Australia*. Geol Surv. W. Aust. Bull. 61.
Jutson, J. T. 1940. The shore platforms of Mount Martha, Port Phillip Bay, Victoria, Australia. *Proc. R. Soc. Vic.* **52**, 164–76.
Kemp, E. M. 1978. Tertiary climatic evolution and vegetation history in the southeast Indian Ocean region. *Palaeogeog. Palaeoclim. Palaeoecol.* **24**, 169–208.
Kessler, D. W., H. Insley and W. H. Sligh 1940. Physical, mineralogical and durability studies on the building and monumental granites of the United States. *J. Res. Nat. Bur. Stand.* **24**, 161–206.
Kidson, C. 1953. The Exmoor storm and the Lynmouth floods. *Geography* **38**, 1–9.
Klaer, W. 1957. "Verkarstungserscheinungen" in Silikatgesteinen. *Abh. Geogr. Inst. Freien Univ. Berlin.* **5**, 21–7.
Lamplugh, G. W. 1902. Calcrete. *Geol Mag.* **9**, 575.
Leith, C. K. 1925. Silicification of erosion surfaces. *Econ. Geol.* **20**, 513–23.
Linton, D. L. 1955. The problem of tors. *Geogrl J.* **121**. 470–87.
Livingstone, D. A. 1963. *Data of geochemistry. (Chemical composition of rivers and lakes)*. US Geol Surv. Prof. Pap. 440-G.
Lofgren, B. E. 1965. Land subsidence due to artesian-head decline in the San Joaquin Valley, California. In *Guidebook for field conference I: northern Great Basin and California*, 104–2. VII Congress INQUA.
Lofgren, B. E. and R. L. Klausing 1969. *Land subsidence due to groundwater withdrawal, Tulare–Wasco area, California*. US Geol Surv. Prof. Pap. 437-B.
Logan, J. R. 1849. The rocks of the Palo Ubin. *Genoots. Kunsten Wetenschappen (Batavia)* **22**, 3–43.
Logan, J. R. 1851. Notices of the geology of the Straits of Singapore. *Q. J. Geol Soc. Lond.* **7**, 310–44.
Loughnan, F. C. 1969. *Chemical weathering of silicate minerals*. London: Elsevier.
Lovering, T. S. 1959. Geological significance of accumulator plants in rock weathering. *Bull. Geol Soc. Am.* **70**, 781–800.
Ludbrook, N. L. 1969. Tertiary Period. In *Handbook of South Australian geology*, L. W. Parkin (ed.), 172–203. Adelaide: Geol Surv. S. Aust.
Mabbutt, J. A. 1961a. "Basal surface" or "weathering front". *Proc. Geol Soc. Lond.* **72**, 357–8.

Mabbutt, J. A. 1961b. A stripped land surface in Western Australia. *Trans Paps Inst. Br. Geogs* **29**, 101–14.

Mabbutt, J. A. 1965. The weathered land surface in central Australia. *Z. Geomorph.* (N.S.) **9**, 82–114.

MacCallien, W. J., B. P. Ruxton and B. J. Walton 1964. Mantle rock tectonics. A study in tropical weathering at Accra, Ghana. *Overseas Geol. Min. Resour.* **9**, 257–94.

MacCulloch, J. 1814. On the granite tors of Cornwall. *Trans Geol Soc.* **2**, 66–78.

Maignien, R. 1966. Review of research on laterites. *Nat. Resour. Res. Ser.* IV, Unesco.

Mann, A. W. and R. C. Horwitz 1979. Groundwater calcrete deposits in Australia: some observations from Western Australia. *J. Geol Soc. Aust.* **26**, 293–303.

Meade, R. H. 1968. *Compaction of sediments underlying areas of subsidence in central California.* US Geol Surv. Prof. Pap. 497-D.

Merrill, G. P. 1897. *Treatise on rocks, weathering and soils.* London: Macmillan.

Merrill, G. P. 1898. *Desert varnish.* US Geol Surv. Bull. 150, 389–91.

Michel, P. 1978. Cuirasses bauxitiques et ferrugineuses d'Afrique occidentale. Apercu chronologique. *Trav. Doc. Géogr. Trop.* **33**, 11–32.

Miller, R. P. 1937. Drainage lines in bas-relief. *J. Geol.* **45**, 432–8.

Moss, A. J. 1973. Fatigue effects in quartz sand grains. *Sed. Geol.* **10**, 239–47.

Moss, A. J. and P. Green 1975. Sand and silt grains: predetermination of their formation and properties by microfractures in quartz. *J. Geol Soc. Aust.* **22**, 485–95.

Moss, A. J., P. H. Walker and J. Hutka 1973. Fragmentation of granitic quartz in water. *Sedimentology* **20**, 489–511.

Nace, R. L. 1960. *Water management, agriculture and groundwater supplies.* US Geol Surv. Circ. 415 1–11.

Netterberg, F. 1971. *Calcrete in road construction.* Nat. Inst. Road Res., Pretoria, Bull. 10.

Ollier, C. D. 1969. *Weathering.* Edinburgh: Oliver & Boyd.

Öpik, A. A. 1954. *Mesozoic plant-bearing sandstones of the Camooweal region and the origin of freshwater quartzite.* Tech. Rep. Bur. Min. Resour.

Peel, R. F. 1941. Denudation landforms of the central Libyan Desert. *J. Geomorph.* **4**, 3–23.

Peel, R. F. 1966. The landscape in aridity. *Trans Inst. Br. Geogs.* **38**, 1–23.

Pouyllau, D. and M. Seurin 1981. Geomorphologie et pseudo-karsts sur grès et quartziteo du Roraima dans la region de la Gran Sabana, (sud est due Vénézuela). *Trav. Doc. Géogr. Trop.* **44**.

Prescott, J. A. and R. L. Pendleton 1952. *Laterite and laterite in soils.* Common. Bur. Soil Sci. Tech. Comm. 47.

Price, W. A. 1925. Caliche and pseudo-anticlines. *Bull. Am. Assoc. Petrol Geols* **9**, 1009–17.

Prider, R. 1966. The lateritized surface of Western Australia. *Aust. J. Sci.* **28**, 443–51.

Pugh, J. C. 1956. Fringing pediments and marginal depressions in the inselberg landscape of Nigeria. *Trans Paps Inst. Br. Geogs* **22**, 15–31.

Ratcliffe, F. N. 1936. *Soil drift in the arid pastoral areas of South Australia.* CSIRO Pamph. 63.

Ratcliffe, F. N. 1937. *Further observations on soil erosion and sand drift with special reference to southwest Queensland.* CSIRO Pamph. 70.

Reiche, P. 1950. *Survey of weathering processes and products.* Albuquerque, NM: University of New Mexico Publications in Geology.

Renault, P. 1953. Caractères généraux des grottes gréseuses du Sahara méridional. *Congr. Int. Spéléol. (Paris) Publn* **2**(1), 1–15.

Riley, F. S. 1970. *Land- surface tilting near Wheeler Ridge, southern San Joaquin Valley, California.* US Geol Surv. Prof. Pap. 497-G.

Russell, R. J. 1958. Geological geomorphology. *Geol Soc. Am. Bull.* **69**, 1–22.

Ruxton, B. P. 1958. Weathering and subsurface erosion in granite at the piedmont angle, Balos, Sudan. *Geol Mag.* **45**, 353–77.

Ruxton, B. P. and L. R. Berry 1961. Notes on faceted slopes, rock fans and domes on granite in the east-central Sudan. *Am. J. Sci.* **259**, 194–206.

Scheffer, F., B. Meyer and E. Kalk 1963. Biologische Ursachen der Wüstenlachbildung – zur Frage der chemischen Verwitterung in den ariden Gebieten. *Z. Geomorph.* **7**, 112–9.

Sharpe, C. F. S. 1941. Geomorphic aspects of normal and accelerated erosion. *Trans Am. Geophys. Un.* 236–40.

Sivarajasingham, S., L. T. Alexander, J. G. Cady and M. G. Cline 1962. Laterite. *Adv. Agron.* **14.**

Springer, M. E. 1958. Desert pavement and vesicular layer of some soils in the Lahontan Basin, Nevada. *Proc. Soil Sci. Soc. Am.* **22**, 63–6.

Smith, B. J. 1978. The origin and geomorphic implications of cliff-foot recesses and tafoni on limestone hamadas in the northwest Sahara. *Z. Geomorph.* **22**, 21–43.

Stephens, C. G. 1964. Silcretes of central Australia. *Nature* **203**, 1407.

Stephens, C. G. 1970. Laterite and silcrete in Australia. *Geoderma* **5**, 5–52.

Szczerban, E., F. Urbani and P. Colvée 1977. Cuevas y simas en cuarcitas y metalimolitas del grupo roraima, meséla de Guaiquinima, Estado Bolivar. *Bol. Soc. Venezolana Espel.* **8**(16), 127–54.

Tanner, V. 1944. *Outlines of the geography, life and customs of Newfoundland – Labrador.* Helsinki: Tilgmann.

Thomas, T. M. 1954. Swallow holes on the Millstone Grit and Carboniferous Limestone of the New South Wales coalfield. *Geogrl J.* **120**, 468–75.

Thomas, T. M. 1963. Solution subsidence in south-east Carmarthenshire and south-west Brencanshire. *Trans Paps Inst. Br. Geogs* **33**, 45-60.

Trendall, A. F. 1962. The formation of "apparent peneplains" by a process of combined lateritisation and surface wash. *Z. Geomorph.* **6**, 51–69.

Tschang, H-L. 1961. The pseudokarren and exfoliation forms of granite on Pulau Ubin, Singapore. *Z. Geomorph.* **5**, 302–12.

Tschang, H-L 1962. Some geomorphological observations in the region of Tampin, southern Malaya. *Z. Geomorph.* **6**, 253–9.

Tucker, M. E. 1978. Gypsum crusts (gypcrete) and patterned ground from northern Iraq. *Z. Geomorph.* **22**, 89–100.

Turner, F. J. and J. Verhoogen 1960. *Igneous and metamorphic petrology.* New York: McGraw-Hill.

Twidale, C. R. 1956. Chronology of denudation in northwest Queensland. *Geol Soc. Am. Bull.* **67**, 867–82.

Twidale, C. R. 1960. Some problems of slope development. *J. Geol Soc. Aust.* **6**, 131–47.

Twidale, C. R. 1962. Steepened margins of inselbergs from north-western Eyre Peninsula, South Australia. *Z. Geomorph.* **6**, 51–69.

Twidale, C. R. 1964. Effect of variations in the rate of sediment accumulation on a bedrock slope at Fromm's Landing, South Australia. *Z. Geomorph.* **5**, 177–91.

Twidale, C. R. 1966. Geomorphology of the Leichhardt–Gilberton area of northwest Queensland. *Land Res. Ser.* **16** (CSIRO).

Twidale, C. R. 1967. Origin of the piedmont angle, as evidenced in South Australia. *J. Geol.* **75**, 373–411.

Twidale, C. R. 1968. Origin of Wave Rock, Hyden, Western Australia. *Trans R. Soc. S. Aust.* **92**, 115–23.

Twidale, C. R. 1971. *Structural landforms.* Canberra: ANU Press.

Twidale, C. R. 1976a. *Analysis of landforms.* Sydney: Wiley.

Twidale, C. R. 1976b. Geomorphological evolution. In *Natural history of the Adelaide region*, C. R. Twidale, M. J. Tyler and B. P. Webb (eds), 43–59. Adelaide: Royal Society South Australia.

Twidale, C. R. 1976c. The origin of some recently initiated exogenetic landforms, with special reference to South Australia. *Env. Geol.* **1**, 227–40.

Twidale, C. R. 1978a. Granite platforms and the pediment problem. In *Landform evolution in Australia*, J. L. Davies and M. A. J. Williams (eds), 288–304. Canberra. ANU Press.

Twidale, C. R. 1978b. On the origin of Ayers Rock, central Australia. *Z. Geomorph. Suppl.* **31**, 177–206.

Twidale, C. R. 1978c. Early explanations of granite boulders. *Rev. Géomorph. Dynam.* **27**, 133–42.

Twidale, C. R. 1980. Origin of minor sandstone landforms. *Erdkunde* **34**, 219–24.

Twidale, C. R. and J. A. Bourne 1975. The subsurface initiation of some minor granite landforms. *J. Geol Soc. Aust.* **22**, 477–84.

Twidale, C. R. and J. A. Bourne 1976a. The shaping and interpretation of large residual granite boulders. *J. Geol Soc. Aust.* **23**, 371–81.

Twidale, C. R. and J. A. Bourne 1976b. Origin and significance of pitting on granite rocks. *Z. Geomorph.* **20**, 405–16.

Twidale, C. R. and J. A. Bourne 1978. A note on cylindrical gnammas or weather pits. *Rev. Géomorph. Dynam.* **26**, 135–7.

Twidale, C. R., J. A. Bourne and D. M. Smith 1974. Reinforcement and stabilisation mechanisms in landform development. *Rev. Géomorph. Dynam.* **23**, 115–25.

Twidale, C. R., J. A. Bourne and D. M. Smith 1976. Age and origin of palaeosurfaces on Eyre Peninsula and in the southern Gawler Ranges, South Australia. *Z. Geomorph.* **20**, 28–55.

Twidale, C. R., J. A. Bourne and N. Twidale 1977. Shore platforms and sealevel changes in the Gulfs region of South Australia. *Trans R. Soc. S. Aust.* **101**, 63–74.

Twidale, C. R. and W. K. Harris 1977. The age of Ayers Rock and the Olgas, central Australia. *Trans R. Soc. S Aust.* **101**, 45–50.

Twidale, C. R., J. M. Lindsay and J. A. Bourne 1978. Age and origin of the Murray River and Gorge in South Australia. *Proc. R. Soc. Victoria* **90**, 27–42.

Twidale, C. R., J. A. Shepherd and R. M. Thomson 1970. Geomorphology of the southern part of the Arcoona Plateau and of the Tent Hill Region west and north of Port Augusta, South Australia. *Trans R. Soc. S. Aust.* **94**, 55–67.

Urbani, F. 1977. Novedades sobre estudios realizados en las formas carsicas y pseudocarsicas del escudo de Guayana. *Bol. Soc. Venezolana Espel.* **8**, 175–97.

Verrall, S. T. 1975. *Origin of minor sandstone landforms, southern Flinders Ranges*, South Australia. BA hons. thesis. University of Adelaide.

Vogt, J. 1953. Érosion des sols et techniques de culture en climat temperé maritime de transition (France et Allemagne). *Rev. Géomorph. Dynam.* **4**, 157–83.

Washburn, A. L. 1956. Classification of patterned ground and review of suggested origins. *Geol Soc. Am. Bull.* **67**, 823–66.

Wayland, E. J. 1934. *Peneplains and some erosional landforms.* Geol Surv. Uganda Ann. Rep. Bull. 1. 77–9.

Wellman, P. W. 1971. *The age and palaeomagnetism of the Australian Cainozoic volcanic rocks.* PhD dissertation. ANU Canberra.

White, S. E. 1973. Is frost action really only hydration shattering? A review. *Arctic Alp. Res.* **8**, 1–6.

White, W. A. 1944. Geomorphic effects of indurated veneers on granites in the southeastern States. *J. Geol.* **52**, 333–41.

White, W. B., G. L. Jefferson and J. F. Haman 1966. Quartzite karst in southeastern Venezuala. *Int. J. Speleol.* **2**, 309–16.

Wilhelmy, H. 1958. *Klimamorphologie der Massengesteine*. Brunswick: Westermann.

Wopfner, H. 1960. On some structural development in the central part of the Great Australian Artesian Basin. *Trans R. Soc. S. Aust.* **83**, 179–94.

Wopfner, H. 1978. Silcretes of northern South Australia and adjacent regions. In *Silcrete in Australia*, T. Langford-Smith (ed.), 93–141. Armidale: University of New England Press.

Wopfner, H., R. Callen and W. K. Harris 1974. The Lower Tertiary Eyre Formation of the southwestern Great Artesian Basin. *J. Geol Soc. Aust.* **21**, 17–51.

Wopfner, H. and C. R. Twidale 1967. Geomorphological history of the Lake Eyre Basin. In *Landform studies from Australia and New Guinea,* J. N. Jennings and J. A. Mabbut (eds), 118–42. Canberra: ANU Press.

Wright, R. L. 1963. Deep weathering and erosion surfaces in the Daly basin, Northern Territory. *J. Geol Soc. Aust.* **10**, 151–64.

6
Potential effects of acid rain on glaciated terrain

William W. Shilts

Introduction

The concern about acid precipitation has prompted a great deal of research into the potential effects of acid loading on the terrain of eastern North America. Ideally, the response of terrain to acid loading is measured in terms of "sensitivity," the potential of the ecosystem to be changed by acid loading. Alternatively, the response is predicted on the basis of the terrain's "buffering capacity," or ability to absorb protons without significantly reducing the natural pH of surface- or groundwaters. In reality, assessments of sensitivity involve the evaluation of a number of terrain components that interact in complex ways. Thus, estimates of sensitivity and methods of estimating sensitivity vary widely among geologists, biologists, pedologists, limnologists, etc.

In most discussions of sensitivity, the terrain is divided into terrestrial and aquatic components, and the effects on each are generally discussed separately. However, groundwater has indirect effects on both terrestrial and aquatic systems and is a critical link between them. The composition of the rock or unconsolidated medium through which groundwater flows is of critical importance in evaluating the ultimate composition of the water as it is used by surface vegetation or as it enters the surface-water (aquatic) drainage net. The composition of the unconsolidated glacial overburden and of the glacial and modern sediment in lakes is particularly critical in evaluating the effects of acid precipitation on terrestrial and aquatic systems.

Effects of glaciation

It is recognized that varying mineralogical properties of bedrock have a strong influence on the chemical and physical properties of its unconsolidated cover or overburden, and various maps depicting the bedrock influence have been prepared (Research Consultation Group 1979, 1980, Hendrey *et al.* 1980, Shilts *et al.* 1981). However, the significance of one important geological factor, the distortion of the bedrock signature on terrestrial and aquatic systems by processes associated with glaciation, has only recently been pointed out

(Coker & Shilts 1979, National Research Council 1981). Thus, the boundary marking the maximum extent of Pleistocene glaciation in North America is a critical dividing line between soils formed by *in situ* weathering of bedrock on the one hand and soils formed on unconsolidated sediments produced by processes associated with glacial erosion and deposition on the other.

Studies of the sub-bottom of arctic and southern Canadian Shield lakes have brought to light another effect of glaciation that may have an important influence on the sensitivity of lakes to acid loading. Using low-frequency sonar profiling equipment, it has been possible to map thicknesses of as much as 25 m of glacial and pro-glacial sediments underlying the modern (post-glacial) sediment of lakes. The modern sediments are variable in thickness, comprising 0–4 m of fill on top of till, pro-glacial varved clay, outwash gravels or marine sediments. These thick, glacial and late-glacial sediments are generally inorganic and fine grained and constitute a body of material by which acid waters may be buffered, either by direct contact at or near the sediment–water interface or by passing through the sediments as groundwater inflow into the lake.

All but about 5% of the land area of Canada, mountainous areas in the western United States, and the northeastern third of the United States has been glaciated. In all but the glaciated regions of the high arctic, the unconsolidated cover over bedrock (henceforth referred to as drift or overburden) is not composed of weathering products resulting from the chemical breakdown of bedrock, but is composed largely of products of physical abrasion, formed by clast-to-clast contact in the ice or by the interaction of bedrock and stones held firmly in the base of the glaciers. This mode of formation of the debris is transported by glaciers and ultimately incorporated into drift or waterlain sediments has two important implications with regard to effects of acid precipitation:

(a) The mineral components of glacial deposits and sediments derived from them include phases that under normal weathering circumstances would be quickly destroyed, their chemical components being redistributed among other, more stable secondary minerals. Many of the labile phases, particularly the primary carbonate minerals, impart buffering capacity; others, such as sulphides or finely divided phyllosilicate minerals, may contain potentially noxious metals that can be readily mobilized because of the labile nature of their host lattices. Finally, because the minerals have been liberated from their rock matrix and crushed to fine sizes, their internal surface area has been greatly augmented, increasing the potential for reaction, under the appropriate eH conditions, with ground- or surface waters containing abnormal concentrations of protons.

(b) The materials produced by glacial erosion were transported one or more times by glaciers for varying distances from their source outcrops. Consequently, in many areas, overburden has mineralogical and chemical characteristics bearing little similarity to those of underlying bedrock. In exceptional cases, such as southwest of Hudson and James bays, these mineralogically foreign deposits may completely "seal off" the effects of underlying bedrock (Coker & Shilts 1979, National Research Council 1981).

Glacial deposits have direct effects on agricultural or forest soils developed on them and on modern lake or stream sediments derived in part from them. In addition, groundwater, which eventually enters lakes or the surface drainage net, must pass through and may be chemically altered by glacial overburden. Because the bulk of the aquatic and terrestrial systems that are under the most immediate threat of damage from acid rain are located in glaciated areas, it is important that the special chemical and mineralogical characteristics of *glacial overburden* be distinguished from the characteristics of *residual soils* that cover most of the area beyond the glacial boundary.

Sensitivity to acid loading of glacial deposits of the Frontenac Arch

In 1980, a sampling project was initiated in the region of the Precambrian Frontenac Arch or Axis of Ontario. Its objectives are: (a) to map those natural chemical and mineralogical characteristics of glacial and derived sediments that are perceived to bear directly on the sensitivity of terrestrial and aquatic systems to acid loading, i.e to man-induced depression of the natural pH of precipitation; and (b) to assess, through sub-bottom profiling of major lakes, the distribution, type and thickness of glacial and pro-glacial overburden as well as post-glacial lake sediment. This area was chosen for a number of reasons, among which were its tourist economy that is strongly dependent on preservation of the natural state of its aquatic systems, its geology, and the ease of access to most of the region.

Bedrock geology

The Frontenac Arch is a southeastward extension of the Grenville Structural Province of the Canadian Shield and connects the main body of the Shield to the Adirondack Mountains across the Thousand Island segment of the St. Lawrence River (Baer *et al.* 1977). Unlike most of the rest of the Grenville and older Precambrian provinces, the Frontenac Arch includes extensive areas of carbonate metasediments. It is also characterized by several massive granitic plutons and less important plutons of mafic affinity. Northeasterly trending belts of metavolcanic and non-calcareous metasedimentary rocks make up most of the rest of the terrane except for a major belt of intrusive alkalic rocks that lies near its northwestern boundary, sub-parallel to its contact with the Algonquin Batholith.

The Frontenac Arch is flanked on its northeast and south sides by relatively flat-lying, unmetamorphosed Paleozoic sedimentary rocks. The Paleozoic formations are predominantly limestones and dolomites except along the northeastern contact where quartzites and orthoquartzites of the Nepean Sandstone (Potsdam Sandstone in New York) form the base of the Paleozoic section. Along both the northeast and south flanks of the Frontenac Arch, it is common to find deeply weathered zones on the Precambrian rocks. Some of these are obvious remnants of the Precambrian–Paleozoic paleosol that underlies Nepean Sandstone, but it is possible that some of this weathered surface may be much younger, perhaps Tertiary or inter-glacial in age.

Surficial geology

The Frontenac Arch was glaciated during the Wisconsinan by ice moving predominantly south to southwest, except late in the last glacial phase when lobes of thinning ice were deflected down major topographic features southeastward along the Ottawa–Bonnechere graben and westward along the St. Lawrence–Lake Ontario lowlands (Gadd 1980, Henderson 1973). The higher parts of the region are mantled by a discontinuous cover of till, the mineralogic and lithologic components of which generally reflect southward or southwestward transport. Major valleys in the higher areas are commonly filled with outwash gravels, eskers and hummocky ice-contact gravels with minor and poorly delineated areas of windblown sand and laminated glaciolacustrine deposits. Extensive deposits of lacustrine and marine silty clay occur below altitudes of about 100 m in areas adjacent to the St. Lawrence River valley. A heavy cover of marine clayey silt and scattered deposits of nearshore sand and gravel occur below ~170 m in the Ottawa River valley from Renfrew eastward.

Field procedures

Samples were collected from natural and artificial exposures along roads, along streams, or in pits in areas easily accessible by road. In most cases, till was sampled because it is thought to represent best the sediment load carried by the last glaciers to cover the area and, therefore, to reflect the parent material from which all other unconsolidated sediments or soils were derived by weathering, or by wind or water sorting. The genesis of the till is difficult to determine in most cases, but in this region, a number of samples are flow or leeside tills. The nature of the environment of deposition of the till sampled is of little significance in this compositional study, however, because in any case, till represents accurately the local composition of glacial debris that was the parent material for all other sediments or soils with the following exception: in some areas, such as the Ottawa valley or St. Lawrence valley, which are covered heavily by marine or lacustrine clayey silt or silty clay, the lacustrine or marine deposits were sampled. In these areas, the water-laid sediments constitute the bulk of the overburden and are composed of detritus derived from glaciers or fluvial erosion and dispersed widely throughout major depositional basins.

At many sites, multiple samples were collected, either to compare compositional parameters among different surficial sediment facies at a single site or to investigate vertical variation in one particular facies, such as till. Vertical variations are caused by variation in sedimentation processes or by alteration by weathering or groundwater flow.

Profiling of lake basins was carried out using a Raytheon RTT–1000A–1 Portable Survey System with a dual low-frequency (3.5 & 7.0 kHz) transducer coupled with a high-frequency (200 kHz) transducer. This system was mounted on a 4.5 m inflatable Zodiak boat. Navigation was accomplished by "dead reckoning" between headlands, and bottom penetration in the 3.5 kHz mode was more than 25 m under optimum conditions. Resolution of sediment structures down to 0.5 m is possible with this system.

Laboratory procedures

Several analytical tests were performed on the samples to provide information on potential buffering capacity and on potential for mobilization of noxious metals by decreasing pH of ground- or surface waters. Calcium carbonate equivalent was determined using a modification of a procedure developed for the Leco Carbon Analyzer, which destroys carbonate minerals by burning the samples at high temperature and detects the amount of carbon in the CO_2 evolved (Foscolos & Barefoot 1970). The samples were burned twice, once untreated and once after dissolution in concentrated HCl. The amount of carbon evolved after the HCl treatment was considered to represent non-carbonate carbon (graphite, Quaternary or modern plant detritus, and other organic compounds). The carbon from the untreated sample less the carbon from the treated sample was considered to be carbonate carbon and was converted into calcium carbonate equivalent, as if all carbonate carbon was evolved from calcite.

To evaluate partitioning of carbonate minerals among various grain sizes in the till, carbonate concentrations were determined in several grain sizes in samples collected in vertical profiles from two, deep, natural exposures of calcareous till near Burnstown, Ontario, and near Thetford Mines, Quebec.

Carbonate equivalent was determined on the $< 64\,\mu m$ (silt and clay) fraction of all till and fine-grained, water-laid sediments and on the fine-sand

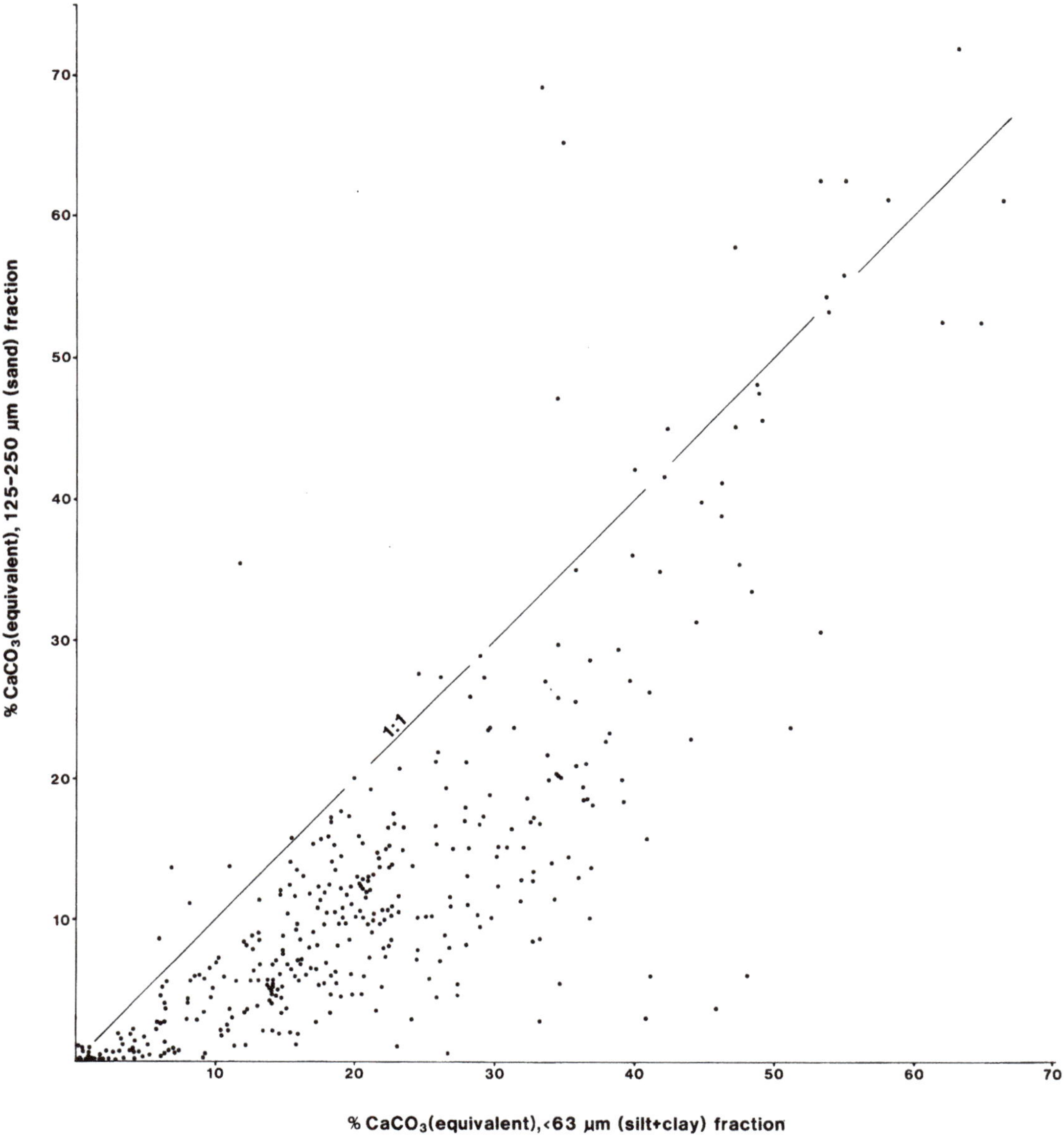

Figure 6.1 Comparison of carbonate content in silt and clay versus sand fractions of till samples from Frontenac Arch.

(125–250 μm) fraction of all till and sand or gravel samples (Fig. 6.1). This duplication was carried out first to see whether the finer fractions were being preferentially leached in the near-surface environment and second to establish the relationships of carbonate concentration between the two fractions. The latter comparison was made to determine if it would be possible to estimate the buffering potential of till from samples of its surrogate, coarse-grained, ice-contact gravel and sand. Ice-contact deposits were collected where till exposures were poor or lacking. On the assumption that the coarser sediments are equivalent to till that has had silt and clay removed by meltwater processes, the carbonate analyses of sand locally should be a good measure of the buffering capacity of nearby till that is poorly exposed for sampling.

Clay-sized (<2 μm) particles were separated from all till and fine-grained, water-laid sediments. Clay fractions were analyzed for selected trace and minor elements. Concentrations of copper, lead, zinc, molybdenum, cadmium, chromium, nickel, iron and manganese were determined by atomic absorption after a hot, one hour, mixed-acid ($HCl-HNO_3$) leach. Arsenic was determined colorimetrically, and uranium was determined using fluorimetric methods. All analyses were done in a commercial laboratory on samples prepared at the Geological Survey of Canada.

These two suites of analyses are thought to yield significant information on natural regional variations in buffering capacity of overburden, and in regional potential for overburden to release toxic metals to the ground- or surface-water system under the stress of significant depression of pH by acid precipitation.

Results

Figure 6.2 illustrates the variation of carbonate concentrations in various grain-size fractions of till from the Burnstown and Thetford Mines sections. In

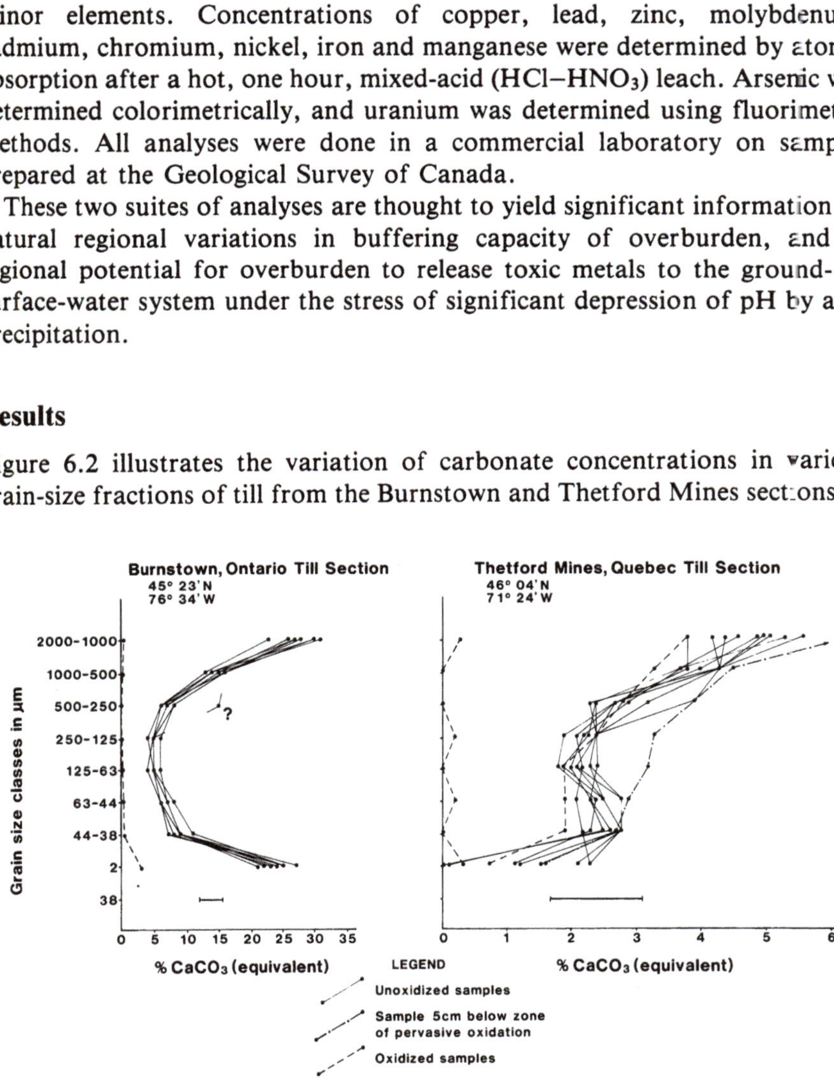

Figure 6.2 Distribution of carbonate (% $CaCO_3$ equivalent) in various size fractions of till from vertical profiles in tills of moderate (Burnstown) and low (Thetford Mines) carbonate concentration. Samples collected at ~3 m vertical intervals in 20 m section at Burnstown; at ~1 m intervals in 10 m section at Thetford Mines.

general, it can be seen that in the Appalachian tills, carbonate concentrations are relatively high in the gravel and silt sizes, very little carbonate being found in the clay sizes. Medium and fine sand-sized grains contain roughly half as much carbonate as do the gravel and silt sizes. These results are predicted by the terminal mode concept of Dreimanis and Vagners (1972) which suggests that glacial crushing and abrasion produce a bimodal size distribution of carbonate minerals, a phenomenon related to the physical characteristics both of the carbonate minerals and of the rocks in which they occur. An interesting feature revealed by Leco analysis of the Appalachian tills is a high concentration of clay-sized, non-carbonate carbon, presumably representing graphite or organic complexes derived from the black slates of the region.

In the till section from the Madawaska River at Burnstown, Ontario, the Appalachian pattern is repeated, except that the carbonate content increases sharply in the clay fraction. Clay separated from these tills by centrifugation was examined with a scanning electron microscope with X-ray energy dispersive capability. Oversized ($> 2 \mu m$) aggregates of clay-sized particles, apparently cemented by calcium carbonate, were observed in the separates examined (Fig. 6.3). It is thought that the anomalous concentrations of carbonate in the clay-sized fraction, at least at this particular site, originate through post- or syndepositional precipitation of carbonate from carbonate-rich groundwaters. As a result, the total carbonate content of these tills is not solely the product of glacial dispersal of clastic carbonate as seems to be the case with Appalachian tills, but includes a secondary component produced by chemical precipitation from groundwater circulating within the glacial sediment during or after deposition. The secondary chemical component is only important in analysis of separates containing the finest size fractions, such as the $< 64 \mu m$ fraction. The carbonate may be precipitated preferentially at the face of the exposure from which the samples were collected, a phenomenon observed on gravel pit faces in the region.

Dispersal of carbonate

Figures 6.4 and 6.5 show carbonate concentrations in the silt–clay and sand sizes, respectively, of till and other overburden samples from the study area. Although the maps show the same general patterns of carbonate dispersal, the map of carbonate concentrations in sand has more sample points because it contains the well-sorted sand and gravel samples from sites where till samples were unobtainable. As discussed above, these ice-contact sands and gravels serve as surrogate samples for till where the latter is poorly exposed and difficult to sample.

(a) (b)

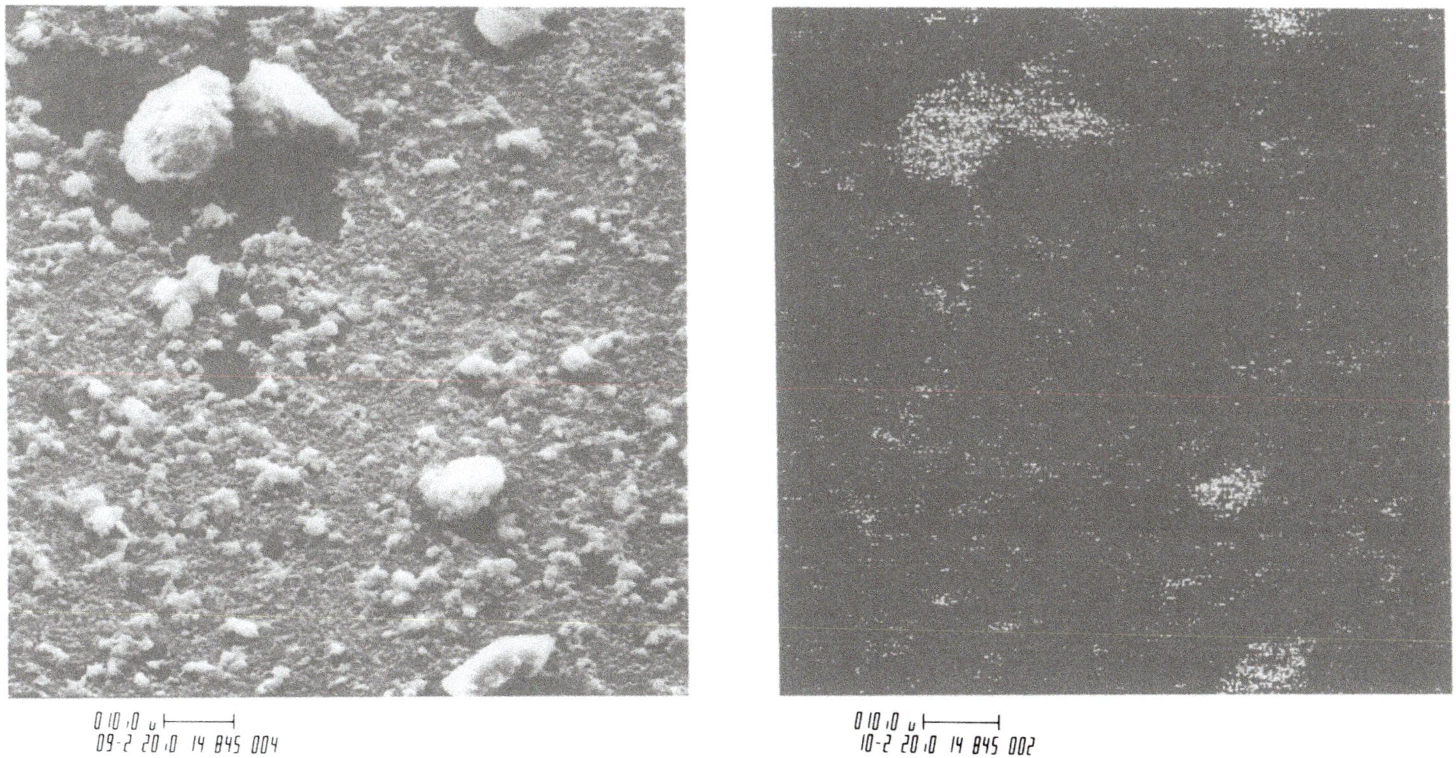

Figure 6.3 Scanning electron micrographs of $< 2\,\mu m$ debris from till in Burnstown, Ontario, section. (a) Secondary image showing $> 2\,\mu m$ aggregates of clay-sized detritus; (b) X-ray map of calcium for image A. Note that calcium is concentrated in aggregates suggesting that they are cemented by $CaCO_3$.

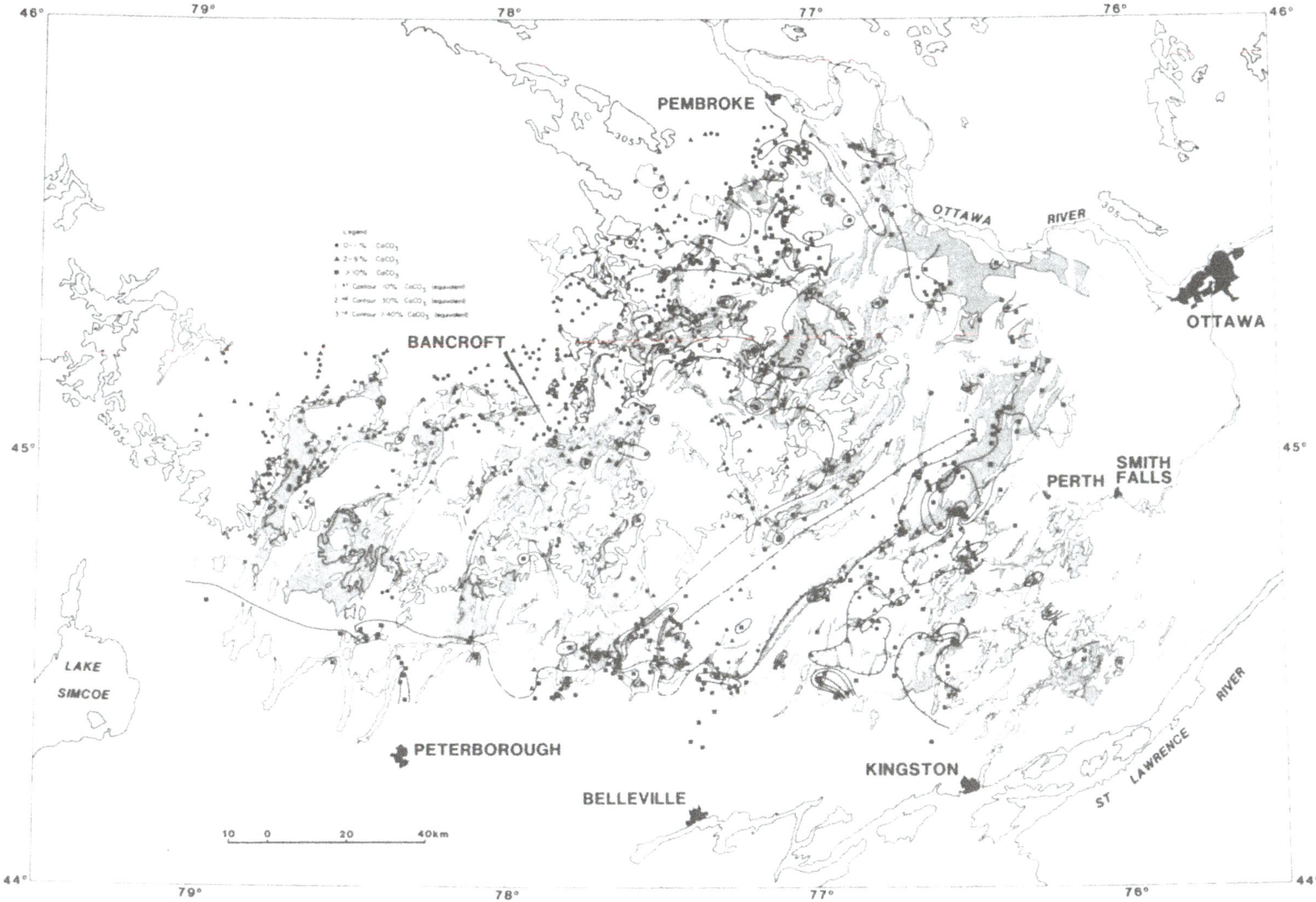

Figure 6.4 Map of carbonate ($CaCO_3$ equivalent) concentrations in silt and clay (< 64 μm) fractions of till and derived glacial sediments

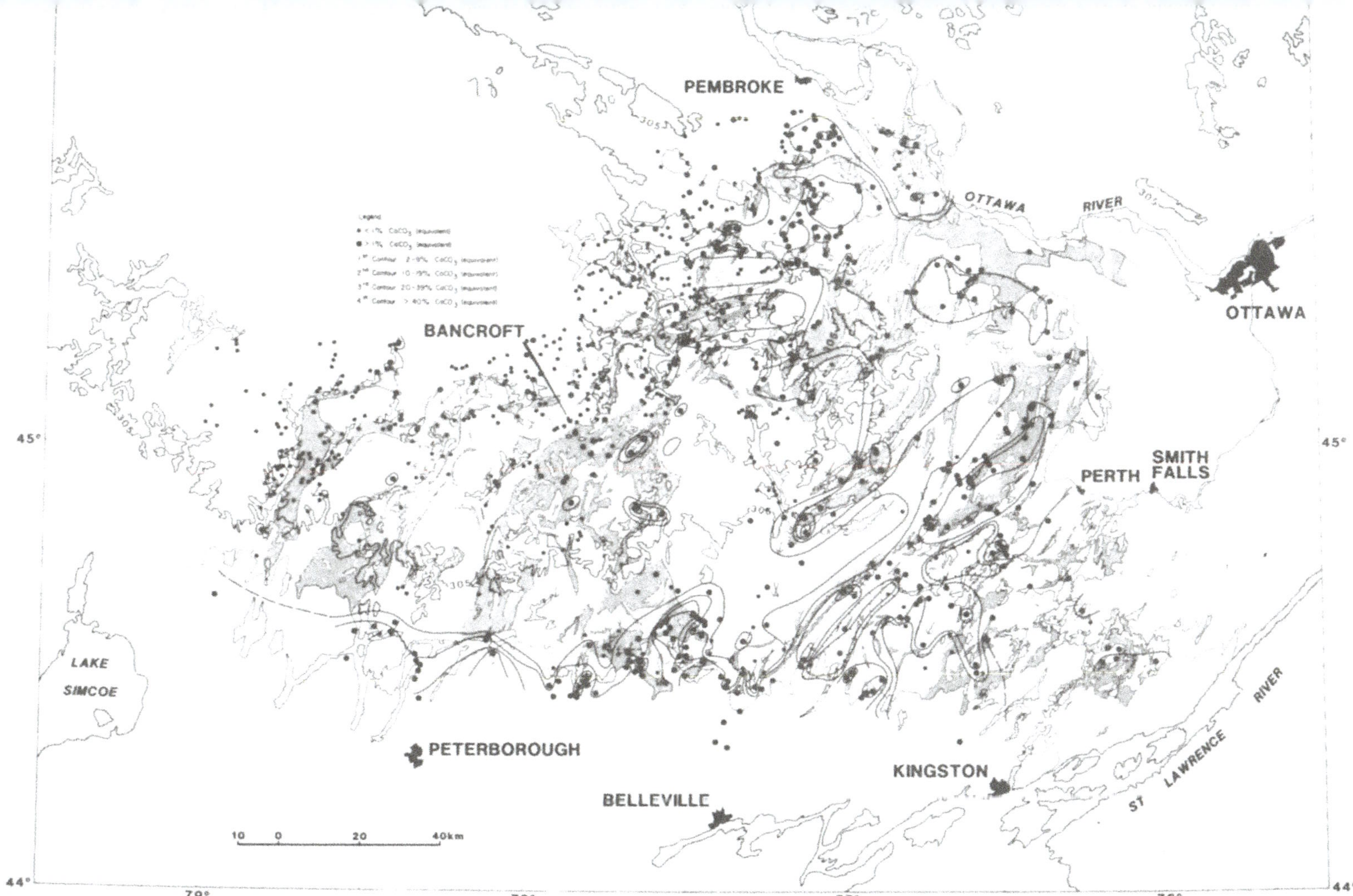

Figure 6.5 Map of carbonate ($CaCO_3$ equivalent) concentrations in sand (125–250 μm) fractions of till and derived glacial sediments.

Table 6.1 shows the comparison of sand from ice-contact gravels and from till in localities where both facies were found at one site. In general, the carbonate content of the two facies is comparable, the greatest discrepancies occurring between flow tills and ice-contact gravels that they cap at two of the three exposures of flow till. This could be caused by weathering in the flow till, which is the uppermost unit in these sections and on which the post-glacial soil was developed. At the other flow till site, where sand compositions are comparable, the samples were collected > 3 m below the post-glacial solum.

The apparent carbonate dispersal patterns for both size fractions may be affected by weathering or leaching, a problem that has long plagued attempts to use regional carbonate contents of glacial sediments for correlation or dispersal studies (Merritt & Muller 1959). Merritt and Muller (1959) have shown that at lower carbonate contents, leaching depths increase radically. In an attempt to assess and control the effects of partial leaching, several profiles were sampled in tills with varying background carbonate contents. Table 6.2 summarizes the results of measurements on some of these profiles and indicates that at depths < 1 m below the original surface, severe leaching effects are noted, regardless of the original carbonate content of the samples. In some cases, as in the particularly sandy till near Packenham, Ontario (SAR 0008, etc.), leaching effects are apparent in the silt–clay fraction at depths of 2 m, but the sand fraction is affected only in the upper meter of the section.

Table 6.1 $CaCO_3$ equivalent in < 250 > 125 μm (sand) fraction of ice-contact and till facies at same site.

Sample no.	Mode of occurrence	%$CaCO_3$ equivalent (sand)
SAR 0004	till beneath sand-gravel	10.1
SAR 0004 A	sand in ice-contact gravel	9.3
SAR 0042	flow till	13.0
SAR 0042 A	flow till	12.9
SAR 0042 B	ice-contact sand	12.5
AR 0005 B	ice-contact sand	7.9
AR 0005 A	unoxidized till below sand	10.4
AR 0005	unoxidized till below sand	10.5
AR 0008 B	till clast in ice-contact gravel	28.6
AR 0008	ice-contact sand	22.9
AR 0025	flow till	14.9
AR 0025 A	ice-contact gravel	18.1
AR 0026	flow till	12.3
AR 0026 B	red flow till	14.1
AR 0026 C	flow till	15.0
AR 0026 A	ice-contact sand	18.6

Table 6.2 Profiles of $CaCO_3$ equivalent through oxidized till.

Till sample no.	Location, notes	%$CaCO_3$ (silt and clay)	%$CaCO_3$ (sand)	Sample depth (m)
SAR 0008 C	Packenham,			
	Ontario;	1.3	3.8	1.0
8 B	8B and 8C oxidized	14.3	12.6	2.0
8 A		23.3	13.3	2.3
8		24.1	13.9	3.0
SAR 0023 A	highly oxidized	1.4	0	0.5
SAR 0023	thin till on			
	weathered marble	16.0	10.2	1.5
SAR 0027 K	section on	0.8	0	1.0
27 J	Madawaska River,	14.8	5.3	1.5
27 I	Burnstown, Ontario;	13.8	5.4	3.0
27 H	secondary carbonate	12.5	5.8	4.5
27 G	along roots in 27I,	13.3	5.9	6.5
27 F	27J; 27K highly	14.8	4.5	8.5
27 E	oxidized	13.8	5.8	10.5
27 D		14.2	5.3	13.0
27 C		14.9	7.9	15.0
27 B		13.3	5.5	18.0
27 A		14.3	4.5	19.7
27		14.8	5.3	21.0
SAR 0033	highly disturbed, oxidized till	13.4	2.3	1.0
SAR 0033 A	till	18.6	6.4	2.0
SAR 0036 B	oxidized till	0.4	0	1.0
SAR 0036 A	till	17.5	10.5	2.0
SAR 0036	till	20.3	10.3	3.0
AR 0001 A	oxidized till	15.9	5.9	1.0
AR 0001 B	till	26.6	6.7	2.0
AR 0001	till	26.3	7.3	2.0
AR 0013 C	oxidized till	13.1	9.1	1.5
AR 0013 B	unoxidized till	21.7	14.5	3.0
AR 0013 A	unoxidized till	21.1	13.1	3.0
AR 0013	unoxidized till	21.8	13.8	3.0
AR 0014 E	oxidized till in	18.9	8.3	~2.0
AR 0014 D	marble quarry	16.3	6.9	~3.0
AR 0014 C	till	16.8	8.3	~4.0
AR 0014 B	till	15.1	6.9	~6.0
AR 0014 A	till	17.3	9.3	~7.0
AR 0014	till on marble	17.4	8.8	~8.0
AR 0023 B	leeside till	25.6	10.2	0.5
AR 0023	leeside till	31.9	12.6	4.0

In many cases, a small amount of carbonate is detected in the silt–clay fraction even though all of the carbonate has been leached from the sand fraction, and the sample appears to be thoroughly weathered. In the zone of complete leaching, secondary calcium carbonate is often observed along roots or root casts, and traces of this finely divided material may supply the small amount of carbonate detected.

Considering the potential for inconsistent carbonate contents caused by leaching that can be expected in samples from < 2 m depth, it is surprising that the patterns of carbonate dispersal are as clear-cut as shown on Figures 6.4 and 6.5. It should be borne in mind that individual low carbonate values or even groups of low values could result from lack of adequately deep sample sites.

The carbonate dispersal patterns shown by both size fractions of till and other overburden indicate that carbonate content of drift that lies on bedrock of the eastern part of the Frontenac Arch far exceeds that of the western part. The northeast–southwest trend of carbonate enrichment partially reflects the similarly trending strike of both the marble outcrops and the topography; it also partially reflects the generally southwestward dispersal of carbonate debris by the last glaciers (Gadd 1980). The lack of significant carbonate content in tills of the western part of the Arch, even where they overlie marble, suggests that much of the carbonate in the eastern part was derived from Paleozoic rocks lying northeast of the Arch in the Ottawa–St. Lawrence Lowlands. The marbles do not seem to have contributed much carbonate to the tills in the west, and the higher carbonate contents of tills lying on marble along the northeast flank of the Frontenac Arch may be largely an accident of topography; marbles generally underlie topographic depressions along which Paleozoic debris was transported preferentially. The possibility of topographic control is further suggested by the effects of the low (300 m) escarpment that forms the north edge of the Madawaska Highlands. This modest escarpment appears effectively to have blocked glacially transported Paleozoic debris from being carried southward or southeastward into the Highlands, except along major river valleys.

Dispersal of trace elements

The distribution of anomalous concentrations of potentially toxic metal in surficial sediments has only been plotted systematically for a few elements. For the purposes of this chapter, I will discuss the dispersal of arsenic, both because of its implications for environmental geochemistry studies and because its provenance can be related easily to geological features of the study area. Similar discussions could be written for all of the other trace and minor metals analyzed. It should be noted that the analyses on which this discussion is based were done on the $< 2\,\mu m$ fraction of till and other samples. Trace element concentrations in this chemically reactive, very fine, grain size are much more important to consider in studies of mobilization than would be analyses of coarser fractions.

Three major areas of arsenic enrichment stand out clearly on Fig. 6.6. The clay fractions of tills in these areas contain five to several hundred times the < 2 ppm background concentration of arsenic present in tills elsewhere in the sample area. Arsenic is absorbed onto or is structurally part of the phyllosilicates or secondary iron–manganese oxides and hydroxides that make up the bulk of the clay-sized fraction. Therefore, it is more likely to be mobilized by hydrolysis or exchange reactions with excess amounts of protons than if it was concentrated in the coarser fractions.

The three major areas of arsenic-enriched till occur within a triangle, the apices of which are Mazinaw Lake, Bob's Lake and the towns of Marmora and Deloro. This triangle encompasses the original Ontario goldmining belt that was the site of active mining in the late 1800s and early 1900s. Gold occurrences were worked along northeast–southwest striking belts of metasedimentary and metavolcanic rocks where gold mineralization was found typically in vein quartz and was accompanied by arsenopyrite and pyrite. The principal

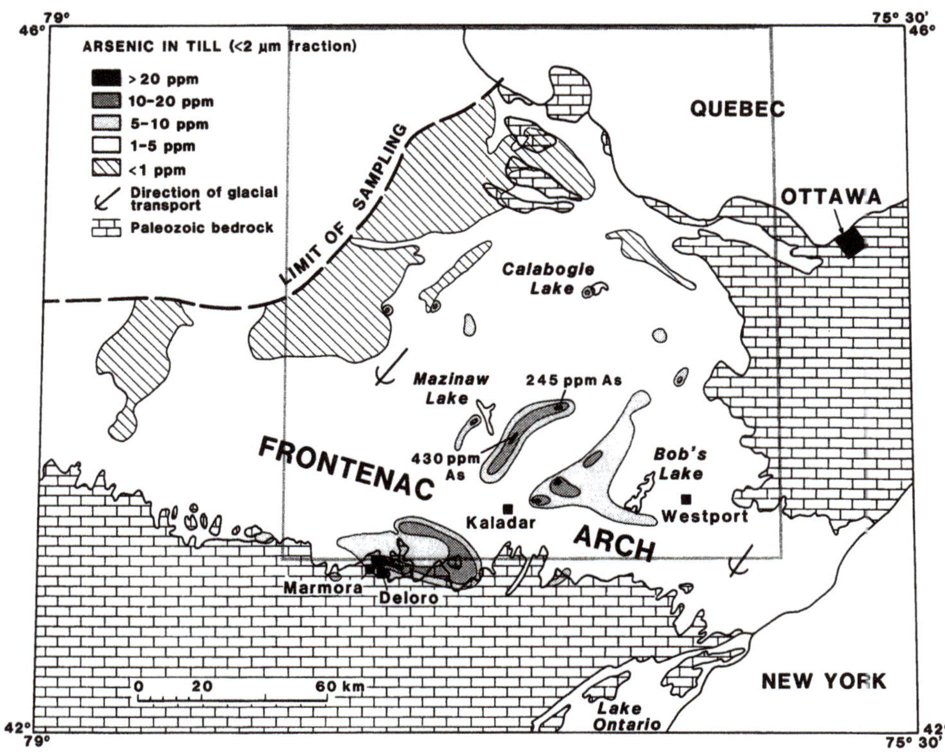

Figure 6.6 Arsenic concentrations in ~850 evenly spaced samples of till and derived sediments of the Frontenac Arch. Shaded box shows location of area depicted on Figure 6.7.

mines in the region were at Deloro and at nearby Cordova (Hewitt & Freeman 1972).

Arsenic concentrations in till faithfully reflect the bedrock belts known to have hosted gold mineralization. Arsenic being a geochemical pathfinder for gold in the study area as well as elsewhere, it is not surprising that the glacial sediments should be enriched in arsenic. It is somewhat surprising, however, to note the magnitude of the arsenic concentrations and the large area over which the high concentrations occur, the typical gold-bearing veins being no wider than a few meters or tens of meters. The relatively large area of arsenic enrichment may result from glacial homogenization of debris from many unknown sources occurring over an area much larger than that covered by the gold occurrences that have been discovered and/or mined.

The patterns of natural concentrations of arsenic in the tills of this region correspond very closely to patterns of arsenic anomalies in lake sediments (Coker 1981) (Fig. 6.7). This correspondence is a clear-cut example of the impact of drift composition on the chemistry of the sediment that settles in lake basins. The sediments and (presumably) waters of the surface- and groundwater drainage net, being derived from and filtered by overburden, faithfully reflect the chemical peculiarities of the drift.

The arsenic dispersal pattern further points out a common problem in

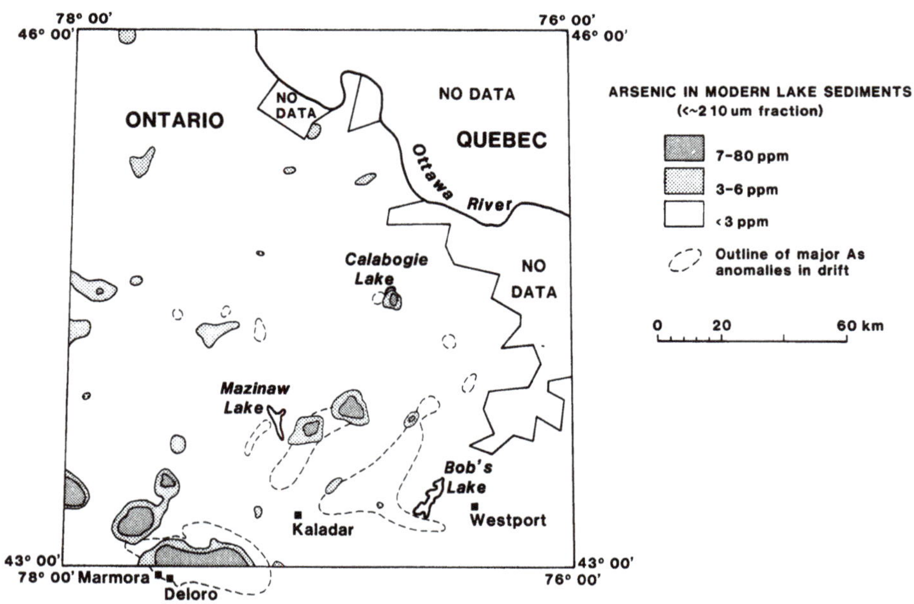

Figure 6.7 Arsenic concentrations in lake-bottom sediments from 10 cm or more below sediment–water interface of 1250 lakes on Frontenac Arch (modified from Coker 1981).

environmental geochemistry, the question of whether a toxic (usually inorganic) substance is naturally or artificially enriched in the environment. Because gold mining and refining were carried out in the vicinity of Deloro over several years, it is possible that arsenic released from the ore or tailings during or after processing could have entered drainage or groundwater systems with accompanying deleterious effects on plant and animal life. It is certain that the overburden of the region around Deloro is naturally enriched to 10–20 times the background levels of arsenic that are found elsewhere on the Frontenac Arch and that sediment derived by natural erosion of this landscape over the past 10 000–12 000 years has been enriched in arsenic, as indicated by the high arsenic levels in samples from well below the sediment–water interface in the lakes of the region (see Coker & Jonasson 1977, p. 69, for details of sampling procedure in lakes). Thus, the legal question of whether natural or anthropogenic processes are responsible for elevated levels of arsenic in the various ecological units of the environment around Deloro would be difficult to resolve. Metal enrichment cannot be attributed casually to man's activities without concerted study of the bedrock and glacial sediment geochemistry, as illustrated by the classic study of Hornbrook and Jonasson (1971).

Finally, the natural arsenic concentrations in tills of this region illustrate our concern for the possible effects of proton loading of the waters passing through and over such metal-enriched drift. The glacial deposits of this and other parts of the Canadian Shield characteristically display significant local metal anomalies (any of the metals analyzed in this project could have been used to make the same point). In the absence of buffering components, such as carbonates, which can consume the excess protons, hydrolysis or cation exchange processes may free such metals from their more labile hosts. Even in the areas of carbonate bedrock and carbonate-rich drift, there is a strong potential for zinc–cadmium mineralization in the marbles and for copper–uranium mineralization in the base of the Paleozoic carbonate section (March Formation) north of the Frontenac Arch. In fact, strong cadmium and zinc anomalies have been found in the drift over some of the marble outcrops. If such metals are present in carbonate phases, they may be released in the course of the buffering reaction. The more destruction of carbonates or silicates that takes place as a result of buffering the increased proton flux caused by increasing acidity of precipitation, the more likely metals are to be mobilized in the aquatic and terrestrial system. Therefore, it seems to the author that in estimating the sensitivity of glaciated terrain to acid loading, some term should be added to account for potential mobilization of toxic trace metals from natural substances (research on pH-induced mobilization of metal in plumbing or water distribution systems has already been initiated, but distinguishing between internal and external sources of metal will be difficult). Presently, leaching studies are concentrating on mobilization of toxic minor elements such as aluminum.

Sub-bottom profiling

Systematic surveys of the sediment fill of major lakes in the northeastern part of the Frontenac Arch were begun in 1981. Figure 6.8, which shows the configuration of the sedimentary fill in Calabogie Lake, illustrates the relative importance of late-glacial and modern, lacustrine fill. Modern, organic-rich, lake sediments are typically almost transparent to the 3.5 kHz acoustic signal and are thus only thought to reach significant thickness (2–4 m) at the east end of the profile. The deformed, laminated sediment that forms a cover > 20 m thick over the rugged Precambrian outcrops or till-mantled outcrops of the eastern two-thirds of the profile makes up the basin fill. The deformation of the laminated sediment is thought to be caused by lateral shortening of the beds of pro-glacial lake sediment as a result of subsidence over masses of buried, melting, glacier ice. Such fill can provide significant buffering capacity both at the sediment–water interface and for groundwaters entering a lake basin through fractures in surrounding bedrock.

Discussion and conclusions

Overburden in glaciated areas has several characteristics that are important to understand in evaluating the potential effects of acid rain. Glacial overburden is largely produced by physical abrasion of rock, and thus contains all of the mineral components that are present in fresh, unweathered bedrock, except in the uppermost meter or two where post-glacial weathering and soil formation have modified the original mineralogy significantly. Many of the unweathered components may be chemically active or labile minerals, such as carbonates, and may provide effective buffering action against precipitation of lower than normal pH. Other labile minerals, such as sulphides or clay-sized phyllosilicates, may release noxious metals into the terrestrial and aquatic environment in the course of their degradation. In contrast, in the areas of residual soil beyond the limit of glaciation, overburden is formed largely by chemical destruction of bedrock and so is largely composed of less reactive or more stable secondary mineral phases. Thus, the glacial boundary is of critical significance in determining what techniques must be adopted to study the effects of acid rain and what the nature and severity of the effects of acid loading are likely to be.

A second important characteristic of glaciated terrain is transportation by glaciers and their associated meltwaters of physically comminuted rock components away from their source outcrops. With respect to acid rain, this phenomenon is most important where glaciers transported a load of debris with abundant buffering components onto bedrock with little potential for producing, through glacial or weathering processes, soils with buffering capacity. Glacial dispersal is important in northwestern Ontario where calcareous drift derived from the Hudson Bay basin and lowlands forms a con-

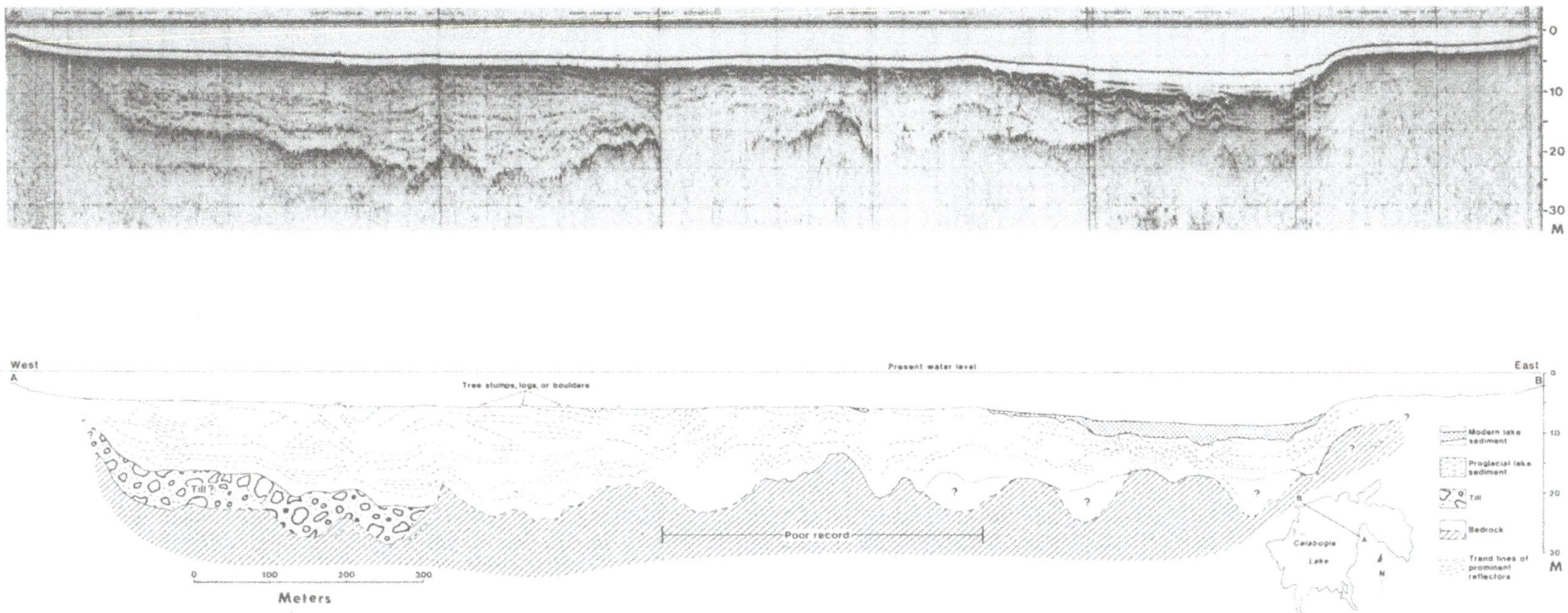

Figure 6.8 Sub-bottom profile and profile interpretation, Calabogie Lake, Calabogie, Ontario. Heavy vertical lines on upper profile are due to chart recorder malfunction. Heavy line above bottom on upper profile is electronically offset 200 kHz record of sediment–water interface; sub-bottom obtained with 3.5 kHz signal. Present lake level estimated to be 2–4 m above natural level as a result of damming.

tinuous cover over granitoid bedrock for several hundred kilometers south and southwest of the carbonate basin, "sealing off" the effects of the underlying, largely granitoid bedrock. Glacial transport can also carry minerals with potentially noxious metal components away from their source outcrops to spread metal-rich drift over an area many times greater than the area of the source outcrops.

Glacially transported debris can also be transported in pro-glacial streams, seas or lakes, often forming thick fill in depressions, such as those occupied by present-day lakes. This fill may be so thinly covered by modern lake sediment that it can react with lake water at or near the sediment–water interface. Groundwater entering a lake may be filtered through it. If these sediments contain significant amounts of fine-grained carbonate minerals, as they commonly do, especially in central Canada, they may provide considerable buffering capacity. The abundant clay-sized particles in marine or glaciolacustrine sediments also provide buffering capacity through hydrolysis.

The relationship of the effects of glaciation to evaluation of potential effects of acid rain is well illustrated by studies of glacial dispersal and pro-glacial lacustrine sedimentation in the Precambrian terrane of southeastern Ontario. Calcareous debris, derived from the Paleozoic limestones and dolomites that occur northeast of the Frontenac Arch, has been transported across the lower parts of the Precambrian terrain, providing an overburden cover with significant buffering capacity, regardless of the lithologies of the underlying bedrock. In areas of higher relief, however, transportation of this material, presumably carried near the base of the glacier, was blocked, and drift composition is much more closely adjusted to local lithologies. The marble belts that occur throughout the Frontenac Arch appear to have provided little carbonate debris to the glacier load and do not impart significant buffering capacity to drift much beyond their outcrops. Although the reasons for the contrast between the response of Precambrian carbonate beds and Paleozoic carbonate beds to glacial erosion is not entirely clear, the author suspects that it is related primarily to the massive, little fractured, coarse-grained nature of the marble outcrops as opposed to the thinly bedded, highly jointed, Paleozoic outcrops. Material was apparently much more easily stripped from the latter terrain by glacial erosion than from the former, providing a greater abundance of carbonate debris per area of outcrop and more highly debris-charged, basal, glacial ice.

In the lake basins studied with low-frequency acoustic, sub-bottom, profiling systems, as much as 25 m of laminated glacial sediment has been detected. Lake bottoms are covered by a discontinuous mantle of modern lake sediment that tends to fill depressions, leaving slopes and much of the flat floor of lakes available for reactions with lake water at the glacial sediment–water interface. Groundwater entering the lake may be filtered through and react with the thick pro-glacial or marine sediments, any excess of protons being consumed by hydrolysis.

Finally, large arsenic anomalies, representing glacial erosion and transportation of debris from arsenic-rich outcrops, occur in till of the east-central part of the Frontenac Arch. The till anomalies correspond closely to arsenic anomalies in lake-bottom sediments, the inorganic components of which were largely derived from glacial overburden. This pattern illustrates (a) the glacial dispersal of a component to cover an area much greater than that of its source outcrops; (b) the degree to which natural variations in metal levels can constitute an environmental menace; (c) the effects of drift composition on sediments derived from it or groundwaters filtered through it; and (d) the potential hazard of metal mobilization in such anomalous areas by abnormally decreasing pH of meteoric waters.

Acknowledgements

My colleague, Inez Kettles, has worked closely with me on the Frontenac Arch project and has carried out much of the sampling on which parts of this chapter are based. Trace-element analyses were carried out by Bondar-Clegg and Co. Ltd. on separates provided by the Drift Composition Lab of the Geological Survey of Canada. Carbonate analyses of the samples were carried out by personnel in this lab under the direction of P. J. Higgins. J. Hornsby, L. Farrell and R. A. Klassen directed the sub-bottom profiling.

References

Baer, A. J., W. H. Poole and B. V. Sanford 1977. *Rivière Gatineau. Geol Surv. Can.* Map 1334A.

Coker, W. R. 1981. *National geochemical reconnaissance, southeastern Ontario.* Geol Surv. Can. 1:2 million colored compilation map series Open File 747.

Coker, W. B. and I. R. Jonasson 1977. Geochemical exploration for uranium in the Grenville province of Ontario. *Can. Inst. Min. Metal. Bull.* **70**(781), 67–75.

Coker, W. B. and W. W. Shilts, 1979. Lacustrine geochemistry around the north shore of Lake Superior: implications for the evaluation of the effects of acid precipitation. In *Current research. Geol Surv. Can.* Pap. 79-1C, 1–15.

Dreimanis, A. and U. J. Vagners 1972. The effect of lithology upon texture of till. In *Research methods in Pleistocene geomorphology*, E. Yatsu and A. Falconer (eds), 2nd Guelph symposium on geomorphology, 1971, University of Guelph. Geogr. Publ 2, 66–82.

Foscolos A. E. and R. R. Barefoot 1970. *A rapid determination of total organic and inorganic carbon in shales and carbonates; a rapid determination of total sulphur in rocks and minerals.* Geol. Surv. Can. Pap. 70-11.

Gadd, N. R. 1980. Late-glacial regional ice-flow patterns in eastern Ontario. *Can. J. Earth Sci.* **17**, 1439–53.

Henderson, E. P. 1973. *Surficial geology of Kingston (north half) map-area, Ontario.* Geol Surv. Can. Pap. 72-48.

Hendry, G. R., J. N. Galloway, S. A. Norton, C. L. Schofield, P. W. Shoffer and D. R. Burns 1980. *Geological and hydrochemical sensitivity of the eastern United States to acid precipitation.* US Envir. Prot. Agency Rep. 600/3-80-024. Corvallis, Oregon: Corvallis Environmental Research Laboratory.

Hewitt, D. F. and E. B. Freeman 1972. *Rocks and minerals of Ontario.* Ont. Dept Mines North. Affairs Geol Cir. 13.

Hornbrook, E. H. W. and I. R. Jonasson 1971. *Mercury in permafrost regions: occurrence and distribution in the Kaminak Lake area, Northwest Territories (55 L)*. Geol Surv. Can. Pap. 71-43.

Merritt, R. S., and E. H. Muller 1959. Depth of leaching in relation to carbonate content of till in central New York state. *Am. J. Sci.* **257**, 465–80.

National Research Council of Canada 1981. Acidification in the Canadian aquatic environment: scientific criteria for assessing the effects of acidic deposition on aquatic ecosystems. In *Environmental Secretariat*, H. H. Harvey (chr.), NRCC Publ. 18475. Ottawa: NRCC Publications.

Research Consultation Group 1979. The LRTAP problem in North America: a preliminary overview. In *Report of the United States–Canada Research Consultation Group on the long-range transport of air pollutants*, A. P. Altshuller and G. A. McBean (cochr.). US State Department & Canada Department of External Affairs.

Research Consultation Group 1980. *Second report of the United States–Canada Research Consultation Group on the long-range transport of air pollutants*. A. P. Altshuller and G. A. McBean (cochr.). US State Department & Canada Department of External Affairs.

Shilts, W. W., K. D. Card, W. H. Poole and B. V. Sanford 1981. *Sensitivity of bedrock to acid precipitation: modification by glacial processes. Sensitivity of bedrock and derived soils to acid precipitation, southcentral and southeastern Canada*. Geol Surv. Can. Pap. 81-14, Maps 1549A, 1550A, 1551A.

7
Hydrologic classification of caves and karst

John E. Mylroie

Introduction

Karst landscapes form where solutional removal of bedrock is the dominant process in landform development. Rock solution is usually manifested by formation of conduits (caves) in the subsurface. The term karst is derived from the classic locality in western Slovenia (Yugoslavia) where the slavic word *krs* means crag or stone (Jennings 1971). The German form of the word, "karst," has come into international usage.

The general public poorly understands karst landforms and processes, as evidenced by its opinion of cave conduits and its use of land in karst areas. Solution caves (as opposed to lava tubes or tectonic caves) are common throughout the world and are universally the source of colorful superstition and folklore. Caves are often regarded as the lair of demons, as gateways to hell or as foreboding, nasty places by people who otherwise consider themselves enlightened and informed. More serious is the actual use of karst areas by inhabitants. Solution conduits, because they focus water flow in the subsurface under conditions remarkably different from those on the surface, are vulnerable and may contribute to a wide variety of environmental problems. Solution conduits can transmit pollutants without filtration over long distances in a very short time. Sinkholes and pits are convenient disposal sites (intentional or otherwise) especially if they are deep and subsiding actively. On the other hand, the term "spring water" has traditionally assumed an almost mystical water-quality connotation. Recent advertisements by a well known distillery laud the superior qualities of its limestone spring water. The public may put great trust in the purity of water from such springs, when in many cases the springs are merely outlets of conduits delivering water from convenient waste sites.

The major experience of the public, and some geologists, with caves is by way of a commercial or show cave. These caves are arranged for public display, and the emphasis is on visual impact and entertainment rather than explanation of processes and products. The outgrowth of this is that the public, geologists included, misunderstand the function and importance of caves. As an example, out of 13 introductory geology texts on the author's shelf, 11 show a cave photograph, each one a picture of secondary calcite deposits (stalactites, stalagmites), primarily from two caves, Carlsbad Caverns, New Mexico, and Luray Caverns, Virginia. This is analogous to discussing plate tectonics and showing a picture of slickensides. Further lack of understanding of caves is demonstrated by the three texts that show a cave map. In each case (Cameron Cave, Missouri; Mark Twain Cave, Missouri; and Anvil Cave, Alabama), the cave shown is a maze cave, an unusual and very unrepresentative type. This commentary about cave photographs and cave maps in geology textbooks points out an ignorance syndrome that has plagued cave geology since the days of the maze and minotaur in ancient Crete.

This chapter will review what karst forms are and what function they perform in the evolution of karst landscapes. Karst landscapes are fluvial landscapes wherein the soluble nature of the rock results in significant chemical weathering and subsurface water flow. Other, more common fluvial processes also occur and are important. However, the fact that much of the karst process takes place, occultly, in the subsurface, has led to many misconceptions about how these processes work. This chapter will organize karst forms by function, and the chapters following will discuss genesis, mechanisms and interpretation.

Classification of karst forms according to function

Karst forms can be classified by their position and function in a hydrological regime, i.e. part of the terrestrial, fluvial component of the hydrologic cycle. The study of many of the complexities of cave formation and the development of karst landforms often leads to a classic "can't see the forest for the trees" situation. The investigator must remember that the karst features under study perform functions. Explanations of mechanisms, processes and genesis must be reconciled with reality.

The classification of karst forms presented here suffers from arbitrary oversimplification that affects all classifications; nonetheless, one can order and label a suite of forms that appear to have some common origin. Some new terminology is introduced and used although the karst literature has quite enough terms already. This is done so the investigator might better understand the natural features being studied and so can unambiguously transmit this understanding to workers elsewhere. The terms offered here are unimportant; the hydrologic environment they attempt to describe is important.

Landforms at any scale that evolve due to solution of bedrock can be separated into two main groups: those formed by surficial solution and those developed through subsurface waterflow. These two groups can be considered to some extent independently of each other. The surficial features form because surface water falls or flows on them. The subsequent destination of the water is not important. Subsurface features can be considered to form independently of surficial features although the nature of the surface can affect the chemistry, volume and input mode of the water. The classification described is outlined as follows:

- I. Surficial karst forms
  - A. Exposed
  - B. Mantled
- II. Interface forms
  - A. Insurgences
    1. Diffuse
    2. Confluent
  - B. Resurgences
    1. Gravity spring
    2. Artesian spring
    3. Overflow spring
  - C. Intersection forms
    1. Vertical
    2. Lateral
- III. Subsurface forms
  - A. Active cave passages
    1. Tributary passage
    2. Master cave passage
    3. Diversion passage
    4. Tapoff passage
    5. Abduction passage
  - B. Abandoned cave passage

Surficial forms

Surficial forms are found on bedrock surfaces altered solutionally by local rainfall, snowfall, condensation and runoff, independent of the eventual flow path of the water (surface or subsurface). Extensive literature describes solutional forms on exposed limestone surfaces (see Jennings 1971, Sweeting 1973, Quinlan 1979), and no attempt is made here to further subdivide or classify these forms except to note that they tend to fall into two broad categories: those formed on exposed bedrock surfaces and those formed on mantled bedrock surfaces. Exposed bedrock surfaces are in equilibrium contact with the atmosphere, and waterflow on and over them is of an unconfined, often mechanically energetic nature. Mantled bedrock surfaces are often not in geochemical equilibrium with the atmosphere, and water flowing on and over the bedrock must pass through a permeable medium (soil) and so is rarely mechanically energetic. Mantled bedrock surfaces may be covered with impervious overburden and not be in significant contact with circulating water. Quinlan (1979) gives an exhaustive review of karst forms and classifies karst in terms of cover rock type, climate, geologic structure, physiography, hydrology, modification during and after karstification, and dominant landforms. It is important to realize that karst forms, regardless of the multitude of factors that control their expression and morphology, are a link in the chain of water transport over and through a soluble material. Function must be clearly understood before morphology and history can be explained.

Interface forms

At the interface where surficial and subsurface environments meet, water may enter the subsurface (insurgences), or be discharged from the subsurface (resurgences). Where the surface environment intersects the subsurface environment due to some weathering process not involving the mass transfer of water, intersection forms occur. Because interface forms vary widely in scale and structure, their surface expression assumes a wide variety of morphologies. Whereas the terms insurgences, resurgences and intersection forms readily describe hydrologic processes by which the surface and subsurface interact, these terms do not describe the configuration that the surface landscape will assume at interface points.

As in the discussion of surficial karst forms, no attempt is made here to describe and classify the morphology of karst depressions other than to say they are all, by definition, interface forms. Differences in morphology among depressions are a measure of the properties of the materials present and the rates at which the karst processes work on them. The basic processes involved, however, are the same – solutional and gravitational removal of material from the surficial environment into the subsurface environment.

Insurgences **Insurgences** are points of input for surface water into subsurface conduits. Great variation can exist in the morphology of an insurgence depending on the structure and lithology of the soluble rock, the local relief, and the volume of water sinking underground. Sweeting (1973) discusses insurgences and their terminology and prefers to call them "swallow holes." She then subdivides these forms into groups based on morphology. The term swallow hole connotes a volume and implies existence of a cavity or hole into which the water goes. The term insurgence is free of size and volume connotations and describes no specific morphological feature. The following subdivisions of Sweeting (1973) are used here with some modification:

(a) *Diffuse insurgence*. Water enters the limestone by percolation (through overburden if it is present) as a series of small, diffuse inputs into small openings in the limestone. The small openings are usually the primary permeability of the carbonate rock or joint and fracture permeability. Given a large expanse of limestone, the water enters in approximately equal volume over the entire area. Water concentration takes place in the subsurface environment in the case of diffuse insurgences, not in the surface environment as with confluent insurgences (see below).

(b) *Confluent insurgence*. Water enters the limestone after it is concentrated into identifiable streams that sink into the limestone at one or more discrete points, localities that can be seen and measured as point inputs. In many cases, growth and evolution of subsurface conduits and interface forms result in abandonment of some confluent insurgences in favor of newly formed ones upstream. The adjective **abandoned** can be used to describe those insurgences no longer in use. Insurgences that are only utilized when insurgences upstream cannot handle the water flow are called **overflow insurgences**.

Where confluent insurgences tend to cluster, they are called a **confluent insurgence complex**. Given a pre-existing subsurface conduit, insurgences may cluster around its course because the conduit is the drain for water entering the surface area above it. Insurgences may also cluster because hydraulic inefficiencies in one insurgence require the use of overflow insurgences in times of flood. Insurgences also cluster because the soluble rock is exposed over a limited area; surface water flows to that area to sink.

Confluent insurgence formation depends on the presence of some capping layer, such as impervious lithology or thick overburden, that provides a surface on which meteoric water can collect and flow as confluent streams. At the actual point of contact of the surface stream with the limestone, the conditions existing there (water volume, lithologies and overburden) control the morphology of the insurgence.

Diffuse and confluent insurgences represent end members of water input into the subsurface.Cave systems that collect insurging water usually gather such water from both diffuse and confluent inputs. Cave systems in areas such as the Bahamas (Mylroie 1978), where the soluble material is almost entirely very porous, eolian calcarenite, come closest to approximating the completely diffuse insurgence condition. Cave systems such as the Friar's Hole Cave System (Medville 1981), where the limestone appears as very small windows in otherwise clastic terrain, approach the completely confluent insurgence condition.

The terms diffuse and confluent are not directly comparable to the terms authigenic (autochthonous) and allogenic (allochthonous). Authigenic input defines water derived from the soluble rock catchment only, whereas allogenic water is collected on non-soluble terrain and delivered to the soluble terrain. As discussed by Palmer (1984) the origin of the insurging water has great geochemical importance. The terms confluent and diffuse describe the volume of water whereas authigenic and allogenic are terms that describe the source of the water. In areas where thick residual soil exists or where the limestone bedrock lacks well-developed primary and secondary porosity, the water insurgence will be authigenic but confluent. In areas where recharge passes through an overlying porous sandstone or ash bed, the water insurgence will be authigenic but diffuse. The equating of confluent water with allogenic water and diffuse water with authigenic water is usually correct as a broad rule

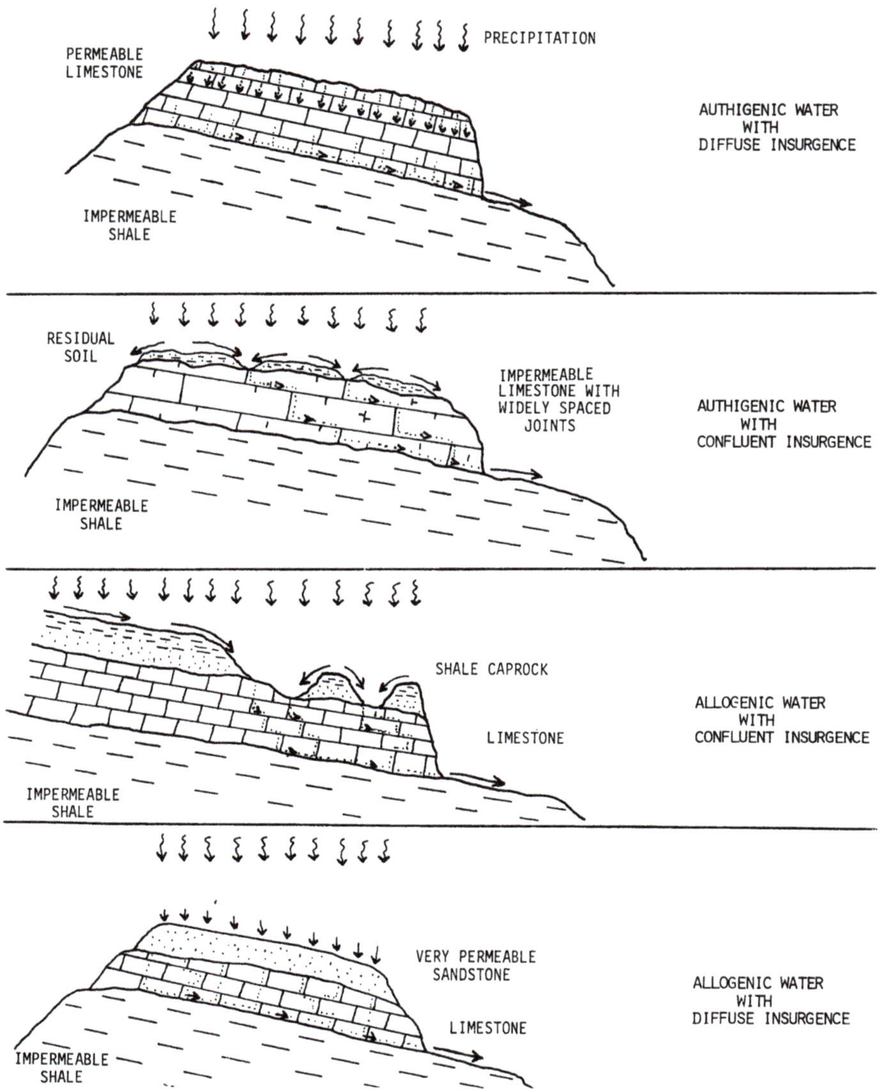

Figure 7.1 Examples of water source and methods of insurgence.

of thumb, but confusion among the terms is possible and avoidable. Figure 7.1 demonstrates the relationships described by these terms.

Resurgences **Resurgences** occur at the interface between surface and subsurface environments where water that is collected at insurgences and transmitted by solution conduits returns to the surficial environment. Resurgences, or karst springs, are the downstream end members of karst terrains. They include the following types of springs:

(a) *Gravity springs.* These are resurgences where water under gravity flow leaves the subterranean conduit much like the flow of a surface stream, following a down-slope gradient. The actual opening to the surface may be totally or partially blocked by collapse or colluvium, obscuring the spring and causing local ponding in the conduit behind the spring.

(b) *Artesian springs.* These springs are those in which water from a confined conduit flows due to a hydrostatic pressure gradient and not a down-slope gradient. Artesian springs include **Vauclusian** springs and **alluviated** springs. The *Vauclusian spring* is named after the type example at the Fontain de Vaucluse in France (Sweeting 1973). Under hydrostatic pressure, water flows up and out of an ascending, solutional bedrock conduit. Ebbing and flowing springs are special cases of Vauclusian springs. *Alluviated springs* were described by Powell (1963). These are springs in which the conduits emerge below local base level into unconsolidated sediment and flush out openings to the surface through which the water escapes. Water is under hydrostatic pressure and is not gravity flow. Alluviation is usually due to changes in regional base level (common in glaciated regions). A special case is the submarine spring where the "alluviation" is produced by ponded water (fresh or salt), and the conduit discharges its water into this larger ponded feature below the surface.

(c) *Overflow springs.* Overflow springs are resurgences that flow only during flood in a cave system when the normal spring is incapable of handling the volume of water transmitted to it. An overflow spring can act as either a gravity spring or an artesian(Vauclusian) spring depending on the morphology of the passages connecting it to the main (low flow) conduit. During flood conditions, what would normally be a gravity spring can take on the internal characteristics of an artesian spring. Overflow springs are commonly associated with the downstream distributary pattern of many large cave systems (see Palmer 1984).

Jennings (1971) and Sweeting (1973) differentiate between **resurgence**, which they define as reappearance of a sinking stream (confluent insurgences) and **exsurgence**, which is the reappearance of local infiltration of water (diffuse insurgence). Both investigators admit the difficulty of demonstrating a spring to be a resurgence only, an exsurgence only, or a combination of both. Jennings states that almost every resurgence must have some exsurgent input. The term exsurgence can be applied as an adjective modifying the term resurgence if exsurgent qualities can be demonstrated to be significant.

As with insurgences, maturation of karst may result in abandonment of resurgences in favor of newer ones. In this case, the resurgence classification is modified with the adjective *abandoned*.

Intersection forms. In many karst areas, connections exist between surface and subsurface environments that do not relate to appreciable water insurgence to, or resurgence from, a subsurface conduit system. These are called **intersection forms**. They are generally caused by weathering phenomena other than solution, although solution may contribute greatly. An intersection form may evolve into an insurgence (or resurgence) by capture of water from nearby areas. **Vertical intersection**, or collapse due to mechanical failure of the roof into a void, is the most common type of intersection form. Because the voids are caused by solution, it can be argued that intersection forms are solutional in origin: however, their final expression is usually due to other weathering processes. **Lateral intersection** occurs when retreating cliff lines and slopes uncover pre-existing solution conduits.

Subsurface forms

Subsurface solution conduits are best studied as cave systems, forms produced during part of the karst hydrologic cycle. Individual caves are segments of an active or abandoned subsurface flow path. Cave systems carry (or carried) water from a point or group of points on the surface to a point or group of points elsewhere on the surface. Solutional cave systems were classified in terms of genesis by Ford (1977). White (1977) looked at cave systems as part of the carbonate aquifer. A functional classification, presented here, sidesteps (with conceptual risk) the genetic aspects of the cave system and treats it as a *fait accompli*. The function of the various parts of the cave system in transmitting water is important to the overall system's interpretation.

Cave conduits are generally only explored as cave segments, as it is rare that an investigator can travel the total distance from insurgence to resurgence. A terminology is developed in this chapter that describes individual caves in terms of their function with regard to the whole cave system. The following terminology does not imply specifically what local conditions (vadose or phreatic) existed at the time of the cave segment formation, or what feature (structure or lithology) controlled its formation. The term **cave system** is used to describe the sum of the subsurface solution conduits that exist in one subsurface drainage basin. Cave systems, being enclosed in rock, may interact and compete in three dimensions. These individual segments are defined as to function.

Active cave passages **Active cave passages** within a cave system carry water from insurgences to resurgences, either permanently or seasonally. They consist of the following types of passages:

(a) A **tributary** *passage* enters the cave system and brings water directly from an insurgence.
(b) The **master** *cave passage* is the primary conduit through which the tributaries discharge water to the resurgence(s).
(c) A **diversion** *passage* carries water in competition with the master cave (see Palmer 1984) and can be a **loop** passage, which is a competing passage that carries water from the same upstream source as another passage to

the same downstream objective within the cave system, or an **overflow** *passage*, which is a passage that during flood or high water carries water past hydraulic constrictions in the cave system to the cave system downstream or to an overflow spring.

(d) A **tapoff** *passage* is a newly formed passage (with respect to the rest of the cave system) in the distal portion of the cave system. It carries water to a new resurgence by the abandonment of, or in competition with, an earlier resurgence.

(e) An **abduction** *passage* forms to remove all or part of the water from one cave system into a neighboring pre-existing cave system.

Abandoned cave passages. **Abandoned cave passages** no longer carry water. Where enough data exist, the former function of the inactive passage can be applied, prefixed with the adjective *abandoned*.

Discussion

This classification is arbitrary to some degree in organization and application. Sweeting (1973) classifies caves into three main types: phreatic, vadose and vertical. The first two types are based on apparent genesis of the passage; the last involves an explorational bias (all caves have a vertical component). None of the three terms describes the primary function of the cave system, which is to carry water through the subsurface from insurgence(s) to resurgence(s). Jennings (1971) classifies caves as inflow, outflow, through and between allowing the limits of exploration to determine the cave's classification. This, too, is an unacceptable means of describing cave passage function.

The cave classification used in this chapter avoids explorational bias and the need to describe the solutional processes that formed the passage. One finds difficulty in deciding, at the junction of two equal-sized cave streams, which one is the tributary. Overflow passages may be difficult to label because they may look like abandoned master caves, utilized only in flood conditions when the cave is inaccessible (an explorational bias). The distinction between tapoff passage and abduction is critical because the former is a common speleological occurrence, but the latter is rare. Tapoff formation is commonly the adjustment of a cave stream in the distal portions of the cave system to local base-level changes. These changes are caused by influences of surface conditions such as increased jointing and joint gapping, blocking of resurgences, and intersection by slopes and scarps. Tapoff formation is not to be confused with a total cave system re-adjustment of passage elevations in response to a regional drop in base level, with the subsequent abandonment of the higher-elevation passages. Abduction is the capture of water from part or all of one cave system by a pre-existing, competing cave system. The term piracy is used to describe this phenomenon on the surface, but the three dimensions seen in limestone solutional conduits make rigid application of this term confusing. In the speleological literature, the term piracy is used to describe the capture of water from one cave passage by another, regardless of whether the captur-

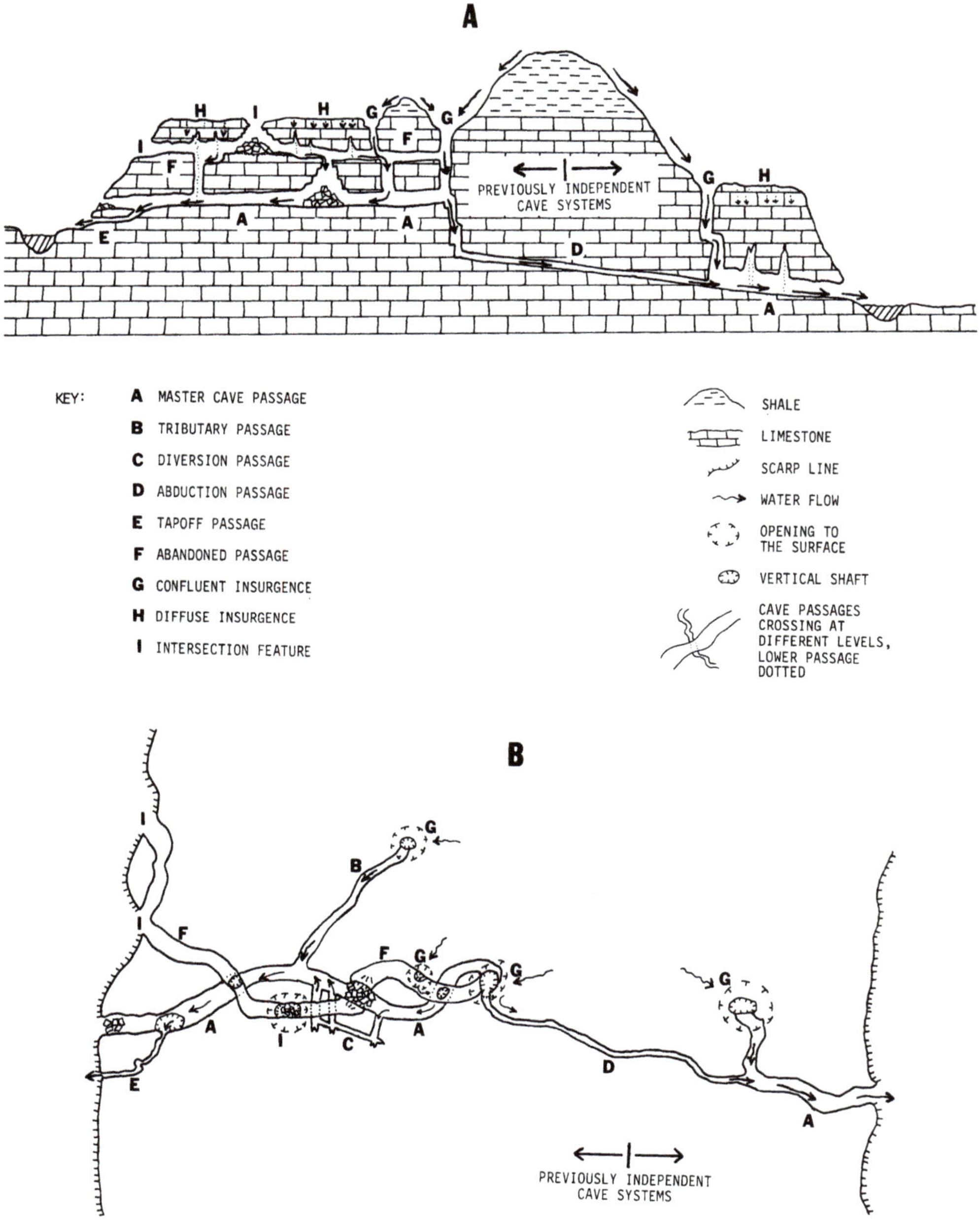

Figure 7.2 (A) Idealized cave systems, shown in cross section, displaying some of the interface and subsurface features described in the text. (B) Continued, idealized cave system shown in plan view, displaying some of the interface and subsurface features described in the text. Note the different amount of information contained on the plan view. Use of both the plan view and the cross section provides the best representation of cave systems.

ing passage was a diversion, tapoff, overflow route or abduction passage. The term abduction is preferred because it has no previous speleologic interpretation and therefore can be strictly defined.

Figure 7.2a is a schematic diagram showing various karst forms. The accompanying cave map (Fig. 7.2b) represents an essential piece of the karst data base. Successful interpretation of cave passages as shown on a cave map can convert a confusing mass of lines into a logical sequence of hydrological events. Figure 7.4 of this chapter, and Figures 8.11, 8.12 and 8.13 in Palmer (1984) are good examples of cave maps that serve as data bases. The various cave components shown on these maps reveal their hydrologic function within the overall cave system.

Karst topography results from solutional sculpturing of bedrock initiated when this soluble material enters the domain of the hydrologic cycle. The philosophy of the foregoing classification presumes that each individual karst form evolves as part of the water transfer chain or because it interacts with parts of the water transfer chain. Any karst landform can be categorized and studied as a surficial, interface or subsurface form. Cave systems function as water transmitters, collecting water from one surface area and delivering it to another. In interpreting karst landscapes, the role of karst forms as participants in the hydrologic cycle must be considered.

Interpretation

Evolution of a karst landscape can be greatly influenced by how groundwater makes its way to a resurgence. Figure 7.3 shows a variety of subsurface flow situations, controlled by lithology, structure and interaction with a master surface stream. Given the proper soluble lithology placed above base level (the A to B distance in Figure 7.3, where B represents base level with respect to water resurgence from the cave system) with an adequate source of recharge, karst landforms develop. Assuming that recharge and soluble material are available, the interaction of the structure with the base-level stream will produce primarily two main flow paths – down the dip or along the strike, or often a combination of both. Figure 7.4 shows the formational history of cave systems in a karst area of east-central New York (Mylroie & LaFleur 1977). Fox Creek has incised through the limestones (dip SSW at 2°) on the down-dip side of the Helderberg Plateau. Cave systems here developed sequentially, as Fox Creek continued headward incision eastward. In the Cobleskill Creek area, incision occurred also on the down-dip side, but the incision of the master stream, Schoharie Creek, provided favorable release for water flowing along the strike, from northwest to southeast. A single, large cave system formed in the Cobleskill Creek basin, with water initially flowing down dip in tributary passages to a master cave oriented along the strike. Continued incision by Cobleskill Creek westward perched the master cave up dip from Cobleskill Creek, resulting in partitioning of the pre-existing cave system into smaller,

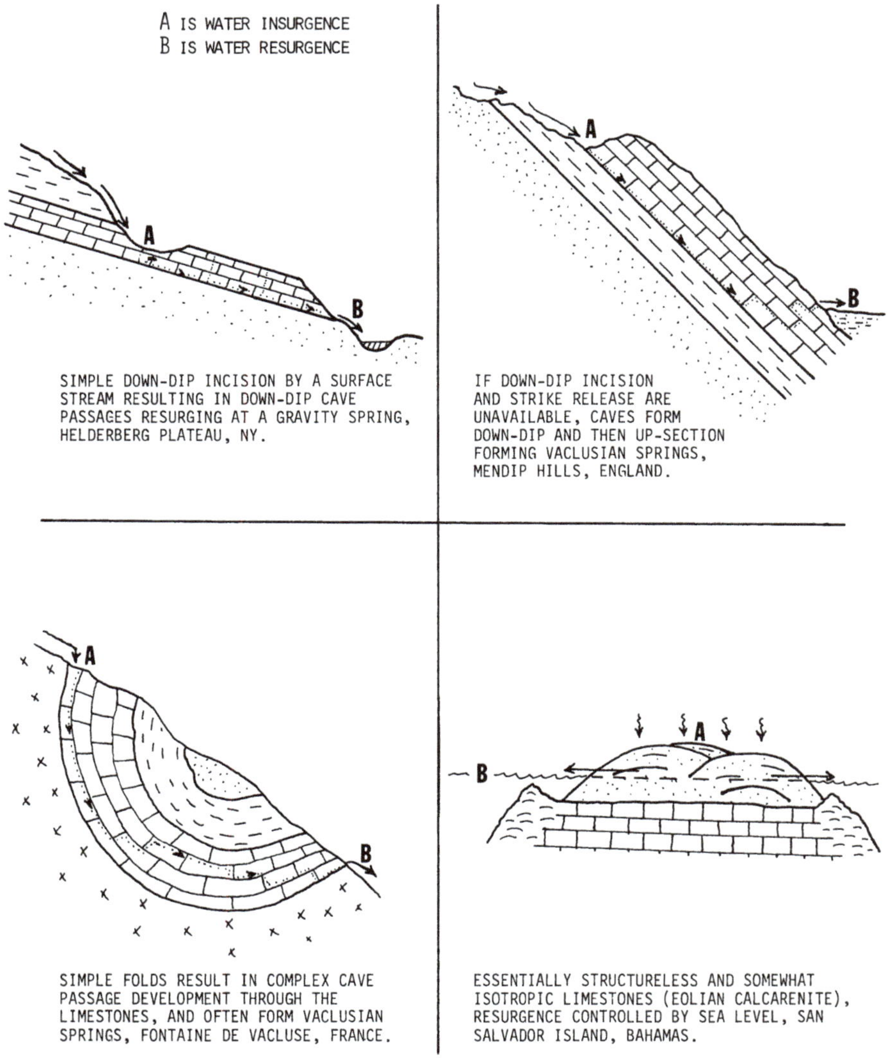

Figure 7.3 Some examples of structural control of solutional conduits.

down dip-oriented systems similar to those of the Fox Creek area. In this area of uniform recharge, structure and lithology, the location of surface streams relative to bedrock structure had a major effect.

The problem of determining the role of structure in cave development can be illustrated by examining joint control of cave-passage formation. Joints obviously are important in the genesis and maturation of solution conduits,

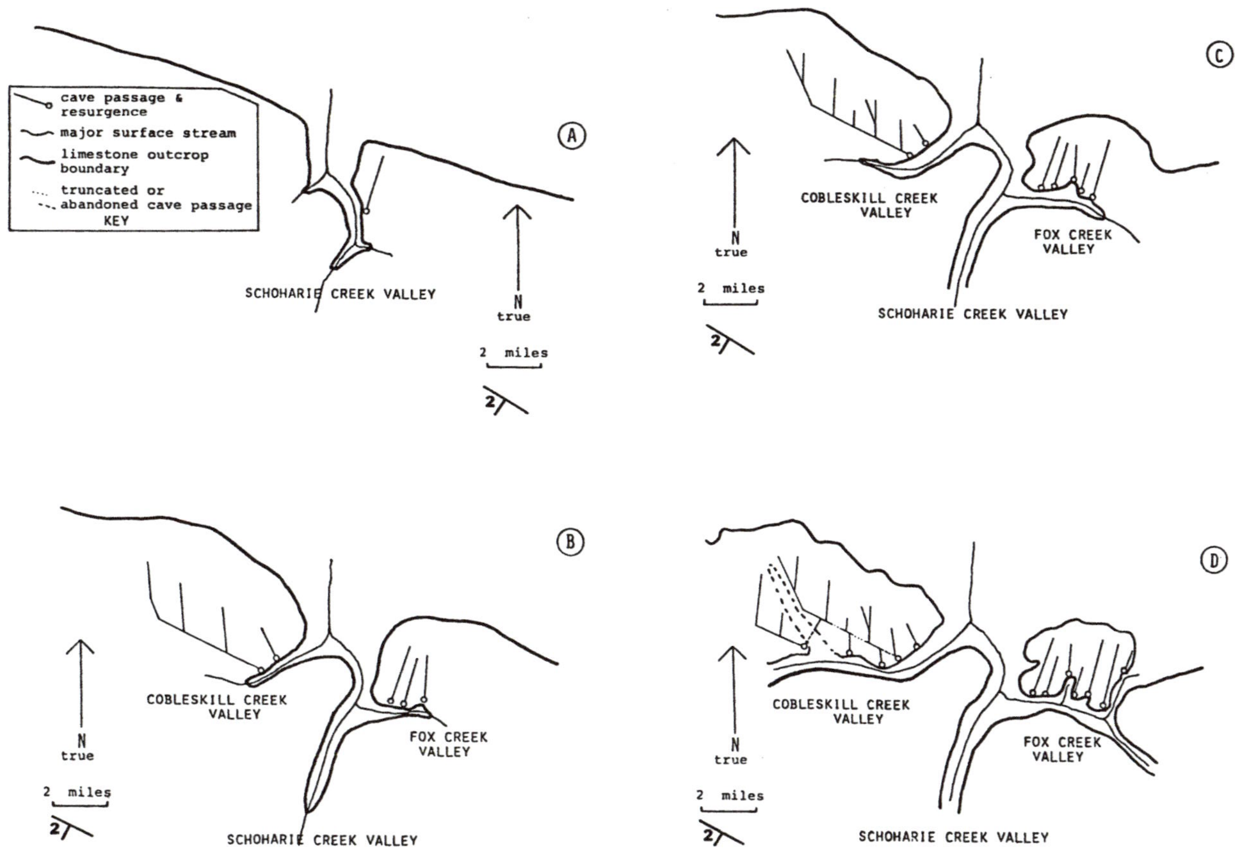

Figure 7.4 Diagrammatic representation of the evolution of the Schoharie Valley and its major cave systems. See the text for a detailed explanation.

but correctly interpreting their effect requires careful observation. Figure 7.5 demonstrates how a meandering passage can appear to be bedding-plane controlled when actually it was joint controlled (Fig. 7.5a vs b). Wall and ceiling collapse into a conduit along joint planes can make a bedding-plane meander look like angular joint control (Fig. 7.5c vs d). The initial controlling conditions of passage genesis must be differentiated from later developing features. Collapse along joint planes, sediment armoring and cyclic flooding can rapidly change the expressed surfaces in a cave passage. Joint planes in a cave wall or ceiling that had no effect on initial passage genesis may be selectively enlarged by the increased discharge of floodwaters as the cave matures.This will lead to passages with well-expressed joint enlargement and a consequent strong impact on the observer. As cave passages enlarge, the increased span of the ceiling causes sagging and the opening of tension cracks, which may or may not have any relation to the regional joint pattern. These cracks may become enlarged solutionally and appear significant genetically.

Cave passages can be confused genetically because of explorational bias. Joint-controlled passages are usually narrow, vertical openings, more easily explored than low, bedding-plane passages that may carry high discharges. This can mislead the cave surveyor to assume joint-controlled passages to be very important or dominant hydrologically, because the observer does not have a complete view of overall, cave-system morphology. Studies identifying structure as a controlling agent can be biased by limited portions of a given cave system available for on-site examination.

Soluble bedrock, usually limestone, may have a geomorphic effect very much out of proportion to its outcrop area or total volume. This is especially true in areas of folded rock, where relatively thin layers of limestone or marble outcropping along hillsides may capture water and deliver it great distances along the strike. This is well documented in West Virginia (Medville 1981) and Norway (Lauritzen 1981), representative of folded and metamorphic areas, respectively. The capture of confluent streams, with subsequent abandonment of their downstream reaches, and the possibility of the water going underground to a different drainage basin (with rejuvenation there because of increased discharge), results in a major geomorphic expression that persists through time.

Karst landforms and cave systems can develop on a variety of scales from a few meters to many kilometers in magnitude. Springs with discharges up to $130\,m^3\,s^{-1}$ and source areas over 30 km away are known in karst areas (Jennings 1971). Karst areas may be quite large in areal extent as in Slovenia (Yugoslavia), which has $20\,350\,km^2$ of karstlands (Sweeting 1973). From these major examples, karst forms can scale down to entities of minor local importance.

The vertical component of groundwater flow is perhaps best shown in karst areas. Unlike surface stream waters, which flow generally in response to a down-sloping valley gradient, karst waters may descend and ascend significant distances within the subsurface, depending on recharge–discharge relation-

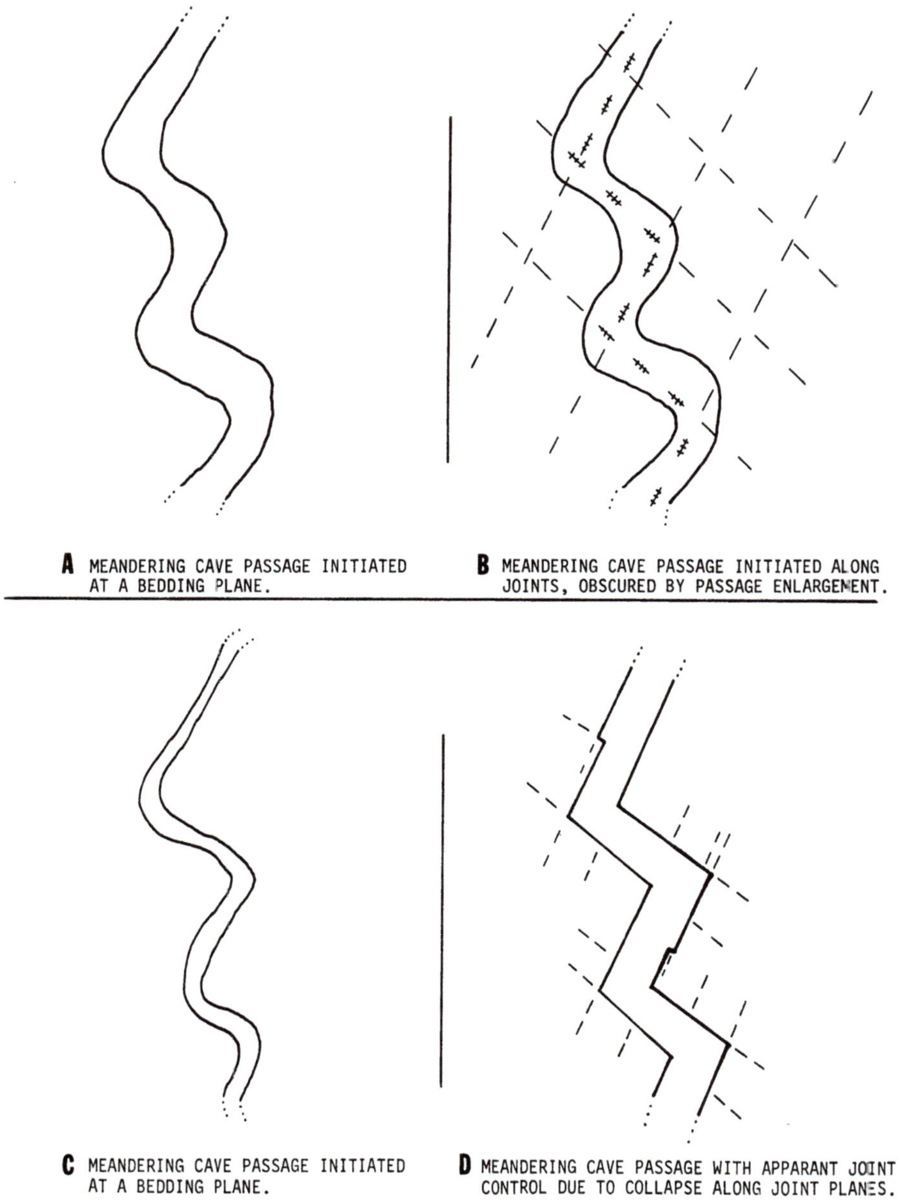

Figure 7.5 Interpreting joint control in cave-passage development.

ships. The cave systems delivering water through the subsurface may not be dendritic, but rather highly distributary (Zotl & Maurin 1959). The degree to which a cave system is distributary or interacts with neighboring cave systems may depend merely on the discharge at any given time.

The complete understanding of cave systems and their effect on surficial geomorphology has been hampered by the relative inability to observe detailed activity, especially as compared to surface fluvial studies. The data base for cave-system studies comes primarily from cave exploration and from dye tracing of underground streams. Dye tracing, as with any remote sensing technique, requires numerous assumptions prior to data gathering, and often complex interpretation afterward. Cave exploration is limited to those areas accessible to the explorer, introducing an explorational bias when progress is halted by small passage size, collapse, or water-filled passages. The advantages of direct exploration are numerous and important. The observation of passage size, flow markings, sediments and size and orientation of abandoned upper levels (unmeasurable by dye tracing) can help unravel the geomorphic history of a karst area. The changes in dye-test interpretation that follow exploration in a karst area are demonstrated by Ewers and Quinlan (1981).

Direct subsurface observation is an extremely arduous field exercise, and difficulty in obtaining complete data on subsurface conditions has meant slow progress in understanding of subsurface processes. Some of this understanding comes from learning the functional roles of karst forms.

References

Ewers, R. O. and J. F. Quinlan 1981. Cavern porosity development in limestone: a low dip model from Mammoth Cave, Kentucky. *Proc. 8th Int. Speleol Congr.* 727–31. Bowling Green, Kentucky.

Ford, D. C. 1977. Genetic classification of solutional cave systems. *Proc. 7th Int. Speleol Congr.*, 189–93. Sheffield, England.

Jennings, J. N. 1971. *Karst.* Cambridge, Mass: MIT Press.

Lauritzen, S-E. 1981. Glaciated karst in Norway. *Proc. 8th Int. Speleol Congr.*, 410–11. Bowling Green, Kentucky.

Medville, D. M. 1981. Geography of the Friars Hole Cave system, U.S.A. *Proc. 8th Int. Speleol Congr.*, 412–14. Bowling Green, Kentucky.

Mylroie, J. E. 1978. Speleogenesis in the Bermuda Islands (Abs.) *Prog. 1978 Nat. Speleol Soc. An. Conv.* 38. New Braunfels, Texas.

Mylroie, J. E. and R. G. LaFleur 1977. Karst geohydrology of the Helderberg Plateau, Schoharie County, New York. *Proc. 1977 Nat. Speleol Soc. An. Conv.*, 27–32. Alpena, Michigan.

Palmer, A. N. 1984. Geomorphic interpretation of karst features. In *Groundwater as a geomorphic agent,* R. G. LaFleur (ed.), 173–209. Boston: Allen & Unwin.

Powell, R. L. 1963. Alluviated cave springs in south-central Indiana. *Proc. Ind. Acad. Sci.* **72**, 182–89.

Quinlan, J. F. 1979. *Types of karst, with emphasis on cover beds in their classification and development.* PhD dissertation. University of Texas at Austin.

Sweeting, M. M. 1973. *Karst landforms.* New York: Columbia University Press.

White, W. G. 1977. Conceptual models for carbonate aquifers revisited: hydrologic problems in karst regions. In *Hydrologic problems in karst regions*, R. R. Dilamarter and S. C. Csallany (eds), 176–87. Bowling Green, Kent.: Western Kentucky University.

Zotl, J. and V. Maurin 1959. Die Unterschung der Zusammenhänge unterirdischer Wasser mit besonderer Berüksichtigung der Karstverhältnisse. *Steier. Beitr. Hydrogeol., Graz.*

8
Geomorphic interpretation of karst features

Arthur N. Palmer

Introduction

One of the major difficulties of geomorphic interpretation is that the processes of weathering and erosion, which play such an integral part in the geomorphic evolution of a region, tend to obscure and obliterate the evidence for previous events at the Earth's surface. In a karst region, however, the record of past events is preserved far longer and in richer detail than in other continental settings, largely because these events are reflected in the pattern of subsurface features such as caves. Their interpretation depends upon a solid understanding of groundwater flow, chemical processes and local geologic setting.

This chapter first outlines the salient hydrologic and chemical processes involved in the origin of karst features, then applies these concepts to the various surface and subsurface environments in which these features form. Finally, several field examples are described that illustrate how karst features can be used to interpret the geomorphic history of the surrounding region.

Karst groundwater systems

A karst landscape is one in which the dominant geomorphic process is the solution of bedrock or semi-lithified sediment. Although many of the largest surface karst features originate partly as the result of mechanical processes such as collapse, subsidence and stream erosion, these processes are triggered by subsurface solution. Karst processes have been explained in detail by Jennings (1971), Sweeting (1973), and Bögli (1980), and karst aquifer types have been categorized by White (1969) and by Mylroie (1984). To supplement the information from these sources, the following paragraphs briefly explain the interrelationship between surface and subsurface karst features in terms of the mode of groundwater recharge to the soluble rock.

Relationship between recharge type and groundwater flow pattern in karst

Field observation and mapping of solutional caves and their related features has shown that the overall pattern of a cave is determined by the mode of groundwater recharge, i.e. whether the water entering the soluble rock is (a) diffuse, (b) scattered among many small but discrete point sources such as dolines, or (c) concentrated at a few major inflow points such as sinking streams (Palmer 1975, 1981a). There is, consequently, an intimate relationship between caves and the surface karst features that evolve with (and to some extent control) the groundwater recharge. The specific direction and elevation of individual cave passages are determined predominantly by the local geologic structure (Ford 1971, Ford & Ewers 1978).

The three major types of groundwater recharge and their associated karst features are illustrated in Figure 8.1. If most of the water enters a soluble rock formation at discrete points, such as dolines (sinkholes), with each point delivering water from only a small catchment area, a dendritic flow pattern and cave system is the dominant type of karst drainage. Each point source of recharge contributes water to a distinct tributary of such a system, and each tributary joins with others to form cave passages of increasing discharge (and usually of increasing size) in the downstream direction. Water emerges at lower altitudes through springs that are far fewer than the recharge points. Dolines and cave passages evolve simultaneously, as described later in this chapter. The dendritic cave pattern is commonly obscured by multiple stages of development or by the inaccessibility of small tributaries. Structural influence may cause this pattern to be far more angular than that of a surface stream. This type of flow pattern is most typical of soil-covered karst areas with authigenic recharge where no perennial surface streams contribute to the groundwater flow. It is also typical beneath surfaces of bare limestone in which fractures dominate over intergranular porosity, as in most alpine settings.

Allogenic recharge is that which collects on a non-karst surface and flows underground at discrete points where soluble rock is encountered. Caves in such a region are usually fed by only one or a few sinking streams, and therefore, the cave patterns are only crudely dendritic. Many consist of a single major conduit with no significant tributaries. Both the discharge and its variation with time are very large compared with other types of infiltration. At any obstruction in a cave fed by this kind of recharge (such as a collapse zone, presence of relatively insoluble rocks, or partial filling of the cave by clastic sediments), water ponds upstream from the obstruction during high flow, often filling parts of the cave under great hydrostatic pressure. Fractures and other openings that are air filled during low flow are exposed to aggressive water with steep hydraulic gradients. As a result, diversion passages are commonly formed around the blockage. These passages may comprise either a sinuous, braided pattern of interconnecting tubular channels (anastomotic pattern) or an irregular network of intersecting fissure passages. This type of

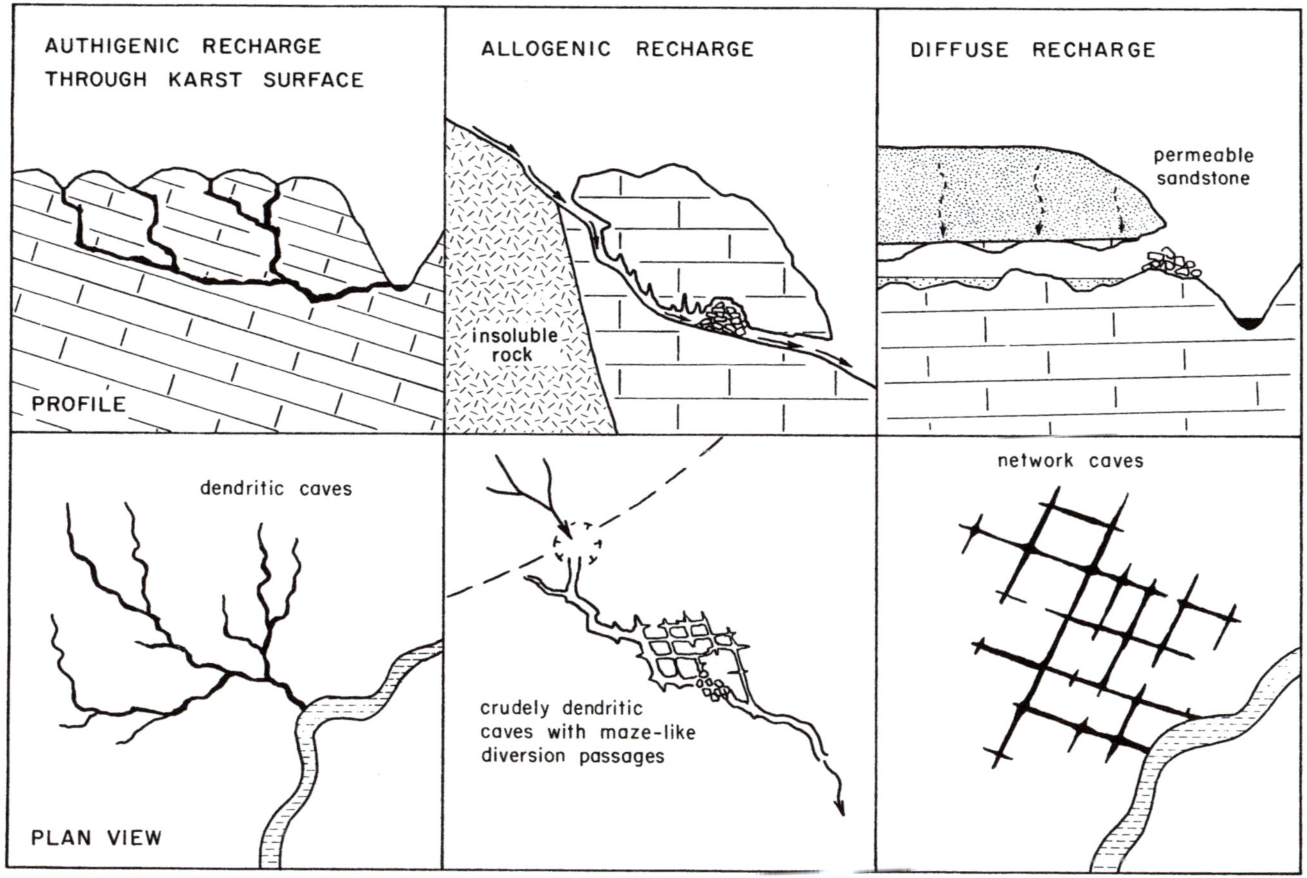

Figure 8.1 Common patterns of solutional caves and their relationship to surface features.

flow system occurs in areas of diverse lithology where insoluble rocks form a large part of the drainage basin.

Where water enters a soluble rock in diffuse form, a network maze of intersecting fissure passages usually develops. This type of recharge is most common where the soluble rock is overlain by a permeable but insoluble rock, such as a thin sandstone. The insoluble rock acts as a governor to the flow, delivering nearly the same amount of water to each major fracture in the soluble rock, regardless of the size of the opening, and allowing them all to enlarge by solution at comparable rates.Diffuse infiltration may also occur in soluble rocks of large intergranular porosity, producing caves that consist of irregular passages and interconnected rooms. In either case, once the initial cave pattern is established, the majority of cave enlargement may be achieved by other sources of water, such as lateral flooding from nearby entrenched surface rivers.

Vadose vs phreatic conditions within a karst aquifer

It is commonly believed that caves originate within the phreatic zone, at or below the water table, and are only modified by invading vadose water after the water table drops. This view is erroneous because vadose and phreatic solution occur simultaneously, often within a single flow path. Along its length between recharge point and spring outlet, an actively forming cave passage normally contains both vadose and phreatic components. The passage conducts water through the vadose zone to the phreatic zone, undergoing solutional enlargement in both zones at the same time. Exceptions to this pattern occur where vadose water exits at perched springs before reaching the phreatic zone, or where the passage is fed by a ponded source such as backflooding from a surface river. Either the phreatic or vadose component may be absent if one of these two hydrologic zones is located entirely within insoluble rock. The distinction between vadose and phreatic cave passages is frequently obscured by solutional and depositional features resulting from temporary flooding of passages that are partly or entirely air filled during low flow. These features are described in detail later in this chapter.

With time, as the water table drops relative to a given cave passage, vadose flow and solution features are superimposed on older phreatic features. Water diverts to lower flow paths, abandoning older passages and forming new ones. Meanwhile, the evolution of such a cave system affects the overlying surface by collapse and subsidence of bedrock and soil, subsurface piracy of streams, and establishment of internal drainage.

Use of the terms "phreatic," "vadose," and "water table" has fallen into disfavor among karst researchers because the complexity of groundwater flow in a karst region cannot readily be described in such a simple way. Nevertheless, their use is fully justified if proper recognition is given to the highly irregular and discontinuous nature of the water table and to the fact that its position varies greatly on a short-term basis according to the amount of groundwater flow. The distinction between vadose and phreatic karst features is essential to the interpretation of the regional landscape history and is discussed in detail later in this chapter.

Chemical reactions

The solutional enlargement of fractures and intergranular pores in a carbonate aquifer requires a source of groundwater acidity, whereas simple dissociation accounts for most of the solution in evaporites. The following chemical reactions are most important to the development of karst features:

(a) *The* CO_2–H_2O–$CaCO_3$ *reaction.* Carbon dioxide from the atmosphere and soil dissolves in water and reacts to form carbonic acid, which dissociates to produce H^+ ions. These react with calcium carbonate to produce the soluble ions Ca^{++} and HCO_3^-. The solution of dolomite takes place by a similar but slower reaction. For a given volume of water, the rate and amount of solution depend mainly on the CO_2 partial pressure of the air in contact with the water. The CO_2 partial pressure is greatest in the soil and in poorly ventilated parts of the vadose zone (as high as 10%). It is low to moderate (typically 0.1%–1%) in cave air, depending on the rate of air exchange with the surface and degree of oxidation of organic material carried in by cave streams. It is lowest in the outside atmosphere (~ 0.035%). If water is exposed to an increasing CO_2 partial pressure in the direction of flow, its solutional aggressivity increases. Conversely, if the CO_2 partial pressure decreases in the direction of flow, CO_2 is given off by the water, and the solutional aggressivity diminishes. In the latter case, water may become supersaturated, and carbonates are precipitated. Major sources of CO_2 in the phreatic zone are rare and localized, including accumulations of natural gas, entrainment of air bubbles by descending water, and biological processes. Normally these sources are of negligible importance, and the CO_2 concentration in phreatic water diminishes with time as equilibrium is approached. In the phreatic zone, the highest initial CO_2 concentrations and solutional aggressivity are located along the flow routes having the largest discharge.

If mixing occurs between solutions of different CO_2 content, the resultant mixture usually has a higher solutional aggressivity than either of the two original solutions (Bögli 1964). Picknett (1977) has shown that the mixing of solutions of different Mg^{++} content can either increase or decrease the solutional aggressivity, depending on the initial concentrations and water in each source. These and similar processes are important in determining the evolution of karst features in areas of converging water sources, or where the chemical environment of the water changes with time or with distance of flow. Further treatment of this subject is provided by Thrailkill (1968) and by Dreybrodt (1981a, 1981b).

(b) *Oxidation of sulfides*. The oxidation of solid sulfides (e.g. pyrite) produces sulfuric acid, which can be of local importance in the solution of carbonates (Morehouse 1968). Oxygenation of rising waters carrying dissolved sulfides, as in hydrothermal areas or in the vicinity of petroleum reservoirs, can also produce sulfuric acid (Egemeier 1973, 1981, Maslyn 1978, Davis 1980). Partial replacement of carbonate by gypsum may be a by-product where the reactions take place rapidly. Oxidation of sulfides is most rapid at or above the water table, but the concentration of dissolved sulfides (e.g. H_2S) is greatest deep in the phreatic zone. Therefore, the production of sulfuric acid from sulfide bearing water must occur almost exclusively where rising water reaches the water table, or where oxygen-rich water from a nearby vadose source converges with rising sulfide-rich water. In addition, the dissociation and hydration of dissolved H_2S can produce acidity even in the absence of free oxygen.

(c) *Biogenic acids*. Solution by organic acids from plants and animals is usually of minor importance, except where bare, soluble bedrock is exposed to frequent wetting, as in coastal areas, or where acid, boggy soil drains into or onto the soluble rock. The jagged, pitted surface of algal phytokarst is a typical example of a biogenic karst surface.

(d) *Dissociation*. The simple dissociation of soluble rock in water is important only in evaporites such as gypsum. This process is nearly independent of the carbon dioxide or oxygen content of the water, and therefore, the proximity of a gaseous phase is not so important as it is in the preceding examples. In a humid region, deep-seated solution by groundwater can remove extensive beds of evaporites from a sedimentary section long before they are exposed above the phreatic zone.

Hydrochemical processes

The development of underground solution conduits and their related surface features involves a selective enlargement of only a small percentage of the initial fractures and pores in a soluble rock formation. The pattern of the resulting karst features depends upon which of the initial openings are enlarged. Karst interpretation not only requires an understanding of the hydrologic and chemical processes involved in bedrock solution, but any quantitative expression of these processes must be generalized enough to explain broad geomorphic trends, rather than applying only to the specific site where field measurements have been made.

The origin of features created by the solutional activity of flowing water must conform to a mass balance in which the mass of rock removed is equal to that carried away in solution. Within any time increment ($\mathrm{d}t$) and incremental distance of flow ($\mathrm{d}L$),

$$\varrho_r \left(\frac{\mathrm{d}V}{\mathrm{d}t}\right)_{\mathrm{d}L} = Q\,\mathrm{d}C \times 10^{-6}\,\mathrm{g\,s^{-1}} \qquad (8.1)$$

where ϱ_r is the bulk density of the rock ($\mathrm{g\,cm^{-3}}$), $\mathrm{d}V/\mathrm{d}t$ is the rate of volume increase of the solutional void ($\mathrm{cm^3\,s^{-1}}$) within the distance $\mathrm{d}L$, Q is the discharge of water ($\mathrm{cm^3\,s^{-1}}$), and $\mathrm{d}C$ is the increase in concentration of dissolved rock over the flow distance $\mathrm{d}L$ ($\mathrm{mg\,l^{-1}}$).

Evaluation of $Q\,\mathrm{d}C$ is essential in determining the rate of solution ($\mathrm{d}V/\mathrm{d}t$) under various conditions, and therefore, in explaining the size and morphology of the resulting features. A solutional cave passage will be used as the most appropriate example. The discharge through a cave passage is governed either by the amount of recharge available (catchment control) or by the efficiency with which the passage transmits water (hydraulic control). In catchment control, which is far more common, the discharge depends upon the catchment area and the amount of rainfall and snow melt that is able to infiltrate into the ground to feed the passage. An example is a doline that collects water within its local drainage area and conducts the water to a vadose passage below. In the vadose zone, the hydraulic gradient that governs the flow is virtually equivalent to the slope of the passage. Where the passage reaches the phreatic zone, the flow is still determined by the amount of recharge from the sinkhole, and the hydraulic gradient within the water-filled section adjusts to the amount of flow and the shape of the conduit. In hydraulic control, however, the passage is a drain for a surface or subsurface reservoir whose capacity is great enough to keep the entire cave passage completely water filled. An example is the seepage of water through a fracture connecting a high-level surface stream to a spring at a lower elevation. As long as the discharge in the stream exceeds the capacity of the fracture to transmit water, the amount of flow within the fracture is determined by the hydraulic gradient and by the width of the fracture. When the fracture enlarges enough by solution that it is able to carry more water than the stream can deliver, the system changes to catchment control, and part or all of the flow in the conduit acquires a free surface.

The initial groundwater flow through an opening in soluble rock is almost always laminar. If the opening enlarges by solution and the discharge increases, the flow eventually becomes turbulent. At the transition from laminar to turbulent flow, there is a change in the flow equations that determine the discharge, as shown below. In addition, wherever the solution rate depends upon the rate of mass transfer of dissolved solids away from the rock surface, the mixing caused by eddies in turbulent flow increases the solution rate. Control of the solution rate by turbulent eddies is of minor importance at the moderate pH values that normally occur in groundwater (Plummer *et al.* 1978), but appears to be significant in the solution of evaporites (Berner 1980, p. 107). For an open channel or water-filled conduit, the conditions that determine whether the flow is laminar or turbulent are expressed by the Reynolds number (N_R):

$$N_R = \frac{\varrho_w Q}{\mu p} \tag{8.2}$$

where ϱ_w and μ are the density and viscosity of water, and p is the wetted perimeter in the channel or conduit. The transition from laminar to turbulent flow takes place at a Reynolds number of roughly 500.

The relationship among discharge, hydraulic gradient and conduit character is given by the Darcy–Weisbach equation, shown here in its most general form:

$$Q = pR^{3/2}\sqrt{\frac{8gi}{f}} \tag{8.3}$$

where R is the hydraulic radius (cross-sectional area of flow divided by p), g is the gravitational field strength, i is the hydraulic gradient, and f is a friction factor that depends on the Reynolds number, wall roughness and cross-sectional shape of the conduit. For laminar flow,

$$f = \frac{\alpha}{N_R} \tag{8.4}$$

where α is a coefficient that varies from 16 to 24 as the conduit cross section varies between a circular and fissure geometry. For turbulent flow in natural solution conduits, f generally falls between 0.03 and 0.05 (see any fluid-mechanics handbook for the evaluation of f).

To evaluate $\mathrm{d}C$ in Equation 8.1, it is useful to convert $\mathrm{d}V/\mathrm{d}t$ into the rate of solutional retreat of the conduit walls.

$$\mathrm{d}V = \mathrm{d}A\,\mathrm{d}L = p\,\mathrm{d}n\,\mathrm{d}L \tag{8.5}$$

where $\mathrm{d}A$ is the change in cross-sectional area caused by solution, and $\mathrm{d}n$ is the distance of solutional retreat of the rock surface. Therefore, Equation 8.1 can be restated as

$$S = \frac{\mathrm{d}n}{\mathrm{d}t} = \frac{Q\,\mathrm{d}C}{\varrho_r p\,\mathrm{d}L} \times 10^{-6}\,\mathrm{cm\,s^{-1}} \tag{8.6}$$

where S is the solution rate.

For the purposes of geomorphic interpretation, in which a more generalized statement is needed, Equation 8.6 can be integrated to give the mean solution rate ($\bar{S}$) within a finite length of passage (L):

$$\bar{S} = \frac{Q(C - C_0)}{\varrho_r pL} \times 10^{-6}\,\mathrm{cm\,s^{-1}} \tag{8.7}$$

where C_0 and C are respectively the mean solute concentrations at the upstream and downstream ends of passage segment L.

The value of C depends both on the reaction kinetics at the rock–water interface and on the rate of transfer of dissolved ions away from the rock walls. Which of these is the major rate-controlling factor depends upon several factors, including pH and reaction type. During enlargement of natural solution conduits, the most common reactions are those of the CO_2–H_2O–$CaCO_3$ system at pH values ~4–6. The solution rate is controlled mainly by the reaction rate under these conditions (Plummer & Wigley 1976, Picknett 1976). At pH values ~4, transfer control predominates.

For reaction control of solution rate in the CO_2–H_2O–$CaCO_3$ system, Plummer and Wigley (1976) have shown that

$$dC = \frac{A'k}{V_w}(C_s - C)^2 dt \times 10^{-3} \tag{8.8}$$

where A' is the area of contact between the wall rock and the solvent water, V_w is the volume of water, C_s is the saturation concentration of the solute ($mg\,l^{-1}$ $CaCO_3$ equivalent), and k is a factor that depends on the character of the bedrock, temperature and concentration of solute. The value of k is $\sim 2 \times 10^{-4}\,cm\,l\,g^{-1}s^{-1}$ for pure, finely crystalline calcite at 15°C (an average groundwater temperature for karst areas in the United States) and slightly less for impure or coarsely crystalline calcite.

Equation 8.8 is valid only at solute concentrations < 80–90% of the saturation values.Beyond this limit, the measured reaction rate decreases more sharply than the equation predicts (Berner & Morse 1974, Plummer & Wigley 1976). More sophisticated experiments and analysis by Plummer *et al.* (1978) show that the solution rate is progressively retarded by the back reaction of dissolved ions to solid. Further insight into the complexities of the solution process are given by Bögli (1964), Curl (1968) and Dreybrodt (1981a). Despite the limited range over which Equation 8.8 applies, however, its simplicity allows several important generalizations to be made about the evolution of karst features.

If Equation 8.8 is integrated,

$$C = C_s\left[1 - C_s\left(\frac{0.001\,pLk}{Q} + \frac{1}{C_s - C_0}\right)^{-1}\right] \tag{8.9}$$

where Q has been substituted for V_w/t and pL has been substituted for A'. This equation provides the value of C needed to solve for the rate of solutional enlargement of a cave passage (Eq. 8.7). The resulting solution rate in $cm\,yr^{-1}$

is shown in Figure 8.2 at various values of discharge, conduit length and hydraulic radius. In this plot, $C_s - C_0 = 150 \mathrm{mg\,l^{-1}}$ $CaCO_3$ equivalent (typical for karst groundwater), and $\varrho_r = 2.7 \mathrm{g\,cm^{-3}}$.

Equations 8.7 and 8.9 account for only the upper part of Figure 8.2 at $C/C_s < 0.9$. The lower part is constructed with the assumption that the groundwater reaches 90% saturation after only a short distance of flow at small values of Q/RL, and that it exits from the conduit nearly saturated. All but the very upstream end of the conduit is enlarged by only the last 10% of the solution, regardless of the exact solution kinetics. This simplification has been vindicated by solving the empirical formulas of Plummer *et al.* (1978) incrementally with computer-aided numerical analysis.

Figure 8.2 reconciles the diverse measurements of solution rate obtained in the laboratory by Howard and Howard (1967) and by Rauch and White (1977), and in the field by Coward (1971). It also agrees with the rates required for the evolution of post-glacial caves (e.g. Palmer 1972). However, it is valid only for carbonate rocks in which the major solvent is carbonic acid.

Figure 8.2 can be used to trace the evolution of solutional caves that form under normal groundwater conditions in limestone (Palmer 1981a). Several conclusions can be drawn from this graph:

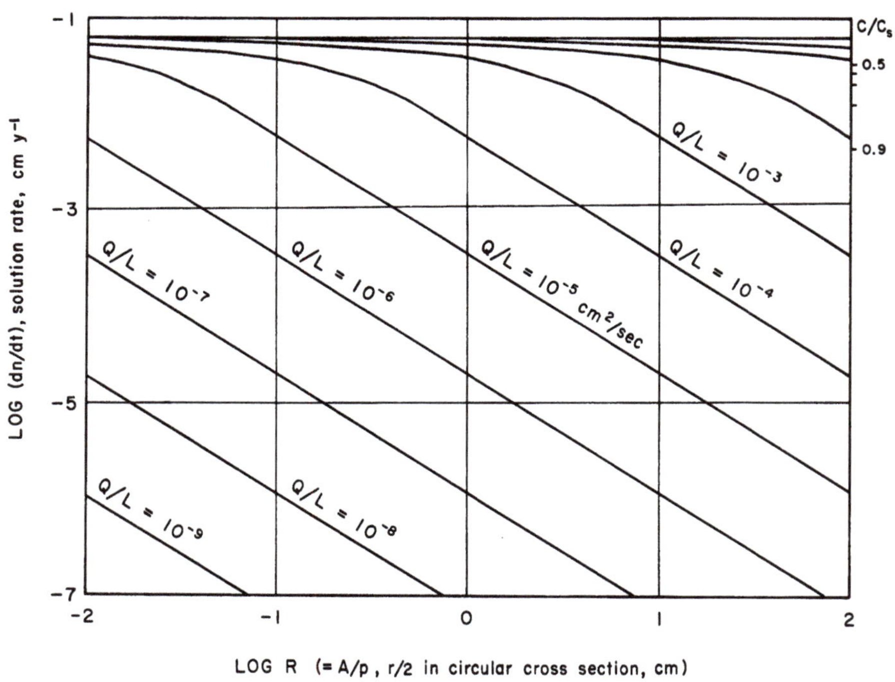

Figure 8.2 Rate of solutional enlargement of karst features as controlled by discharge, length of flow and hydraulic radius, at 15°C. $C_s = 150 \mathrm{mg\,l^{-1}}$, $C_0 = 0$, $k = 2 \times 10^{-4}$ $\mathrm{cm\,l\,g^{-1}\,s^{-1}}$ (see Eqs 8.7 & 9). C/C_s is the degree of saturation for the water at the end of the flow distance L.

(a) The solution rate rises with increasing discharge, but levels off at a maximum roughly on the order of one millimeter per year, depending on the values of C_0, C_s and k.
(b) Passages containing water that is nearly saturated with dissolved limestone possess a great variety of solution rates. Therefore, only those openings that increase their discharge with time are able to grow into major cave passages.
(c) At values of Q/RL greater than ~ 0.001 and C/C_s less than ~ 0.5, the solution rate is nearly independent of hydraulic radius, and all conduits meeting this condition enlarge at almost identical rates.

The three major patterns of karst groundwater flow systems outlined earlier in this chapter can be explained by the hydrochemical processes summarized in Figure 8.2. During the early stages of groundwater solution in a limestone aquifer, the groundwater is nearly saturated with dissolved $CaCO_3$. The various flow paths have a great disparity in solution rate, as shown in the lower part of the figure. Only those paths that acquire sufficient discharge to increase their solution rate with time are able to grow into major cave passages. Because of the finite amount of available recharge, only a few of the initial flow paths are able to do so, and a dendritic cave pattern develops having relatively few passages. Passages join as tributaries because of intersections determined by geologic structure and because major phreatic passages normally possess a relatively low hydraulic head and attract water from surrounding parts of the aquifer.

Caves fed by sinking streams are subject to frequent and severe flooding, particularly if the major stream passages contain local constrictions. As a result, any openings in the limestone in the areas of flooding are injected with aggressive water having steep hydraulic gradients. All openings above a certain minimum size ($R \sim 0.01$ cm) enlarge at the same maximum rate shown at the top of Figure 8.2, commonly resulting in a maze of diversion passages that bypass constrictions.

Water that infiltrates in a diffuse manner into limestone through a permeable, insoluble caprock is highly aggressive and is able to enlarge all openings at comparable rates. Because there is very little increase in discharge to a given opening as it enlarges, all the openings grow at competitive rates.

Recognition of the genetic environment of karst features

Relict landforms are abundant in karst areas and provide clues to the geomorphic history and hydrologic evolution of the surrounding region. The principles established in the preceding sections can be used to interpret the environment in which these features originated. The most easily recognized geomorphic settings are outlined below with descriptions of the specific karst features, processes and characteristics by which they can be identified.

Surficial features formed on exposed bedrock

Karst features on exposed bedrock are formed mainly by surface water from rainfall and snow melt. These features, described in detail elsewhere (Bögli 1980), include solutional rills, canyons, widened fractures and related features (Fig. 8.3). At points of converging runoff, these features may coalesce into shafts or steep, funnel-shaped dolines. A landscape dominated by the solution of bare bedrock is known as karren topography. Lengthy development of such a landscape under high-relief conditions can produce a surface of sharp pinnacles, bedrock blades and jagged fissures that is almost impenetrable to travelers. In addition, biogenic processes, especially in warm, moist climates, can produce a rough, pitted surface on soluble rock. An example is shown in Figure 8.4.

All solutional features produced by surface runoff show the strong influence of gravity, which is the only signficant force affecting the flow. Water moves

Figure 8.3 Solution of a bare limestone surface, Bödmerenalp, Switzerland.

Figure 8.4 Algal phytokarst in the supratidal zone along the Bermuda coast.

directly down slopes, concentrating its flow in areas of slope convergence. On gentle slopes, the flow direction is influenced by minor irregularities in the bedrock surface, and meandering rills or canyons are usually formed. On steep slopes, the influence of surface irregularities is greatly reduced, and nearly straight rills are formed. The most distinctive characteristic of surficial karst features is a strong tendency for the bedrock edges, divides between rills and breaks in slope to be very sharp and abrupt. A jagged, rough-textured surface is common. These characteristics are largely the result of the rapid concentration of runoff into channels, which causes a wide disparity in the rate and duration of solution on the rock surface over short distances. Nevertheless, wherever sheet flow occurs without being channelized, the exposed bedrock can undergo very uniform solutional lowering, which produces smooth, flat surfaces. In many places, steep rills merge into such a surface where the water spreads out at the break in slope at the base of the rills.

Much of the solution on bare bedrock takes place at or near the maximum possible solution rate, because the runoff tends to sink underground before C/C_s reaches a high value (see Fig. 8.2). The presence of broad, nearly flat, bedrock surfaces that have been lowered uniformly by sheet wash supports the idea that the solution takes place at or near the top of the graph in Figure 8.2, where differences in water depth, velocity and length of flow have no significant effect on solution rate. The average atmospheric CO_2 partial pressure is 0.035%. C_0 is nearly zero, and C_s is roughly 60–80 mg l^{-1}. During periods of rainfall, these conditions allow a maximum solution rate approximately equivalent to 0.01–0.02 cm yr^{-1}, according to Equations 8.7 and 8.9.

Subsoil karst features

Water that infiltrates through soil to the underlying bedrock surface produces solutional and depositional features that are usually easy to distinguish from those that form on bare bedrock. Movement of water in the soil is governed both by gravity and by capillary potential. During each wet period, a diffuse wave of water descends into the soil, followed by a period of drying. Whenever the soil has a moisture content less than the maximum that can be held by capillary forces (field capacity), water moves only by capillary potential from moist to dry areas. Where the moisture content exceeds the field capacity, excess water drains downward by gravity to the bedrock surface. Saturation of the soil just above the bedrock surface frequently occurs, allowing water to flow laterally along the contact to openings in the bedrock where it can descend farther into the vadose zone. Perched water of this kind forms many features similar to those on exposed bedrock (including downslope canyons and rills), but with smooth, rounded contours. Gams (1976) describes many of these features. Capillary forces distribute suspended water throughout the soil, drawing water from moist areas to dry areas, and bringing moisture in contact with all parts of the bedrock surface. As a result, solution at the bedrock surface is rather evenly distributed, and the sharp divides and breaks in slope so typical of exposed bedrock are generally absent or much less prominent. The high CO_2 content of the soil atmosphere provides the soil water with very high values of C_s. However, only a small portion of the infiltrating soil water reaches the bedrock, and that which does may contain much dissolved material acquired from the soil. Therefore, the solution rate at the bedrock surface is not as great as might be expected. Trudgill (1976) has shown that the solution rate at the bedrock surface depends on the carbonate content of the soil. In peaty, acid soils containing no carbonate, solution rates for the underlying limestone reach values as high as 0.5 cm yr^{-1}, whereas in carbonate-rich soils, the bedrock solution rate is as little as 10^{-5} cm yr^{-1}.

Water passing through the soil into fractures in the underlying bedrock tends to enlarge the most prominent fractures by solution, forming widened fissures (grikes). Soil subsides into these openings as they grow, as shown in Figure 8.5. Eventually, closed depressions (dolines) develop over the largest sub-soil openings, which causes increasing amounts of water to infiltrate at these specific points. As the dolines expand, the largest few openings in the bedrock acquire a great enough flow to enlarge into cave passages. Lesser openings are robbed of most of their water by the funneling of recharge toward the large ones. With time, entrances to the underlying cave system appear when soil plugs in the bottoms of the largest dolines subside entirely into the growing cave, a process aided by erosional activity of streams running into the dolines during wet weather. Collapse of bedrock into the cave is also an important process in the enlargement of most dolines, and in some cases is responsible for initiating dolines.

The scattered point sources of recharge that develop concurrently with the dolines are also responsible for the origin of dendritic caves. Recharge points, surface depressions and cave passages all form together in a close association with one another.

Vadose cave features

Because solution conduits transmit water so efficiently, the water table (or its equivalent) within a cave passage is located only a small distance above the

Figure 8.5 Subsoil solution features exposed in the sawed face of a quarry in the Salem Limestone, Indiana.

level of the nearest perennial surface stream, except where a stratigraphic or structural barrier causes the groundwater to be perched at a higher level. Therefore, the passages in most cave systems provide a faithful record of past changes in the erosional base level of the region. For this reason the distinction between vadose and phreatic features is one of the most significant aspects of the geomorphic interpretation of a karst area.

Bretz (1942) described solutional cave features in terms of either a vadose or a phreatic origin. However, he assumed that vadose solution merely modifies pre-existing phreatic features and is not responsible for creating caves by itself. He also did not recognize the importance of seasonal fluctuations in the water table as a cause for many of the features that he assumed were of phreatic origin.

All major vadose solution features are formed by gravitational flow. Pressure differences do not affect the direction of flow, as the water pressure is essentially atmospheric throughout vadose solution conduits, except where there is minor ponding. A discrete point of recharge into the soluble rock is required to form a vadose cave passage, because water that is diffused throughout minor openings approaches saturation very quickly. Thrailkill (1968) emphasizes this distinction by use of the term "vadose flow" for discrete trickles or streams and "vadose seepage" for diffuse water.

Gravitational vadose flow follows the steepest possible paths downward through the bedrock. In many places, the steepest openings are pores and fractures too narrow to transmit all of the available water, so the excess water overflows into less steep openings (for instance, dipping bedding-plane partings). Cave streams in vadose passages do not represent the water table. They are influent streams losing water through the more steeply sloping but narrower openings in the floor. These steeper openings may eventually enlarge enough by solution to divert the water to a lower path, causing the original passage to be abandoned.

Passages formed in the vadose zone can be recognized by their continuous downward trend. Examination of the major fracture or parting that once guided the initial flow in a vadose passage should nearly always reveal that the water closely followed the local dip of the fracture or parting. Minor deviations from the dip direction are caused by relatively narrow zones within the initial opening, or by local blockage of the opening with sediment or precipitates. Entrenchment of passage floors is typical of the vadose zone, and canyon-like passages enlarged downward below the initial openings are common (Fig. 8.6). Little solution occurs upward from the initial opening, so in cross section, a typical canyon passage that originated in the vadose zone has a flat or only slightly arched ceiling. The initial path of water flow and the structural feature that controlled it are normally located at the ceiling level. The width of a canyon passage is proportional to the passage slope, unless the slope is so steep that the water becomes fragmented into spray.

Figure 8.6 A canyon passage of vadose origin in Mammoth Cave, Kentucky.

Where a resistant bed forms the floor of a vadose passage, it may acquire a tubular shape with a lenticular cross section elongated along the bedding. Such a passage may be misinterpreted as phreatic, except that a consistent down-dip orientation identifies it as vadose.

Where the water follows a steeply dipping or vertical fracture, a vadose shaft is formed, resembling a well, with a lenticular or rounded cross section and nearly vertical walls (Fig. 8.7). Shaft walls contain vertical rills (flutes) that somewhat resemble those formed on exposed bedrock at the surface. The water spreads out in a thin film or spray, enlarging the shaft to a diameter much greater than the width of a canyon that would be produced by the same amount of flow.

The solution rate in an active vadose cave passage of traversible size is usually close to the theoretical maximum of roughly one millimeter per year (depending on the values of C_0, C_s and k). This solution rate has been verified in direct field measurements by Coward (1971). The length of flow is usually short enough that C/C_s remains well below 0.9 once the passage acquires sufficient discharge for Q/RL to exceed 0.001. The time required for a canyon to be entrenched below the initial openings can be crudely estimated. However, this is not generally possible in a canyon formed by headward retreat of a waterfall rather than by uniform entrenchment. Flutes in walls of the canyon indicate an origin of this type.

Several types of deposits are formed only in the vadose zone. Nearly

Figure 8.7 Vertical shaft formed by descending vadose water in a New York cave.

saturated water that seeps into an air-filled cave through narrow openings loses CO_2 to the cave atmosphere and deposits travertine. Dripping or flowing water produces travertine deposits with a distinct gravitational influence (e.g. vertical stalactites). Where the volume of seepage is small enough to be affected only by adhesive and cohesive forces, eccentric growths (helictites) and similar forms are produced that have no systematic vertical orientation. Evaporites such as gypsum are carried into dry caves that receive only capillary water drawn toward the passages from the moist surrounding limestone. The widespread superposition of one type of vadose deposit on another (e.g helictites and nodular growths on older gravitationally oriented travertine) indicates a change in the amount of infiltration entering the cave, which in turn may represent a change in climate or soil cover.

Uranium–thorium dating of travertine is currently the most precise method for dating stages of cave development and related geomorphic history (Harmon *et al.* 1975). The upper limit of this technique is 350 000 years.

Phreatic cave features

Within the phreatic zone, gravity is offset to varying degrees by the downward increase in hydrostatic pressure. In contrast with the vadose zone, water has no inherent tendency to follow the steepest available paths. Instead, it follows the most efficient routes from the points where it enters the phreatic zone to the available spring outlets. Although there are many alternate flow paths, most of the flow and solution is concentrated along a few especially favorable openings. Equations 8.3 and 8.4 show the strong dependence of discharge on the size of the original opening. (For instance, in a conduit of circular cross section, Q is proportional to r^4 in laminar flow and $\sim r^{5/2}$ in turbulent flow). Hydraulic gradient also controls the discharge through any given opening, but only linearly. Therefore, the paths of phreatic groundwater flow most likely to enlarge into cave passages are rarely the most direct. A roundabout route affording a wide initial-opening size will often be the dominant flow path, despite the fact that its hydraulic gradient is far less than that of the more direct but narrower openings.

However sinuous a phreatic passage is, its optimum location is usually at or near the top of the phreatic zone, because the initial openings in the bedrock tend to diminish in size and frequency with depth. This is particularly true of regions with little structural deformation. Additional but subordinate reasons why deep-phreatic flow is not so common as shallow-phreatic flow are that the hydraulic gradients of shallow paths are usually steeper, and the access to CO_2 in the vadose zone is more direct. In prominently bedded rocks where the water flow is highly concordant to the strata, the most favorable paths of flow in the phreatic zone are generally nearly parallel to the local strike of the beds at or near the water table. In steeply dipping rocks, this tendency imparts a strongly linear pattern to phreatic cave passages. Where an efficient outlet is not available in a strike oriented direction, water may follow deeper paths that loop below the water table and cut discordantly across the beds (Ford 1971).

Deep flow and discordance to the strata are most common in highly deformed rocks, in which the tendency for fractures to diminish in size with depth is commonly disrupted.

Phreatic cave passages have a tubular, fissure-like or irregular shape that shows evidence that solution has taken place upward from the initial openings (Fig. 8.8). Solutional enlargement normally takes place over the entire perimeter of such a passage although clastic sediment may partly or entirely shield the floor from solution. Presently active phreatic passages are accessible only by diving, so most of the evidence for phreatic cave processes is observed in passages that have been partly or completely drained of water because of a drop in the water table. Overall passage gradients are very gentle, no more than a few meters per kilometer, although they may contain individual segments of large upward or downward slope. Many phreatic passages were formed at the level of the water table, in which case the passage was partly filled with water during dry periods and completely filled only during high flow. More commonly, their profiles undulate along their lengths, showing that they descended below the water table to varying depths while forming. To verify a phreatic passage origin, it must be shown that the passage does not descend along the steepest available structural openings, and that any undulations in the profile are not caused by partial collapse or other modification of the original passage shape.

Phreatic passages formed by slow-moving flow from diffuse recharge can be distinguished in several ways from those formed by discrete recharge points

Figure 8.8 Tubular cave passage formed in the phreatic zone, Slovenia, Yugoslavia.

at the surface. Diffuse recharge does not produce distinct connecting routes between inputs and springs, but instead forms a network or irregular pattern having no clearly distinguishable points where water entered and exited (Fig. 8.1). Solutional scallops, which are asymmetrical hollows in the bedrock formed by moderate- to high-velocity turbulent flow, are absent. No coarse-grained sediment is present unless it has been carried in by agents other than the original cave-forming water.

In a presently dry cave passage, to determine the former location of the water table (and therefore, the approximate fluvial base level) that existed at the time the passage formed, the entire passage length must be examined for the following evidence: (a) a change in passage shape, particularly from canyon to tube; (b) a change in direction from a down-dip to a strike orientation or to an unsystematic relationship with respect to the dip; (c) a reduction in gradient or change from a consistent downward slope to an undulatory profile. These changes in passage character are illustrated in Figure 8.9. The depth of phreatic cave development below base level can be estimated by com-

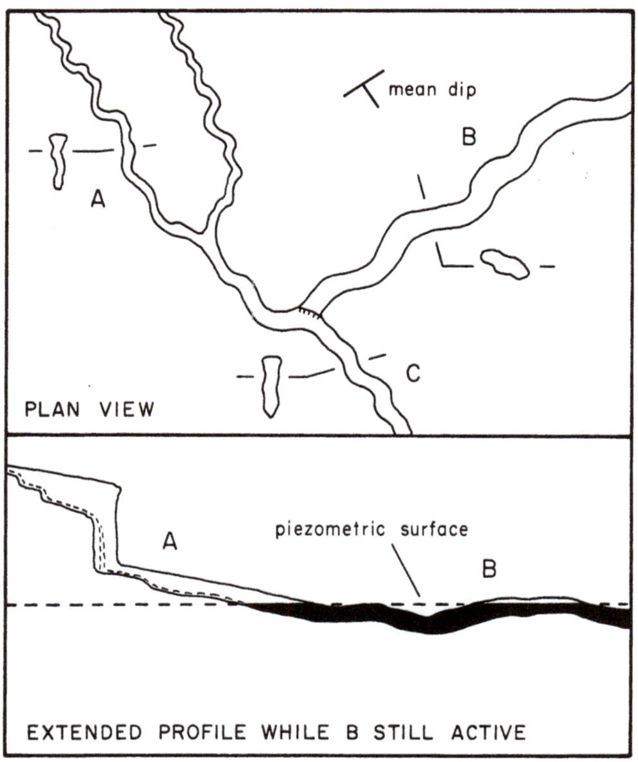

Figure 8.9 Typical changes in the character of a cave passage at the vadose–phreatic transition. A = down-dip canyon passages formed in the vadose zone; B = nearly strike-oriented tubular passage formed in the phreatic zone as the downstream continuation of A; C = down-dip canyon formed by vadose water, which bypassed passage B after the piezometric surface dropped below the level of B.

paring the elevation of the vadose–phreatic transition with that of the lowest solutional ceiling in the phreatic section of the passage.

Where groundwater is locally perched, there can be a change from phreatic to vadose conditions in the downstream direction. However, it is common for vadose entrenchment to breach the perched zone before phreatic features become well developed.

The precision with which the vadose–phreatic transition can be located is remarkable considering the magnitude of seasonal fluctuations in discharge and river level (Palmer 1977). Apparently, the zone of transition from vadose to phreatic features is determined either by the mean annual discharge or by the base flow. Seasonal variations in water table elevation only modify the passage character, but do not determine its overall pattern.

The clarity of the vadose–phreatic transition zone within a cave passage is probably indicative of the length of time the local base level was stable at that elevation during the erosional history of the region. It is necessary to correlate this zone with others at the same elevation throughout the cave region before broad generalizations can be made about base-level history and stability of the region. The time required for the phreatic part of a passage to develop can be crudely estimated from Figure 8.2 although the original C_0, C_s and size of openings, which are critical for determining the first stages of passage evolution, are generally not known.

Phreatic carbonate deposits are limited mainly to sparry calcite, which accumulates as linings in pools of cave water or as pore fillings within the bedrock. A clear record of late-Pleistocene fluctuations in relative land–sea-level elevations has been revealed from cave deposits in Mallorca by Ginés *et al.* (1981).

Cave features formed during alternating vadose and phreatic conditions

A common flaw in the analysis of cave origin is to misinterpret the effects of seasonal or other short-term fluctuations of groundwater flow. For example, sediments deposited by seasonal floods are frequently interpreted as evidence for a single and distinct fill stage in the evolutionary history of the cave. Also, many solutional features that are used almost universally as criteria for phreatic cave origin are actually more commonly formed by periodic floodwaters above the dry-season level of the water table.

The clearest examples of karst features formed within alternating vadose and phreatic conditions occur in caves where great variations in the recharge rate cause rapid and high-amplitude fluctuations in the water table. These features are most pronounced in caves that are fed by sinking streams. It has already been shown how network and anastomotic diversion mazes can form around obstructions in an active stream passage. Associated with these diversion mazes, and often occurring as the result of flooding in other types of caves, are many solution features that share the same floodwater origin. Three of these are described below.

(a) Blind, joint-controlled fissures in the ceiling and walls of a cave passage are most commonly formed in this way. At levels just above the phreatic zone, highly aggressive water is injected into the joints whenever flooding occurs in the cave passage. This water drains out of the joints after the flood, and therefore, it is possible for the next flood pulse to inject freshwater into them under the steepest possible hydraulic gradients. In contrast, within the phreatic zone, joints that intersect cave passages are filled at all times with nearly static water that is close to saturation in its dissolved carbonate content. When a phreatic passage experiences a periodic rise in pressure during peak flow, the water-filled, communicating joints do not enlarge into fissures as easily, because the hydraulic gradients and exchange of water in the joints are much less than in an alternating vadose–phreatic environment.

(b) Bedding-plane anastomoses are mazes of small tubes that intercommunicate in a braided pattern (Ewers 1966). In contrast to a normal anastomotic maze, these tubes are generally too small for humans to enter. Remnants of anastomoses commonly occur along bedding-plane partings in cave walls (Fig. 8.10). Most cave researchers consider them to be the original, phreatic solution channels that eventually coalesce to form cave passages. However, this interpretation does not explain how the many alternate flow routes can enlarge competitively along a narrow parting far away from the source of recharge. Under these conditions, the

Figure 8.10 Anastomoses exposed in the wall of an Indiana cave.

water would be nearly saturated and would occupy a position, near the bottom of Figure 8.2, where there is a great diversity of solution rates. On the contrary, if the bedding planes adjoin a passage capable of periodically delivering high-gradient, aggressive water, their solution rates would be at a maximum, at or near the top of Figure 8.2, and many routes would enlarge simultaneously, regardless of variations in Q, L and R. The difference between anastomoses and anastomotic mazes is probably just one of scale, determined in part by the configuration of initial openings along the bedding plane and whether the passages function as major diversion routes to a new outlet or around an obstruction in the main passage.

(c) Another feature that can be attributed to periodic fluctuations in the water table is the distributary pattern of cave passages that occurs in the vicinity of many large karst springs (Fig. 8.11). During low flow, the phreatic passages in a cave represent zones of low head, and water flows into them from the surrounding areas. During high flow, however, a cave passage responds much more rapidly and intensely to flooding than the smaller openings around it, and the pressure in the passage increases enough that its water has a higher head than that in the narrow, surrounding openings. As a result, during high flow, water follows alternate routes away from the main passage, eventually forming distributary passages that exit at overflow springs. This effect is most pronounced in the downstream ends of underground flow systems where variations in discharge are the greatest. Blockage of spring outlets by collapse also contributes to ponding and increase in hydrostatic pressure in the passage during floods.

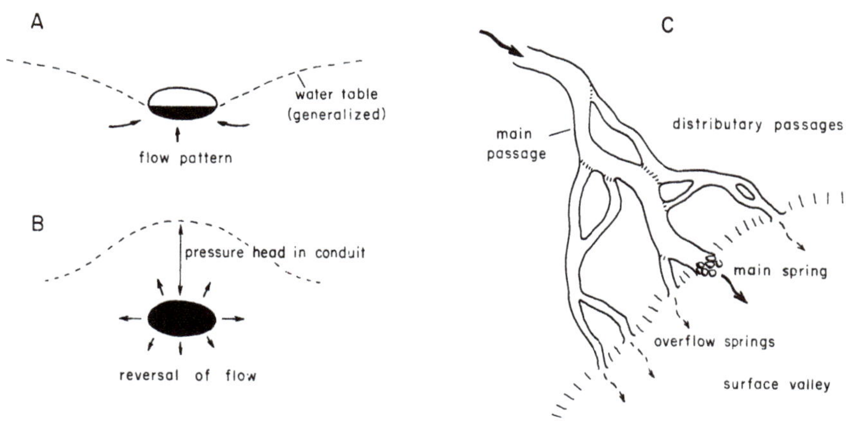

Figure 8.11 Origin and pattern of distributary cave passages in the vicinity of a karst spring. (A) Low-flow pattern of groundwater movement, when the head in the main cave passage is lower than in the smaller surrounding openings. (B) Flow pattern during high flow, when the head in the main cave passage is higher than in the surrounding openings. Water is forced into these openings under steep hydraulic gradients. (C) Resulting distributary pattern of cave passages.

Evidence for long term rises in base level

Thick alluvial sediment in present river valleys is common evidence for a rise in base level. Whether such valley fill represents a rise in sea level or depressed continental crust, or the more subtle influence of changes in climate, stream regimen and land use, can be determined only by extending the geomorphic study over a large region. Regardless of the interpretation, however, such episodes of alluvial filling represent important stages in regional geomorphic history.

In river valleys containing thick alluvial fill, much of the water that once emerged in karst springs at the former (deeper) river level now rises upward through the alluvium at artesian springs. Evidence for more ancient long-term rises in base level, during a previous geomorphic stage, is clear but infrequent. Deep sediment fill in cave passages can indicate a rise in base level following passage development (Miotke & Palmer 1972). However, cave sediments can accumulate for many other reasons, including blockage of an active stream passage by collapse, periodic overflow of floodwaters into otherwise dry passages from active stream passages at lower levels, or settling of sediment to the floors of phreatic passages that are enlarging upward by solution. Cave sediments, no matter how thick, should be interpreted as evidence for a regional rise in base level only if the deposits occur in all or most of the passages at the same elevation and if it can be demonstrated that the fill occupies former vadose passages, such as dip-oriented canyons. Correlation with surface features may be ambiguous because most old alluvial deposits on the surface have been modified by mass wasting and erosion.

One of the most striking phenomena showing evidence for past rises in base level is paleokarst, in which caves and surface karst features have been filled with now-lithified weathering products or with marine or continental sediments. The extensive paleokarst at the top of the Mississippian limestones of west-central North America is an example in which marine sedimentary rocks of Pennsylvanian age fill solutional voids in the limestone (Campbell 1977). Some of the clearest exposures of these paleokarst features and their sediment fill can be seen in more recent caves that intersect them (Palmer 1981b). Environmental interpretation of the fill material in paleokarst is necessary to establish the exact conditions that led to burial of the karst features.

More ambiguous evidence for long-term rises in base level includes phreatic solution superimposed on vadose features. For example, the walls of vadose canyon passages are normally covered with solutional scallops as a result of the high-velocity flow that formed them, but some canyons contain smooth walls devoid of scallops as the result of flooding with low-velocity water.

Field examples

A few generalized examples are sufficient to show how karst features may be used in the geomorphic interpretation of a region. The following paragraphs are not intended to give definitive evaluations of each region, but merely to show the potential for using karst features as clues to geomorphic history.

Pleistocene sea level changes: Bermuda caves

Bermuda consists of a chain of islands on a stable volcanic seamount. Nearly all of the volcanic rock lies below present sea level. The islands are composed of Pleistocene dune calcarenites fringed with bedded marine facies (Land *et al.* 1967). Caves with abundant travertine occur throughout the limestones, particularly in the older, more-indurated rocks. Infiltration into the limestone is diffuse because of the high intergranular porosity, and this water loses nearly all its solutional aggressivity after only a few meters of seepage. The seawater is also saturated with respect to calcite. In the noncavernous areas, a thin lens of freshwater overlies salt water in the bedrock pores. This lens is reduced to a briny mixture of fresh- and saltwater in sea-level caves, with a minimum Cl^- content, usually about 30% that of seawater. Only in this fresh or briny lens of mixing is the water solutionally aggressive enough to form significant caves. Where the lens consists of brine, the water is aggressive only during periods of high infiltration, corresponding to periods of high P_{CO_2} in the groundwater (Palmer *et al.* 1977). The relationships among P_{CO_2}, solutional aggressivity, and saltwater content are shown in Figure 8.12. During periods of little infiltration, the water is supersaturated with respect to calcite, but the high Mg^{++} content of the water prevents calcite from precipitating at that time.

Because major caves can form only in the zone of mixing at or near sea level, the occurrence of sub-horizontal caves at various altitudes is an indication of past stands of sea level. The highest of these caves are located only about 5 m above present sea level, which suggests that the maximum height of sea level during at least the late Pleistocene has been no greater than this. The relative stability of the Bermuda platform makes this hypothesis plausible.

Most of the air-filled caves were created by the collapse of bedrock into solutional voids below present sea level. The evolution of a typical Bermuda cave is shown diagrammatically in Figure 8.13. Major caves may have formed at the limestone/basalt contact during low stands of sea level, but their origin would have depended on a similar mixing phenomenon.

Vadose travertine to depths of 30 m has been dated radiometrically to give a chronology of past sea level changes in response to glacial episodes elsewhere in the world (Harmon *et al.* 1975, Gascoyne *et al.* 1979). Their evidence shows that for the last 200 000 years sea level has probably been higher than that of today only about 5% of the time, with the +5 m stand dated at about 120 000 years BP.

History of changes in erosional base level: Mammoth Cave, Kentucky

Few topics in geomorphology have been debated at such length as that of the significance, interpretation and even existence of a regional base level. The

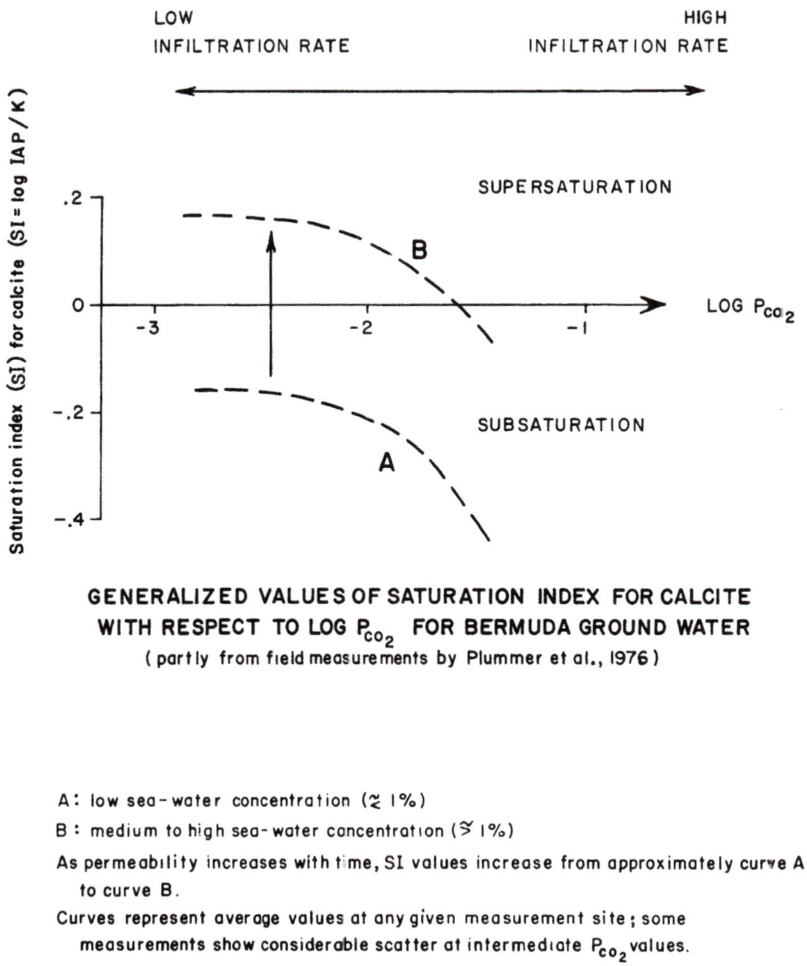

Figure 8.12 Chemical quality of Bermuda karst groundwater at various rates of infiltration. Curve A – conditions within small pores before significant enlargement by solution; water is fresh or slightly brackish and most is solutionally aggressive at all times. Curve B – conditions within solutionally enlarged pores and caves; water is brackish and is solutionally aggressive only during periods of high infiltration. Field data from Palmer *et al.* (1977) and Plummer *et al.* (1976).

character of passages and sediments in a multi-level cave can provide evidence for past changes in local river levels far more precisely than surface features and can help to resolve many of the questions surrounding the concept of base level. The Mammoth Cave System in the Mississippian limestones of west-central Kentucky is a well-documented example (Miotke & Palmer 1972, Palmer 1981c).

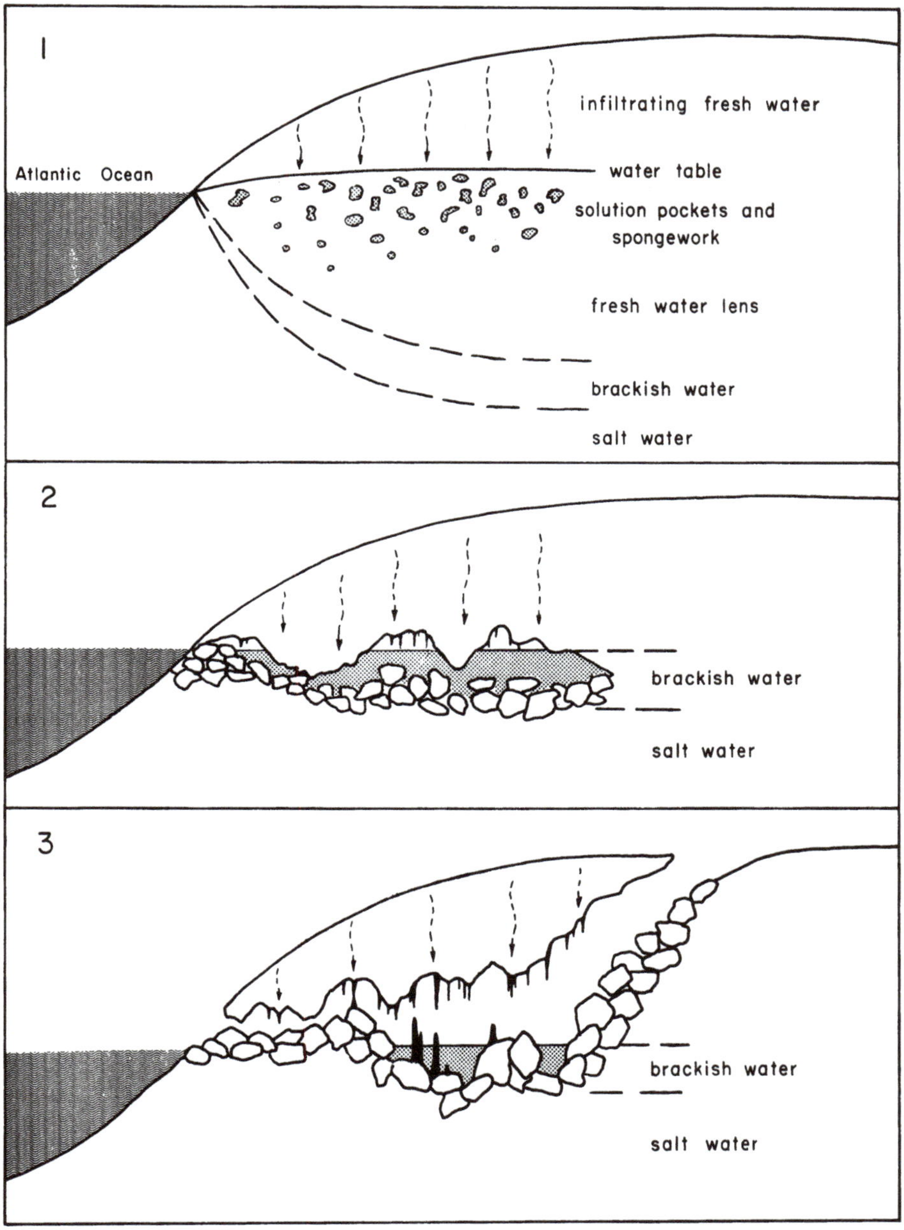

Figure 8.13 Sequence in which most Bermuda caves have formed. (1) Origin of initial solution pockets in the freshwater zone. (2) Disappearance of the freshwater zone as the permeability increases. (3) Modification of caves by collapse and travertine. Many travertine deposits of vadose origin are now under water because of the late-Pleistocene rise in sea level.

Mammoth Cave has a dendritic pattern complicated by the superposition of many levels of development and stages of groundwater diversion. Its main source of recharge consists of sinkholes and other discrete inputs. A protective caprock of resistant detrital rocks has preserved many of the older passages in high ridges whereas elsewhere (e.g. in a neighboring low-relief karst plain called the Pennyroyal Plateau where the caprock is absent) the oldest passages have been obliterated by erosion. Major passages are graded to present and past elevations of the Green River, now at an altitude of 130 m, which is the major entrenched river in the region (Fig. 8.14). The uppermost levels are large tubes and broad canyons up to 25 m in vertical extent at altitudes of 175–210 m. In their downstream ends, they all contain bedded deposits of gravel, sand and silt that reach nearly or completely to the passage ceilings, except where the fill material has been partly removed by later subsurface stream erosion or by subsidence into lower levels. In places, the sediment shows cut-and-fill structures and other features indicative of shallow, open-channel flow during the fill stage.

Below this upper level, nearly all the major passages are tubular in shape, indicating that the original solution took place at or below the water table (and therefore, below the level of the Green River). Passages are smaller in cross section than those at the 175–210 m level, and sediment filling is sparse and thin. Major passages cluster at altitudes of 168 m and 152 m, implying that the Green River was relatively static at those levels for lengthy periods. A few of these passages contain broad, U-shaped loops in their profiles that extend over a vertical range as great as 15 m but the position of the water table at the time they were formed can be ascertained from the elevation at which the vadose–phreatic transition occurs.

The differences between the upper and lower levels of the cave indicate a major change in the erosional–depositional characteristics of the region with time. The wide, partly sediment-filled upper levels correlate in elevation with nearby parts of the Pennyroyal Plateau, and it can be inferred that not only the upper-level cave passages but also the Pennyroyal Plateau formed under conditions of very slow fluvial erosion alternating with periods of aggradation. In contrast, the lower-level passages correlate with narrow, deeply entrenched river valleys and indicate cave development during comparatively rapid erosional dissection of the region. Major rearrangements of the Ohio River drainage resulting from continental glaciation during the early Pleistocene is thought to have caused the accelerated rate of erosion throughout the east-central United States, and therefore, also the change in cave morphology (Miotke & Palmer 1972). The upper levels apparently formed during the late Tertiary period and the two lower levels during the Quaternary period.

Data from the cave can provide a great amount of detail superimposed on the generalized history described above. For instance, the history of regional dissection during the Pleistocene is now being deciphered by dating cave deposits in the various levels and by correlating the passages with surface terraces. Miotke (in Miotke & Palmer 1972) assigns a Yarmouthian age to a

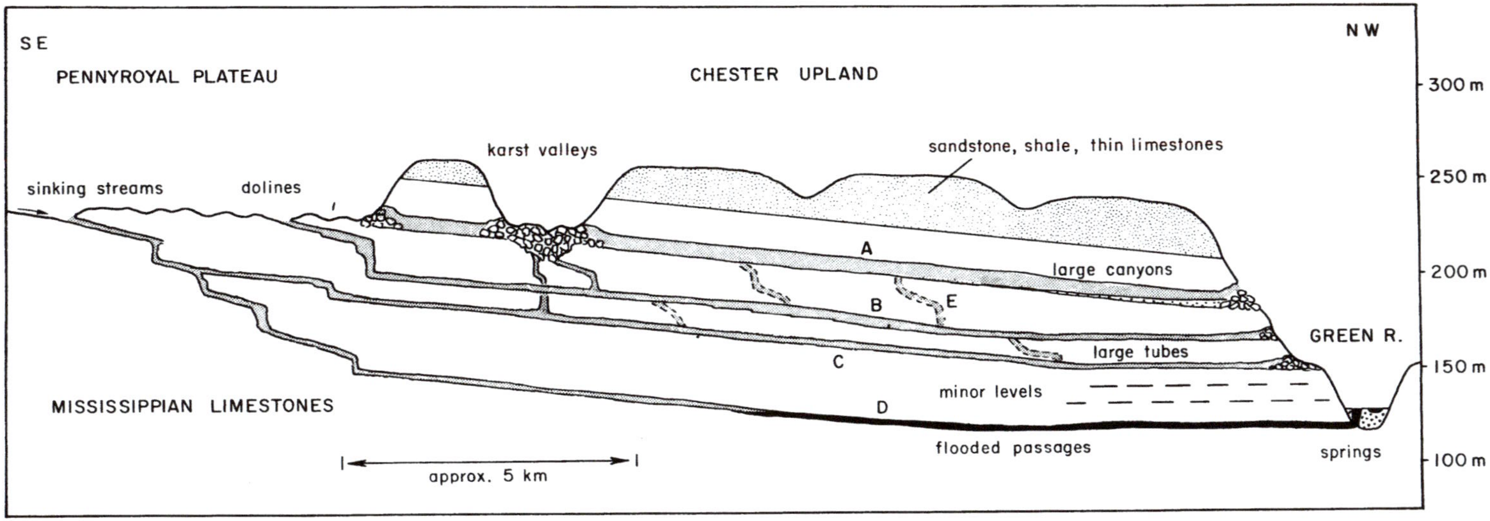

Figure 8.14 Passage levels in Mammoth Cave, Kentucky. A, late-Tertiary canyon passages partly filled with clastic sediment; B, early-Pleistocene tubular passages; C, mid-Pleistocene tubular passages, which correlate with remnants of a major terrace in nearby river valleys; D, late-Pleistocene passages, now partly flooded by a Wisconsinan rise in base level; E, minor canyons and shafts formed in the vadose zone.

set of terraces in the Green River valley that correspond with the 168 m and 152 m levels. Vadose travertine from the 152 m levels has been dated at >350 000 years by Hess and Harmon (1981). Paleomagnetic analysis of cave sediment is now under investigation by Victor Schmidt of the University of Pittsburgh.

The 152 m terrace is well developed, but the 168 m terrace is very fragmentary. However, the number and size of passages at 168 m is greater than that at 152 m, implying that the pause in dissection, or relatively slow change in base level, that produced the upper of these two levels was of greater duration. The largest of the passages at the altitude of 168 m has an average radius of 5 m. Although the chemical properties of the original water flow can only be guessed, the largest cave stream entering that part of the cave today at lower levels has an average C value of 80 mg l^{-1} $CaCO_3$ equivalent and a saturation value (C_s) of ~ 150 mg l^{-1}. From Equations 8.7 and 8.9, the maximum solution rate should have been roughly 0.01 cm yr^{-1}, and therefore, it would have taken, at the very least, 50 000 years for this passage at the 168 m level to enlarge to its present size under these chemical conditions. The passage formed essentially at the level of the water table because throughout its length it maintains the same elevation as its vadose–phreatic transition from canyon to tube and is almost perfectly parallel to the local strike of the beds. Therefore, the Green River must have paused in its downcutting at or near that level for several tens of thousands of years at some time in the mid-Pleistocene.

Evolution of karst drainage and topography in a limestone plateau: Blue Spring Cave, Indiana

Blue Spring Cave is a dendritic system of solution conduits underlying a low-relief, karsted plateau called the Mitchell Plain, which is correlative with the Pennyroyal Plateau of Kentucky (Powell 1964, Palmer & Palmer 1975). The cave is a subsurface tributary of the East Fork of White River and contains 52 active stream tributaries as well as truncated segments of upper-level passages. Although much of the evidence for previous drainage patterns is fragmentary and not fully documented, it is possible to reconstruct the evolution of drainage from the original surface streams to the present underground streams and some of the concurrent changes that have taken place in the land surface.

The surface over the cave is a nearly flat plain pitted with numerous dolines up to 30 m deep that feed water into the limestone. The plain, which lies at an altitude of 185–200 m, owes part of its origin to fluvial erosion. However, all but the largest streams have diverted to underground routes, and only ephemeral remnants of their upstream ends remain at the surface. The general pattern of the former surface drainage can be inferred from the relative altitudes of the divides between dolines and by the location of remnant surface tributaries.

The surface of the Mitchell Plain probably reached its lowest relief late in the Tertiary period. Quaternary glaciers diverted much of the Appalachian drainage into the Ohio River, causing rapid entrenchment of the Ohio and its tributaries, including the White River. As a result, minor streams became perched above the water table. Most of the streams perched on limestone were diverted underground through solution conduits formed by the aggressive influent water. The uppermost passages in Blue Spring Cave that show evidence for phreatic origin occur at ~ 170 m above sea level, which suggests that the surface streams were perched no more than 15 m above the East Fork of White River when diversion took place.

Most of the cave passages that formed as the result of this diversion have quite different trends from the original surface drainage, as shown in Figure 8.15. Groundwater is controlled very strictly by local small-scale geologic structures whereas the pattern of surface streams is adjusted to much broader structures. Also, the perennial streams in the cave are far more numerous than the original surface streams by a factor of at least ten. In addition to dolines that may have existed in interfluvial areas prior to the regional stream entrenchment, many new dolines formed in and around the former stream valleys as the surface drainage was diverted underground. Each doline provides a small but discrete dribble of water that feeds a separate tributary stream passage.

Passages of phreatic origin at an altitude of 150 m indicate a pause in the erosional deepening of the valley of the East Fork of White River. Rather rapid entrenchment below that level is indicated by canyons in the floors of the phreatic passages. In some places, the water abandoned its original flow path and diverted to entirely new, independent routes, leaving dry upper-level passages as evidence of the former drainage pattern. Several, large closed loops were formed where the water diverged from its original path and rejoined it farther downstream. Further evidence for changes in flow pattern is given by several abandoned spring alcoves in the walls of the river valley, which indicate former outlets for the cave streams.

The East Fork of White River is now located at an altitude of 143 m and is underlain by 25 m of alluvial fill. The fact that the streams in the upstream parts of the cave have only entrenched 5–7 m below the 150 m level suggests that the deepening and alluviation of the river valley must have occurred rapidly and rather late in the Pleistocene. Only the farthest downstream kilometer of the cave, which is now almost entirely water filled, shows evidence of having been graded to a level of the East Fork of White River lower than its present altitude.

Near the center of the cave system at a junction with a major tributary, the main passage of the cave became large enough that its roof collapsed, causing intense ponding of water upstream from this point during floods. A joint-controlled network maze bypasses the blockage today. Passages in the maze have been surveyed to a height of 8 m above the present level of the entrenched cave streams and have been plumbed to a depth of 9 m below. The maximum size and frequency of passages is at or slightly above the level of deepest

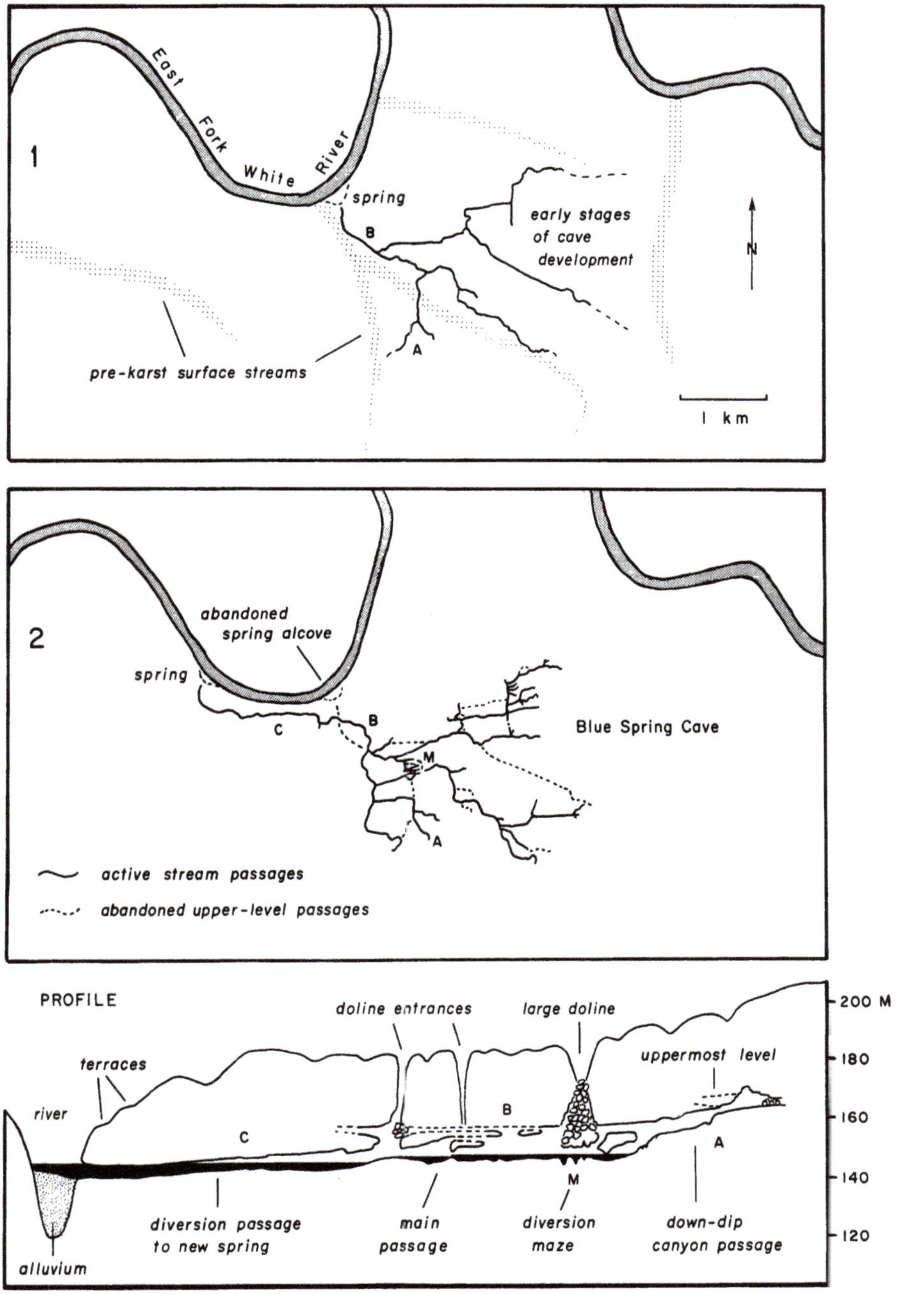

Figure 8.15 Changes in drainage pattern throughout the evolution of Blue Spring Cave and its overlying doline karst. A, down-dip canyon passage; B, main passage; C, diversion passage to new spring; M, diversion maze.

vadose entrenchment in that part of the cave. Therefore, the maze is among the most recent stages of passage development.

A large compound doline is located over the collapse area. This karst depression may have pre-dated the major ceiling collapse in the cave, but most of its present size is attributable to the collapse and to the removal of detrital sediment by the cave stream. As the maze enlarges, further collapse takes place, and the doline widens. The volume of the doline is 50 times greater than the estimated volume of fill in the underlying passages, which indicates the importance of erosional removal of collapse material and soil by the cave stream. Without the erosional removal of this material from below, this and most other dolines would be very shallow and would perhaps have no surface expression at all.

Enlargement of dolines by such processes has obscured the pattern of Tertiary surface drainage to the extent that only the most general drainage lines can be discerned. In contrast, the record of Quaternary drainage patterns and flow conditions is clearly preserved in the underlying cave (Fig. 8.15).

Current research

Recent years have seen a sharp increase in the range and sophistication of geomorphic techniques used in karst studies. This final section describes some of the more promising trends not mentioned in the preceding field examples.

Detailed mapping and description of karst features has increased greatly over the past few decades, so it has been possible to formulate comprehensive interpretations based on a wide variety of field data. International expeditions are mounted each year to remote karst regions, such as those in the Tropics and at high latitudes. Some of these areas are now being explored for the first time by karst scientists. From the simple portrayal of geomorphic features, mapping has evolved to the point where all relevant geologic and hydrographic characteristics are included. Detailed stratigraphic and structural mapping of caves has revealed the often subtle influence of geologic setting on patterns of groundwater flow. Mapping and analysis of cave sediments have helped in the interpretation of former flow patterns and hydraulic regimen.

Hydrologic and chemical measurements have been used to determine denudation rates in the various karst regions in relation to climate, rock and soil type, and topography and have provided information on the geochemical evolution of karst waters. The nature of inaccessible underground solution features is beginning to be understood from the hydraulic and statistical analysis of spring hydrographs.

With the increasing refinement of laboratory techniques, particularly radiometric dating, paleomagnetism, and isotope analysis, great strides have been made in geochronology and the interpretation of past environments and climates. Aside from the estimates of past changes in sea level and base level described earlier, the dating of cave features is beginning to provide information about tectonic and isostatic movements in the crust in various parts of the world. Laboratory simulation has provided insight into the factors that control the origin of solutional features.

Finally, it has been shown in recent years that the carbonic acid reaction in meteoric waters may be of only secondary importance in the origin of karst features in certain areas. The significance of hydrothermal waters and of chemical reactions involving sulfides and sulfates is now recognized in many of the limestones in the western United States and in the lead–zinc mining districts of the Mississippi valley. Little is known about the interactions between hydrothermal fluids and host rock, and the analysis of karst features should provide much of the future information about this subject.

It is apparent that karst studies are transcending the traditional boundaries of geomorphology. Methods from nearly every field of science are employed to provide a comprehensive view of the many diverse processes and events responsible for the origin of karst features.

References

Berner, R. A. 1980. *Early diagenesis: a theoretical approach*. Princeton, NJ: Princton University Press.

Berner, R. A. and J. W. Morse 1974. Dissolution kinetics of calcium carbonate in sea water. IV. Theory of calcite dissolution *Am. J. Sci.* **274**, 108–34.

Bögli, A. 1964. Mischungskorrosion – Ein Beitrag zum Verkarstungs-Problem. *Erdkunde* **18**, 83–92.

Bögli A. 1980. *Karst hydrology and physical speleology*. New York: Springer.

Bretz, J. H. 1942. Vadose and phreatic features of limestone caverns. *J. Geol.* **50**, 675–811.

Campbell, N. 1977. Possible exhumed fossil caverns in the Madison Group (Mississippian) of the northern Rocky Mountains. *Nat. Speleol Soc. Bull.* **39**, 43–54.

Coward, J. 1971. Direct measure of erosion in a streambed of a West Virginia Cave (Abs.). *Caves Karst* **13**, 39.

Curl, R. L. 1968. Solution kinetics of calcite. *Proc. 4th Int. Speleol Congr.*, 61–6. Ljubljana, Yugoslavia.

Davies, D. G.1980. Cave development in the Guadalupe Mountains: a critical review of recent hypotheses. *Nat. Speleol Soc. Bull.* **42**, 42–8.

Dreybrodt, W. 1981a. Mixing corrosion in $CaCO_3$–CO_2–H_2O systems and its role in the karstification of limestone areas. *Chem. Geol.* **32**, 221–36.

Dreybrodt, W. 1981b. Kinetics of the dissolution of calcite and its applications to karstification. Chem. Geol. **31**, 245–69.

Egemeier, S. J. 1973. *Cavern development by thermal waters with a possible bearing on ore deposition*. PhD dissertation. Stanford University.

Egemeier, S. J. 1981. Cavern development by thermal waters. *Nat. Speleol Soc. Bull.* **43**, 31–51.

Ewers, R. O. 1966. Bedding plane anastomoses and their relation to cavern passages. *Nat. Speleol Soc. Bull.* **28**, 133–40.

Ford, D. C. 1971. Geologic structure and a new explanation of limestone cavern genesis. *Trans. Cave Res. Grp* **13**, 81–94.

Ford, D. C. and R. O. Ewers 1978. The development of limestone cave systems in the dimension of length and depth. *Can. J. Earth Sci.* **15**, 1783.

Gams, I. 1976. Forms of subsoil karst. *Proc. 6th Int. Speleol Congr.* 169–79. Prague, Czechoslovakia: Academia Press.

Gascoyne, M., G. J. Benjamin and H. P. Schwarcz 1979. Sea-level lowering during the Illinoian glaciation: evidence from a Bahama "Blue Hole." *Science* **205**, 806–8.

Ginés, J., A. Ginés and L. Pomar 1981. Morphological and mineralogical features of phreatic speleothems occurring in coastal caves of Majorca (Spain). *Proc. 8th Int. Speleol Congr.*, 529–32. Bowling Green, Kentucky.

Harmon, R. S., P. Thompson, H. P. Schwarcz and D. C. Ford 1975. Uranium-series dating of speleothems. *Nat. Speleol Soc. Bull.* **37**, 21–33.

Hess, J. W. and R. S. Harmon 1981. Geochronology of speleothems from the Flint Ridge Mammoth Cave System, Kentucky, USA *Proc. 8th Int. Speleol Congr.*, 433–36. Bowling Green, Kentucky.

Howard, A. D. and B. Y. Howard 1967. Solution of limestone under laminar flow between parallel boundaries. *Caves Karst* **9**, 25–38.

Jennings, J. N. 1971. *Karst.* Cambridge, Mass. MIT Press.

Land, L. S., F. T. Mackenzie and S. J. Gould 1967. The Pleistocene history of Bermuda. *Geol Soc. Am. Bull.* **78**, 993–1006.

Maslyn, R. M. 1978. Cavern development via H_2S dissolved in hot spring and natural gas field waters (Abs.). *Nat. Speleol Soc. Bull.* **41**, 115.

Miotke, F-D. and A. N. Palmer 1972. *Genetic relationship between caves and landforms in the Mammoth Cave National Park area.* Würtzburg: Böhler.

Morehouse, D. F. 1968. Cave development via the sulfuric acid reaction. *Nat. Speleol Soc. Bull.* **30**, 1–10.

Mylroie, J. E. 1984. Hydrologic classification of caves and karst. In *Groundwater as a geomorphic agent,* R. G. LaFleur (ed.), 157–72. Boston: Allen & Unwin.

Palmer, A. N. 1972. Dynamics of a sinking stream system. Onesquethaw Cave, N.Y. *Nat. Speleol Soc. Bull.* **34**, 89–110.

Palmer, A. N. 1975. The origin of maze caves. *Nat. Speleol Soc. Bull.* **37**, 56–76.

Palmer, A. N. 1977. Influence of geologic structure on groundwater flow and cave development in Mammoth Cave National Park, U.S.A., in *Karst hydrology*, J. S. Tolson and F. L. Doyle (eds), 405–14. Int. Assoc. Hydrogeols. 12th Mem.

Palmer, A. N. 1981a. Hydrochemical controls in the origin of limestone caves. *Proc. 8th Int. Speleol Congr.*, 120–2. Bowling Green, Kentucky.

Palmer A. N. 1981b. *The geology of Wind Cave.* Hot Springs, SD: Wind Cave Natural History Association.

Palmer, A. N. 1981c. *A geological guide to Mammoth Cave National Park.* Teaneck, NJ: Zephyrus Press.

Palmer, A. N., M. V. Palmer and J. M. Queen 1977. Geology and origin of the caves of Bermuda. *Proc. 7th Int. Speleol Congr.*, 336–9. Sheffield, England.

Palmer, M. V. and A. N. Palmer 1975. Landscape development in the Mitchell Plain of southern Indiana. *Z. Geomorph.* **19**, 1–39.

Picknett, R. G. 1976. The chemistry of cave waters. In *The science of the speleology*, T. D. Ford and C. H. D. Cullingford (eds), 225–48. London: Academic Press.

Picknett, R. G. 1977. Rejuvenation of aggressiveness in calcium carbonate solutions by means of magnesium carbonate. *Proc. 7th Int. Speleol Congr.*, 346–8. Sheffield, England.

Plummer, L. N. and T. M. L. Wigley 1976. The dissolution of calcite in CO_2 saturated solutions at 25°C and 1 atmosphere total pressure. *Geochim. Cosmochim. Acta* **40**, 191–202.

Plummer, L. N., T. M. L. Wigley and D. L. Parkhurst 1978. The kinetics of calcite dissolution in CO_2 water systems at 5° to 60°C and 0.0 to 1.0 atm CO_2 *Am. J. Sci.* **278**, 179–216.

Plummer, L. N., H. L. Vacher, F. T. MacKenzie, O. P. Bricker and L.S. Land 1976. Hydrogeochemistry of Bermuda: a case history of ground-water diagenesis of biocalcarenites. *Geol Soc. Am. Bull.* **87**, 1301–16.

Powell, R. L. 1964. Origin of the Mitchell Plain in south-central Indiana. *Proc. Ind. Acad. Sci.* **3**, 177–82.

Rauch, H. W. and W. B. White 1977. Dissolution kenetics of carbonate rocks. *Water Resour. Res.* **13**, 381–94.
Sweeting, M. M. 1973. *Karst landforms*. New York: Columbia University Press.
Thrailkill, J. 1968. Chemical and hydrologic factors in the excavation of limestone caves *Geol Soc. Am. Bull.* **79**, 19–45.
Trudgill, S. T. 1976. Limestone erosion under soil. *Proc. 6th Int. Speleol Congr*. 409–22. Prague, Czechoslovakia: Academia Press.
White, W. B. 1969. Conceptual models for carbonate aquifers. *Ground Water* **7**, 15–21.

9
Theory and model for global carbonate solution by groundwater

John J. Drake

Introduction

It is axiomatic that groundwater is the primary geomorphic agent in karst terrains. There are other terrains in which groundwater plays a profound role and other agents that act on karst terrains, but the geomorphic effects of groundwater have been most fully and widely studied in the world's karstlands. The sequence of development of a particular cave system can now be reconstructed and dated using general conceptual models such as that of Ford and Ewers (1978) and speleothem dating methods, and an individual void space may be inferred to be, e.g. a dip tube, a joint-controlled vadose passage, or a collapsed sinkhole.

Despite this intensity of effort, there remains a lack of explanation of landform at other than the individual phenomenological level; there is no explanation of landform at the more general level in the sense that known parameters of process rate, structure and time can be combined to yield a parameterized description of void-space distribution in a karst terrain. Here, I present an illustration of the consistency of karst erosion rates and a discussion of the possible links between them and the various types of landform.

Three main research directions in karst studies can be distinguished. First, there is the research concerned with hydrodynamics and solution dynamics, which has built upon the early work of Frear and Johnstone (1929) and Trombe (1952). Second, there is the work concerned with erosion rates, which was perhaps begun by Goodchild's (1895) measurements on tombstones and translated into the basin context by Corbel (1956): and third, there is the landform description–evolution work that is the continuation of the early geomorphological investigations. Many studies combine elements of more than one of these directions, but process studies are rarely at the terrain scale, erosion rate studies rarely address the evolution of landform, and landform studies rarely invoke process explanation in quantitative terms.

Landform explanation and scale

Two main styles of landform "explanation" are common in karst studies. The first, as suggested above, is a description of the evolution of a particular feature by reference to some general concepts of development. The second, and more recent, is the use of statistical methods to describe an agglomeration of features in terms of parameters whose values can then be tested against some prior hypotheses. In many cases, the testing is rudimentary in that the hypothesized values are not derived from theory, but there are rigorous examples, such as the scallop studies of Goodchild and Ford (1971) and Blumberg and Curl (1974), which have led to the ability to hindcast flow velocities.

The concept of scale is inherent to explanation. For analysis of the structure of variability of karst water chemistry, a three-way classification of time and space has been valuable (Ford & Drake 1982). The same scheme, shown in Figure 9.1, is applicable to the explanation of landform. Ford (1980) has presented a classification of karst forms by linear extent, but it is the classification of space by size that is of importance here. At the micro-scale of space, the features to be explained are a multitude of similar small forms such as karren or scallops. At the meso-scale of space, an individual cave system may be considered, and at the macro-scale of the regional aquifer, there are aggregates of meso-scale features. The relationship of the human scale to the scale of landform is evidently important in determining whether a series of single features are explained individually in phenomenological terms or together as an agglomeration of samples from an infinite population.

The short and medium time scales are of limited importance to landform evolution by chemical groundwater processes. In carbonate karsts, these processes are slow and, it has been suggested, are not subject to dynamic thresholds (Ford 1980). The amount of solution over a period long enough to be geomorphically significant is, therefore, the integral of a continuous

		TIME	
	short up to years	medium centuries	long millennia up
SPACE: micro e.g. scallop bed	dynamic equilibrium	external controls change	space replaced
SPACE: meso e.g. local flow system	may be threshold events	dynamic equilibrium	external controls change, e.g. base level
SPACE: macro e.g. regional aquifer	infinitesimal events	may be threshold events	dynamic equilibrium

Figure 9.1 Three-way classification of space and time for the analysis of karst landform evolution by groundwater solution.

function of time. This is not true, however, of all groundwater action. Groundwater conduit flow does have threshold values, such as the flow regime transition discussed by White (1980) and so does the sediment flux in conduits; Wolfe (1973) has shown that massive fills that appear to be the result of a long fill phase may, in fact, be emplaced or removed by a single large event. The history of a particular feature such as a cave passage, or a collapse, or a suffosion sinkhole may be determined by a threshold event, and there may be significant change in the feature at the short time scale. At the macro-scale of the regional aquifer, however, the collapse of a single sinkhole is an infinitesimal event and does not disturb the equilibrium state. The human time scale is such that we can understand threshold events that happen to meso-scale landforms because they occur as short events that separate medium time periods of dynamic equilibrium. At the aquifer scale, periods of dynamic equilibrium are long-time events, and if there are threshold events, they may be medium-time events that appear to us to be periods of highly dynamic, dynamic equilibrium. The continuous loss of soil from the surface of the Burren, Eire, has recently been shown by Drew (1983) to be such an event caused by people. Future rates of erosion of the Burren aquifer will differ from past rates because of the lack of soil-generated CO_2 in infiltrating groundwater.

It is usually assumed that hypotheses of landform evolution may be tested by comparing expected and actual outcomes in other time periods in the same space, or in other spaces in the same time period. This is the assumption of ergodicity that is common to all geomorphology, but it is not easily shown to hold in the karst setting because of the slowness of groundwater as a geomorphic agent at the aquifer scale. It is not clear whether the macro-scale karst landforms are different because they represent different stages of one evolution stem or because they represent different stems.

The macro–long scale of karst landform evolution

At the macro-scale, the landforms that have received the most attention are the various assemblages of closed depressions and remnant hills that include the very different morphologies of sinkhole plains, cockpit karst and other polygonal karsts, and tower karst. Factors that have been suggested as basic to differentiating among these include those associated with the rock (including fracture density, hydrogeological properties and mechanical strength), with climate (including annual precipitation and temperature and rainfall intensity), and with time. Ford (1980) presents a review of this literature in which he shows that in most studies a single threshold value of a single controlling factor has been postulated, and also that in each such case refuting evidence is available from other studies. A more detailed analysis of possible links between erosion rates and landform is possible in the light of catastrophe theory and is presented following a demonstration of the relative constancy of erosion rates.

Erosion rate variations

In the case of basin erosion rates and groundwater solute concentrations, the lack of threshold values for gross climatic factors has been substantiated by studies of regional variations (e.g. Smith & Atkinson 1976). These studies have shown that in carbonate karsts erosion-rate variations depend linearly and primarily on variations in annual runoff because variations in groundwater solute concentrations are minimal in comparison. No regional discontinuities in erosion rates can, therefore, exist because annual runoff values form a world-wide continuum.

The relative invariance of groundwater solute concentrations and the dependence of erosion rates on runoff regime is shown by a comparison of some of the karst regions of Canada that have been described in summary by Ford (1983). Alkalinity (as equivalent mg l^{-1} HCO_3^-) and temperature for carbonate groundwater data sets identified in Table 9.1 and

Table 9.1 Characteristics of selected carbonate groundwaters in Canada.

Set	Location	HCO_3^- concentration (mg ℓ^{-1}) n	mean	sd	Mean annual temperature
• Wells in carbonate aquifers					
1	Red R. valley, Man.	44	361	96	6.5
2	Selkirk area, Man.	60	407	67	6.5
3	Montreal Is., Que.	160	360	95	6.5 (MAT)
4	Russell Co., Ont.	23	336	107	6.7
5	Schefferville, Que.	35	97	7	5.4
6	Haldimand-Norfolk, Ont.	15	294	84	7.7 (MAT)
7	Hatt Is., Albany R., Ont.	5	402	13	0.5
▲ Springs in carbonate aquifers (diffuse)					
8	Bruce Penn., Ont.	21	301	28	8.5
9	Castleguard, Alta.	—	≈ 120	—	5.0
10	Vendom Fj., Ellesmere Is., NWT	54	425	70	1.0
△ Springs in carbonate aquifers (undifferentiated)					
11	Nahanni R., NWT	10	≈ 285	≈ 79	6.4
12	Anticosti Is., Que.	7	145	21	6.8
○ Wells in carbonate-cemented aquifers, or carbonate tills					
13	Prince Edward Is.	90	158	48	5.8 (MAT)
14	Cochrane, Ont.	9	407	7	5.0
15	Cypress Hills Fm., Sask.	13	298	93	2.3

MAT: Mean annual air temperature.

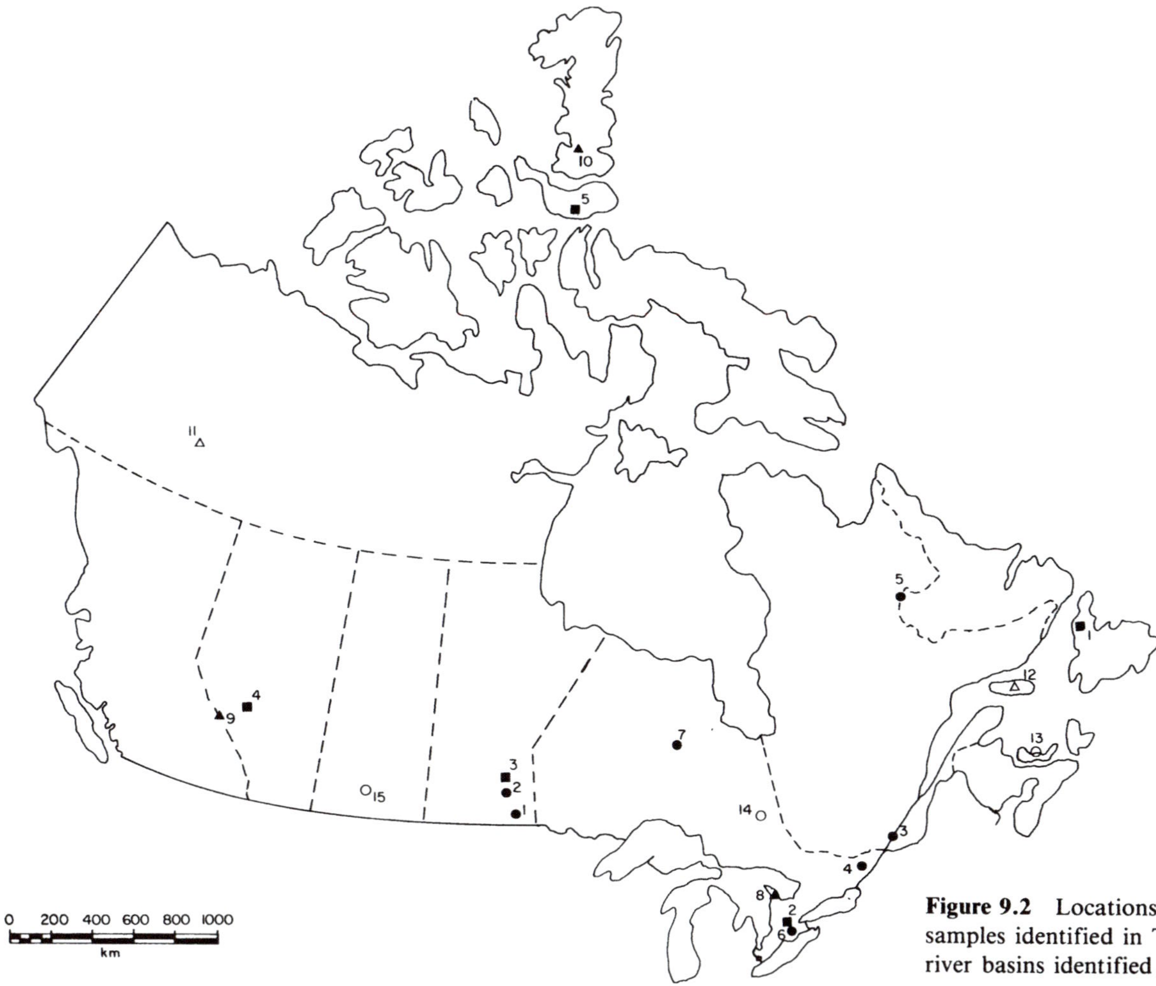

Figure 9.2 Locations of the groundwater samples identified in Table 9.1 and of the river basins identified in Table 9.2.

Figure 9.2 are shown in Figure 9.3. Details are given in Drake (1982). There is a tendency for each set to fall close to one or another of the equilibrium lines of the model described by Drake (1983a), and four groups can be distinguished by a t-test for the mean value at $p < 0.05$: (a) sets that are not different from the coincident system equilibrium but are from the sequential (1, 2, 3, 4, 6, 8, 10, 15), (b) sets that are not different from the sequential system equilibrium but are from the coincident (12, 13), (c) sets that are different from both equilibria (5, 7, 14), and (d) the set that is not different from both (11).

Set 9 cannot be classified in this manner because too few samples exist to calculate a variance. Of the sets in group (c), 5 can be assigned to the closed system group by reaction-path modelling (Drake 1983b), and sets 7 and 14 are small samples taken over very short summer periods from northern Ontario. More representative sets would show greater dispersion and would fall into group (a). Set 11 represents undifferentiated springs in the South Nahanni River area and includes members from a wide range of hydrogeological situations. The analysis also shows that waters from carbonate aquifers, carbonate-cemented aquifers and carbonate tills behave in a similar manner. Apparent differences among reaction systems may be due to recharge through saturated soils rather than to the existence of different system conditions (Drake 1983a). This explanation is particularly likely for the sets in northern Canada where much of the annual recharge occurs immediately following snowmelt. Despite the considerable differences in aquifer and climatic characteristics among the

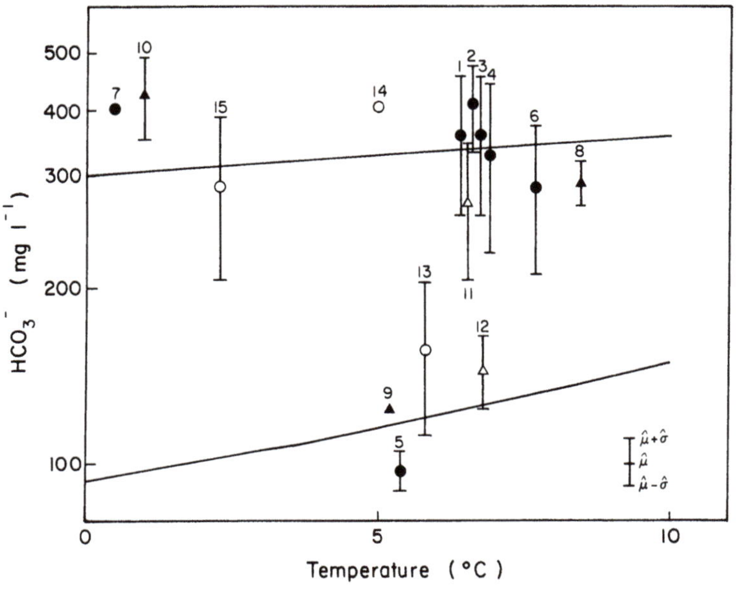

Figure 9.3 Alkalinity and temperature for the groundwater data sets identified in Table 9.1 and Figure 9.2. Upper line is the coincident system equilibrium line from Drake (1983a), and the lower line, the sequential system equilibrium.

various areas, the differences in groundwater solute concentrations are small; there is a factor of about three among the mean values.

In contrast to the relative seasonal and inter-regional stability of groundwater solute concentration values, the hydrologic regime in Canada is extremely variable. Annual runoff varies from over 3000 mm to under 25 mm (Hydrological Atlas of Canada, Plate 24), and high–to–low flow ratios for streams are often infinite because of winter cessation of flow. The conceptually simplest mixing model of a constant base flow of groundwater in equilibrium with a carbonate aquifer mixing with a variable surface flow of lesser solute concentration (Drake & Ford 1974) is capable of describing the hydrochemistry of many streams in karst terrains. Figure 9.4 shows the mean annual cycle of runoff and HCO_3^- and SO_4^{2-} concentrations for the five streams identified in Table 9.2 and Figure 9.2. The choice of streams was severely restricted by the following conditions: a Water Survey of Canada gauging station and a Water Quality Division quality station (or equivalents) had to be present and at the same location; data had to be available for a number of years; the stream had to drain a sufficient area that local spatial and temporal disturbances of the regime were smoothed out; and the basin had to be wholly within one hydrologic and hydrochemical zone. Insofar as was possible, the five streams met these criteria. Most of the karst regions of

Table 9.2 Characteristics of selected karst drainage basins in Canada.

	Data source[1] discharge quality	Area (km^2)	Runoff ($mm\ a^{-1}$)	Carbonate erosion rate[2] ($\mu m\ a^{-1}$)	Carbonate load[2] ($t\ km^{-2}\ a^{-1}$) limestone	Sulphate load[2] ($t\ km^{-2}\ a^{-1}$) gypsum
■ 1. Ste. Genevieve NFLD	WSC 00NF02YA001 WQD 00NF02YA0001	306	936	16.9	44.0	15.4
2. Nith ONT	WSC 00N002GA010 OWRC 16-984-009-02	1031	329	23.9	62.2	55.5
3. Fisher MAN	WSC 00MA05SD003 WQD 00MA05SD0001	1357	47	4.9	12.7	3.45
4. Athabasca ALTA	WSC 00AL07AD002 WQD 00AL07AD0001	9790	573	18.4	47.7	35.5
5. Meecham[3] NWT	Cogley (1976) Cogley (1976)	97.7	296	8.1	21.1	~0

[1]WSC = Water Survey of Canada gauging station number.
WQD = Water Quality Division, Environment Canada, quality station number.
OWRC = Ontario Water Resources Commission quality station number.
[2]Assuming contributing mineral area of 100%.
[3]Data for 1971 only.

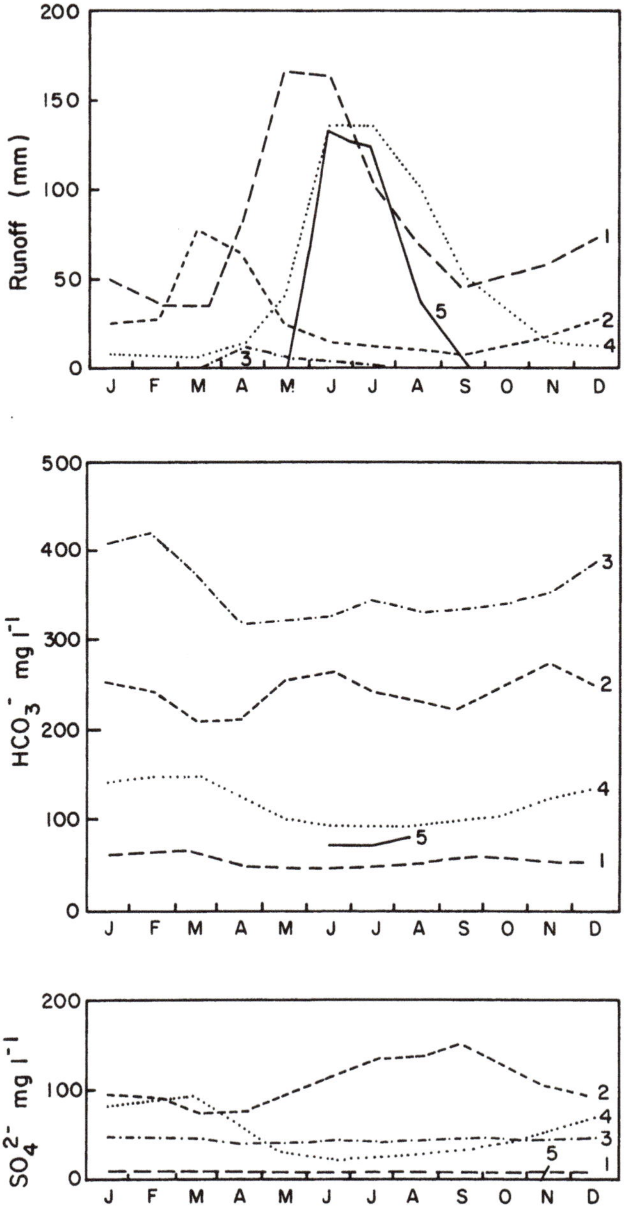

Figure 9.4 Mean annual cycle of runoff and HCO_3^- and SO_4^{2-} for the five rivers identified in Table 9.2 and Figure 9.2.

Canada are represented. Figure 9.5 shows the extremely close concordance of the hydrologic and hydrochemical regimes in terms of the magnitude – frequency distributions of fluxes. Despite the considerable variations in annual runoff, the annual carbonate erosion rate shows a range of no more than a factor of four.

Although the calculation of erosion rates is fraught with problems, it is generally found to be a relatively conservative quantity (e.g. Smith & Atkinson 1976). The high correlations reported between runoff and erosion rate (or dissolved-material transport rates) reported by Pulina (1972), and others are spurious (Benson 1965). Those relationships, therefore, cannot be used for prediction. The lack of any threshold or systematic control other than annual runoff on erosion rate, and the evidence against a simple runoff control of macro-scale karst landform together show convincingly that the explanation must lie elsewhere.

Landform description

The description of karst landform at the macro-scale has largely been derived from analogy between closed depressions and either point processes (stemming from the ecological literature) or fluvial drainage basins. The former approach has led to various inferences of the governing process, but the indeterminacy of process underlying point-pattern distributions make these uncertain. In addition, there is no body of statistical theory that is appropriate to the selection of the best among a series of null hypotheses. The latter morphometric approach has resulted in the same problems as it did in the original fluvial setting. The relationships, e.g., of depression order and number are consequent upon the ordering scheme. There has been little attempt to incorporate the more successful applications of topological concepts.

There are three types of result in these studies: simple or mixed independent random processes (e.g. Haan & Johnson 1967, McConnell & Horn 1972), contagious or multi-generation random processes (e.g. Drake & Ford 1972, Kemmerly 1982), and random processes governed by factors such as competition (Williams 1972), fractures or pre-existing stream networks. A number of such geological controls have been examined by Palmquist (1979).

None of the studies of closed depressions have contained an element of evolution. The parameters of the distribution have been assumed to represent the present situation without addressing the question of change through time, despite a significant early paper by Curl (1960) that did contain an evolutionary mechanism for cave-passage distributions. This omission can be seen either (a) as the assumption that the macroscale, closed-depression landforms are currently in dynamic equilibrium with present processes so that although individual depressions may appear or disappear, the aggregate number and distributions remain constant or (b) as the assumption that the equilibrium state is changing so slowly that they appear to do so. There appears to have been no test of either assumption, and it is difficult to see how one could be performed. Similar difficulties have not prevented the inclusion of

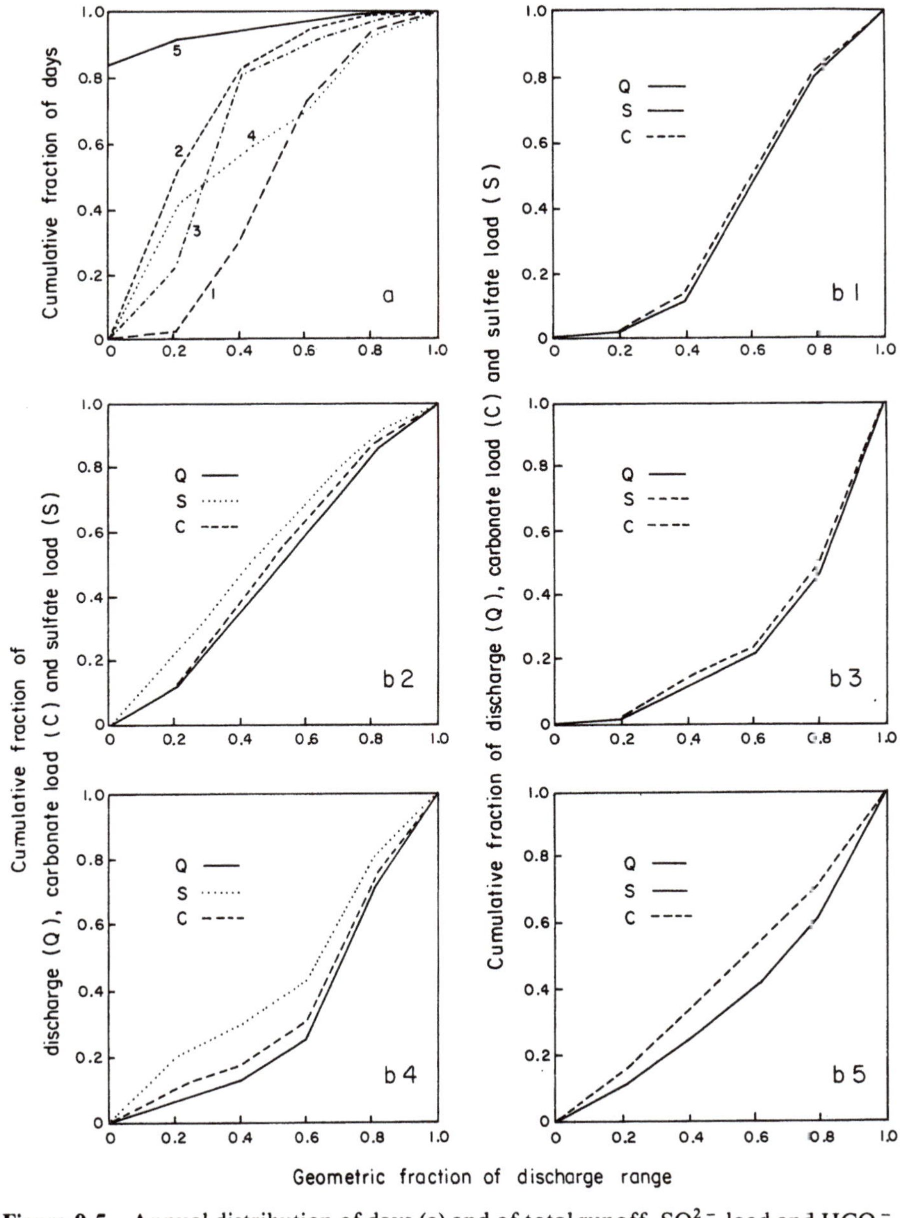

Figure 9.5 Annual distribution of days (a) and of total runoff, SO_4^{2-} load and HCO_3^- load (b) for the five rivers in Table 9.2 and Figure 9.2. Figure 9.5b1 refers to river 1 of Table 9.2, Figure 9.5b2 to river 2, etc.

evolutionary mechanisms in other terrains – Thornes and Brunsden (1977) have considered the question of time at the macro-scale in some detail.

Landform evolution

The lack of threshold values in the karst erosion process or in erosion rates does not imply that there are no thresholds or discontinuities in landform response. The concept of stable and unstable characteristic forms has been reviewed by Thornes (1982), and it is evidently a geomorphic application of catastrophe theory. In this section, I provide a speculative analysis of the evolution and stability of some macro-scale karst landforms.

An alternative to the point-pattern description of the structure of a sinkhole or cockpit terrain is the use of the void space–altitude distribution. After removal of secular space trends of altitude in a representative holokarst area, the range of altitudes is between the highest inter-depression point and the lowest intra-depression point, and the void space–altitude distribution is the inverse of the hypsometric curve. In the appendix to this chapter, it is shown that this may have a number of stable or unstable shapes for a particular erosion pattern. The particular pattern selected here,

$$d\mu/dt = (\mu - 0.5)^2 - \varkappa$$

where μ is the mean reduced altitude, is not derived from observation or from theory, nor is it scaled to time. It is taken as an illustration, but it does preserve some of the properties of groundwater solution distribution:

(a) If the mean altitude is high ($\mu \sim 1.0$), then the rate of surface lowering is small because high hydraulic gradients distribute solution through a large volume of rock.
(b) If the mean altitude is low ($\mu \sim 0$), then the rate of lowering is small because low hydraulic gradients raise water tables and divert precipitation to surface runoff. In both (a) and (b), the dispersion of altitudes is necessarily small, and there is little spatial concentration of flow.
(c) The rate of lowering reaches a maximum for some μ, here arbitrarily chosen as 0.5.
(d) The overall rate of surface lowering may be negative because of the relative lowering of either the maximum or minimum altitudes.

Figure 9.6 shows that $\varkappa < 0.25$ implies that some relative raising of altitude is implied, and that $\varkappa < 0$ implies that no relative lowering exists.

Figure 9.7 shows the rate of change of shape as a function of equilibrium shape at $\mu = 0.5$ ($S = 1$) for all $\varkappa$, but it is stable only for $\varkappa < 0$. In general, residual landscapes have small S, cockpit-type landscapes have S of unity, and sinkhole plains have large S. The suggestion here is that the cockpit landscape

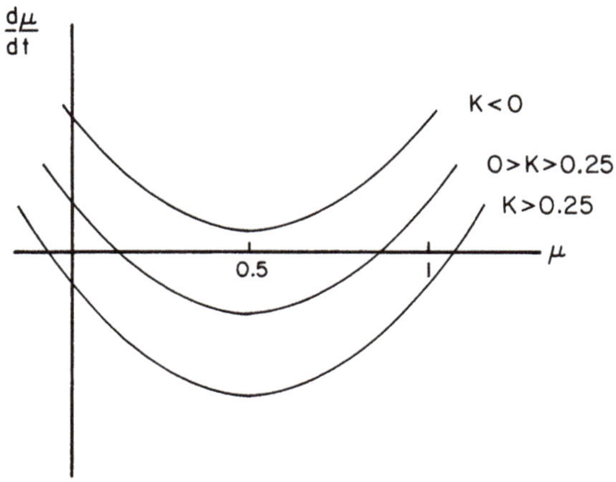

Figure 9.6 Overall lowering rate as a function of mean reduced altitude for various values of $\varkappa$. Negative values of the rate indicate lowering of the area with respect to the maximum and minimum elevations; positive values indicate that the extremes are lowering with respect to the general reduced altitude.

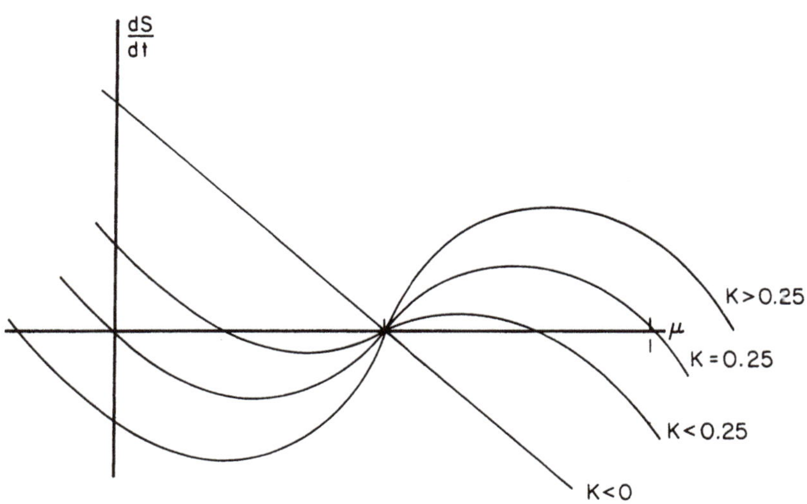

Figure 9.7 Rate of change of shape as a function of mean reduced altitude. Equilibria of shape exists when $dS/dt = 0$, and the presence and stability of equilibria depend on the value of $\varkappa$. Details are given in the appendix to this chapter.

is the stable one in conditions where high points and/or depression bottoms are rapidly lowered with respect to the bulk of the surface. For $\varkappa > 0$, cockpits are an unstable equilibrium form although the rate of change of form for small $\varkappa$ is slow. The considerably different types of cockpit and other polygonal karst that are found may be a reflection of this absolute instability of this shape under conditions of overall relative surface lowering. The polygonal form in a $\varkappa > 0$ environment can only be an inheritance from a previous $\varkappa < 0$ state.

For $0 < \varkappa < 0.25$, there are also two stable equilibria within the range $0 < \mu < 1.0$. Higher μ (higher S) corresponds to the sinkhole plain, which is the stable shape where initial localized lowering has caused generally high relative altitudes. Low μ (low S) corresponds to the residual landscape, which is the stable shape where there is a fixed minimum altitude that forces low relative altitudes. These two complementary shapes appear to be expected where there is general lowering together with an increased lowering of lows if the general altitude is high (such as a strong concentration of flow towards a main flow outlet), and an increased lowering of highs if the general altitude is low (such as an increased erosion rate due to the parallel retreat of several sides of a tower).

For large and ubiquitous rates of lowering ($\varkappa > 0.25$), there are no stable shapes in $0 < \mu < 1.0$, and the tendency is for S to approach zero or unity. This corresponds to an eventual extinction of relief, which is intuitively correct.

Conclusion

The analysis of the geomorphic effect of groundwater in a karst terrain is dependent upon the scales of time and space that are chosen. At the long-term, large-area scale, there is no satisfactory explanation of landscape, in contrast to the comprehensive models of meso-scale form development that have been developed for, e.g., cave systems.

The long-term rate of erosion of a karst terrain taken as a whole is primarily a function of runoff, is relatively invariant between regions, and is stable within a region. Erosion rate differences are not connected in any consistent manner to landform at the regional scale, and there have only been individual phenomenological explanations of landform development.

The lack of thresholds in the karst erosion rate does not imply that there are no thresholds in the landform process. The geomorphic concept of thresholds is a particular example of catastrophe theory, and the conditions under which landform shapes are stable can be derived theoretically for a particular type of distribution of erosion within a karst terrain. In this chapter a particular erosion distribution is assumed, and its implications for the development of macro-scale surface landforms are examined. No claim is made that this particular analysis represents the definitive work on the explanation of surface karst form; rather, it demonstrates that the development of distinct forms can be explained without recourse to a particular control for each case, and suggests what may be the variables that must be quantified to permit such work.

Appendix

The stability of hypsometric form

Assume that a drainage basin can be represented as a set of points whose elevations (H) lie between some externally fixed lower limit (H_{min}, the outlet elevation) and an upper limit (H_{max}, the highest point of the drainage divide), which is also fixed. Then, the distribution of a random sampling of points of the bounded reduced altitude, h_i

$$h_i = \frac{H_i - H_{min}}{H_{max} - H_{min}} \tag{9.A1}$$

follows a *beta* distribution with probability density

$$f(h) = \begin{cases} \dfrac{h^{\alpha-1}(1-h)^{\beta-1}}{B(\alpha,\beta)} & 0 \leqslant h \leqslant 1 \\ 0 & \text{otherwise} \end{cases} \tag{9.A2}$$

where $B(\alpha, \beta)$ is the *beta* function. The parameters α and β can be estimated as

$$\hat{\alpha} = \hat{\mu}\left\{\frac{\hat{\mu}(1-\hat{\mu})}{\hat{\sigma}^2} - 1\right\} \tag{9.A3a}$$

$$\hat{\beta} = (1-\hat{\mu})\left\{\frac{\hat{\mu}(1-\hat{\mu})}{\hat{\sigma}^2} - 1\right\} \tag{9.A3b}$$

where $\hat{\mu}$ and $\hat{\sigma}^2$ are the mean and variance of the random sample h_i. Figure 9.8 shows

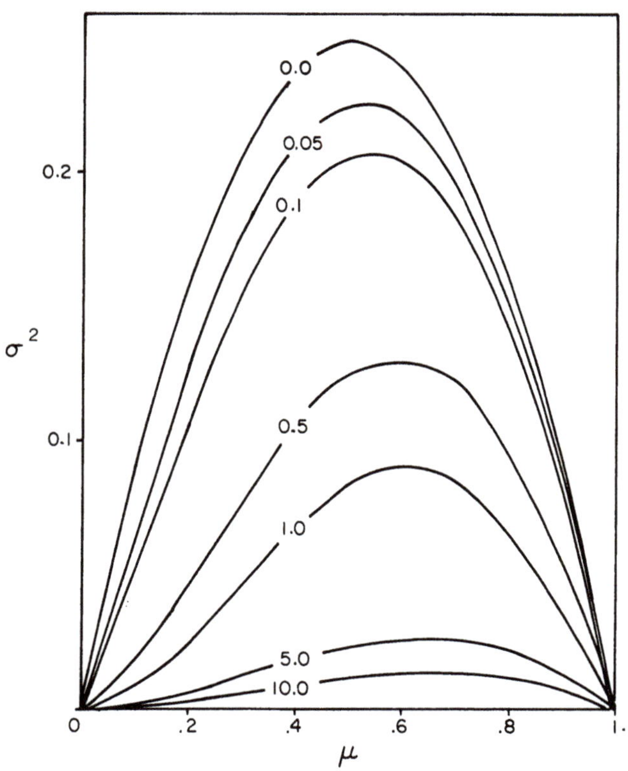

Figure 9.8 Values of α for different means and variances of the reduced altitude. Because this is a bounded variable (between 0 and 1.0), the variance is a function of the mean such that the region below $\alpha = 0$ represents the only possible combinations of mean and variance.

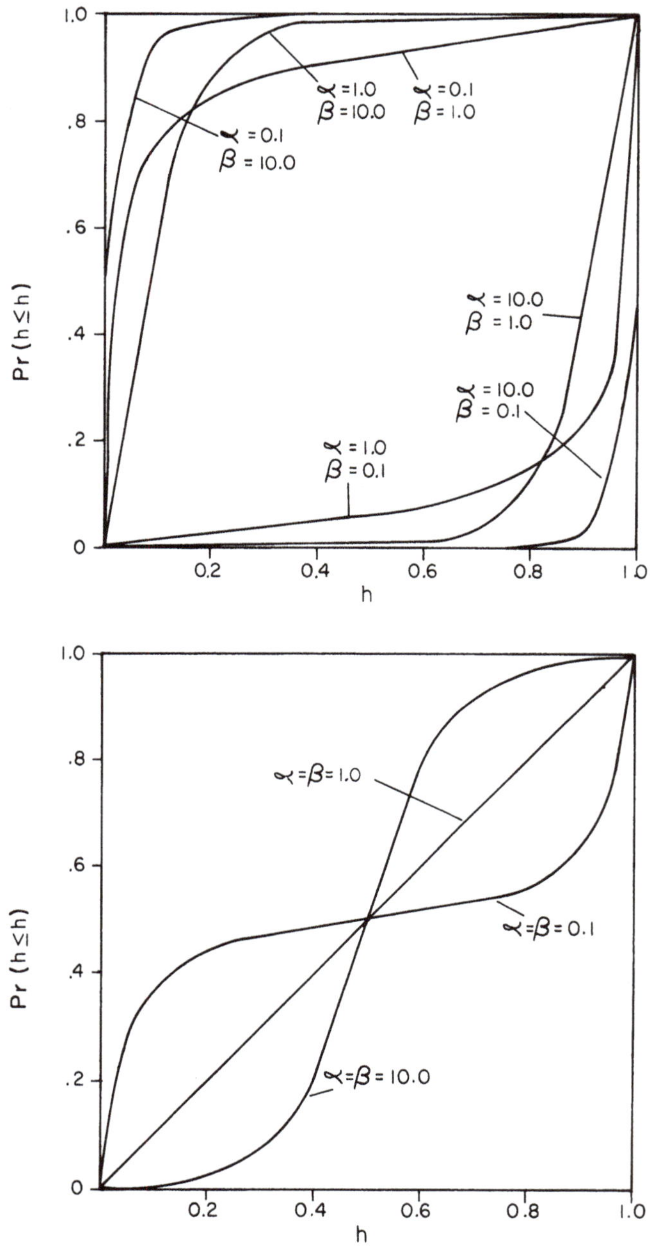

Figure 9.9 Cumulative *beta* distribution for various values of the parameters α and β. Top – asymmetric shapes resulting from values of S $(=\alpha/\beta)$ removed from unity. Bottom – symmetric shapes resulting from $S = 1.0$. If flipped bottom right to top left, these diagrams represent dimensionless hypsometric curves with reduced altitude on the ordinate and fraction of area below on the abscissa.

values of $\hat{\alpha}$ for various values of $\hat{\mu}$ and $\hat{\sigma}^2$. Because $0 \leqslant h \leqslant 1$, only combinations of μ and σ^2 for which $\alpha \geqslant 0$ are possible.

The cumulative distribution function

$$F(h) = Pr(h \leqslant h) = \int_0^h f(h) \quad (9.A4)$$

then provides a representation of the dimensionless hypsometric curve for the basin with reduced altitude as the abscissa and fraction of area as the ordinate. Figure 9.9 shows cumulative distribution functions for various values of α and β. A shape factor, S, can be defined as

$$S = \alpha/\beta \quad (9.A5)$$

which relates to the relative distribution of area above and below the basin mid-altitude $(H_{\max} + H_{\min})/2$.

From (9.A5) and (9.A3)

$$\hat{S} = \hat{\mu}/(1 - \hat{\mu}) \quad \text{or} \quad \hat{\mu} = S/(1 + \hat{S}) \quad (9.A6)$$

and the evolution of S in time is then given by

$$\frac{dS}{dt} = \frac{dS}{d\mu} \cdot \frac{d\mu}{dt} = \frac{(1 - 2\mu)}{(1 - \mu)^2} \frac{d\mu}{dt} \quad (9.A7)$$

Assuming $d\mu/dt$ to be a convex function of μ, for example

$$d\mu/dt = (\mu - 0.5)^2 - x \quad (9.A8)$$

we have

$$\frac{dS}{dt} = \frac{(1 - 2\mu)}{(1 - \mu)^2} (\mu^2 - \mu + 0.25 - x) \quad (9.A9)$$

Equilibrium of shape occurs for $dS/dt = 0$, which occurs at the roots of (9.A9):

$$1 - 2\mu = 0 \quad (9.A10a)$$

$$\mu^2 - \mu + 0.25 - x = 0 \quad (9.A10b)$$

giving $\mu = 0.5$ (9.A10a) and $\mu = 0.5 \pm \sqrt{x}$ (9.A10b).

The general features of the relationship between shape evolution and mean reduced altitude are shown in Figure 9.7. The central equilibrium is unstable and the others stable.

References

Benson, M. F. 1965. Spurious correlation in hydraulics and hydrology. *J. Hydraul. Div. Am. Soc. Civil Engrs* **91** (HY4), 35–42.

Blumberg, P. N. and R. L. Curl 1974. Experimental and theoretical studies of dissolution roughness. *J. Fluid Mech.* **65**, 735–51.

Cogley, J. G. 1976. *Properties of surface runoff in the High Arctic*. PhD dissertation. McMaster University.

Corbel, J. 1956. A new method of study for limestone regions. *Rev. Can. Geog.* **10**(4), 240–2.

Curl, R. L. 1960. Stochastic models of cavern development. *Nat. Speleol Soc. Bull.* **22**(1), 66–76.

Drake, J. J. 1983a. The effects of geomorphology and seasonality on the chemistry of carbonate groundwaters. *J. Hydrol.* **61**, 223–36.

Drake, J. J. 1983b. Groundwater chemistry in the Schefferville, Quebec iron deposits. *Catena*, **14**, 159–68.

Drake, J. J. in press. Karst solution. In *Karst of Canada*. D. C. Ford (ed.), Geological Survey of Canada, Monograph.

Drake, J. J. and D. C. Ford 1972. The analysis of growth patterns of two-generation populations: the example of karst sinkholes. *Can. Geogr.* **16**(4), 381–4.

Drake, J. J. and D. C. Ford 1974. Hydrochemistry of the Athabasca and North Saskatchewan rivers in the Rocky Mountains of Canada. *Water Resour. Res.* **10**(6), 1192–8.

Drew, D. P. 1983. Accelerated soil erosion in a karst area: the Burren, western Ireland. *J. Hydrol.* **61**, 113–26.

Ford, D. C. 1980. Threshold and limit effects in karst geomorphology. In *Thresholds in geomorphology*, D. R. Coates and J. D. Vitek (eds), 345–62. London: George Allen & Unwin.

Ford, D. C. 1983. Effects of glaciations upon karst aquifers in Canada. *J. Hydrol.* **61**, 149–58.

Ford, D. C. and J. J. Drake 1982. Spatial and temporal variations in karst solution rates: the structure of variability. In *Space and time in geomorphology*, C. E. Thorn (ed.), 147–70. London: George Allen & Unwin.

Ford, D. C. and R. O. Ewers 1978. The development of limestone cave systems in the dimensions of length and breadth. *Can. J. Earth Sci.* **15**(11), 1783–98.

Frear, G. L. and J. Johnstone 1929. The solubility of calcium carbonate (calcite) in certain aqueous solutions at 25C. *J. Am. Chem. Soc.* **51**, 2082–93.

Goodchild, J. G. 1895. The glacial phenomena of the Eden Valley and the west part of the Yorkshire Dales district. *Q. J. Geol Soc.* **31**, 55–99.

Goodchild, M. F. and D. C. Ford 1971. Analysis of scallop patterns by simulation under controlled conditions. *J. Geol.* **79**(1), 52–62.

Haan, C. T. and H. P. Johnson 1967. Geometrical properties of depressions in north-central Iowa. *Iowa St. J. Sci.* **42**(2), 149–62.

Hydrological Atlas of Canada 1978. Ministry of Supply and Services Canada, Ottawa.

Kemmerly, P. R. 1982. *Spatial analysis of a karst depression population: clues to genesis.* Unpublished paper, Department of Geology, Austin Peay State University, Clarksville, TN.

McConnell, H. and J. M. Horn 1972. Probabilities of surface karst. In *Spatial analysis in geomorphology*, R. J. Chorley (ed.), 111–33. New York: Harper & Row.

Palmquist, R. C. 1979. Geological controls on doline characteristics in mantled karst. *Z. Geomorph.* **32**, 92–106.

Pulina, M. 1972. Observations on the chemical denudation of some karst areas of Europe and Asia. *Stud. Geomorph. Carpatho-Balcanica* **5**, 79–91.

Smith, D. I. and T. C. Atkinson 1976. Process, landforms and climate in limestone regions. In *Geomorphology and climate*, E. Derbyshire (ed.), 367–409. New York: Wiley.

Thornes, J. B. 1982. Problems in the identification of stability and structure from temporal data series. In *Space and time in geomorphology*, C. E. Thorn (ed.), 327–53. London: George Allen & Unwin.

Thornes, J. B. and D. Brunsden 1977. *Geomorphology and time*. London: Methuen.

Trombe, F. 1952. *Traite de speleologie*. Paris: Payot.

White, W. B. 1980. Kinetic and mass transport effects in the development of karstic drainage systems: the transition from fracture to conduit permeability. *Geol Soc. Am. Abs. Prog.* **12**(7), 548.

Williams, P. W. 1972. The analysis of spatial characteristics of karst terrains. In *Spatial analysis in geomorphology*, R. J. Chorley (ed.), 136–66. New York: Harper & Row.

Wolfe, T. E. 1973. *Sedimentation in karst drainage basins along the Allegheny Escarpment in southeastern West Virginia, U.S.A.* PhD dissertation. McMaster University.

10
Rate processes: chemical kinetics and karst landform development

William B. White

Introduction

Karst landscapes are the result of the dissolution of the bedrock, usually limestone or dolomite, by runoff, by infiltration and by deep circulating groundwaters. To qualify as karst, it is generally taken that mass transport in solution is of greater importance than mass transport by other processes. Karst terrains evolve through time. There is a gradual lowering of the surface, and there is differential solution that forms such characteristic surface forms as dolines, cutters, and deranged drainage. The overall rate processes in the system have a chemical part and a hydrodynamic part. The chemistry of carbonate rock dissolution in turn has an equilibrium part and a kinetic part. Environmental factors determine the maximum amount of dissolved carbonates that can be carried by a given volume of water, and this combined with the flow rate determines the maximum rate at which a karst landscape will develop. However, the characteristic time necessary for limestones to reach chemical equilibrium with the water is about the same as the time necessary for the water to flow through a well-developed karst system. Both times are on the order of a few days. As a result some karst waters are always out of equilibrium with the carbonate rocks and carbonate dissolution kinetics must also be incorporated into the mechanism.

The evolution of a karst landscape is seen, then, as a set of coupled rate processes, some mass transport rates and some dissolution kinetics rates. The boundary conditions for the rate processes are set by the karst rocks, their structure and stratigraphy, and their placement in the drainage basin. The environmental controlling factors on the rate processes are often lumped together as "climate." There has been an intense debate for more than 20 years concerning the role of climate and whether there are identifiable sets of landforms that can be called "tropical karst" or "alpine karst" or "temperate karst." Climate must be factored into chemically definable terms, and the main ones seem to be temperature, precipitation, and concentration of CO_2 all of which enter into both equilibrium and kinetics.

The present chapter has the following objectives:

(a) To summarize the present state of the art concerning the dissolution kinetics of calcite.
(b) To apply geochemical arguments to one of the oldest of karst rate problems, that of the overall denudation of carbonate terrains as a function of various climatic factors.
(c) To discuss some aspects of differential solution and the formation of karst landforms.
(d) To consider the dissolution kinetics of dolomite and how the geochemical models account for some of the obvious differences between karst developed on limestone and karst developed on dolomite.

Equilibria

The dissolution chemistry of calcite and dolomite has been worked out in detail and has been applied to problems of karst landform development, carbonate groundwater geochemistry, cave development and the deposition of secondary minerals in caves. See for example Ford and Cullingford (1976) and Bögli (1980). Carbon dioxide dissolves in water to form weak carbonic acid, which ionizes and reacts with the carbonate minerals. Equilibrium demands that the following reactions be satisfied simultaneously:

$$\left.\begin{array}{l} CO_2(\text{gas}) \rightleftharpoons CO_2(\text{aqueous}) \\ H_2O + CO_2(\text{aqueous}) \rightleftharpoons H_2CO_3 \end{array}\right\} \quad K_{CO_2} \qquad (10.1)$$

$$H_2CO_3 \rightleftharpoons HCO_3^- + H^+ \qquad K_1 \qquad (10.2)$$

$$HCO_3^- \rightleftharpoons CO_3^{-2} + H^+ \qquad K_2 \qquad (10.3)$$

$$CaCO_3 \rightleftharpoons Ca^{++} + CO_3^{-2} \qquad K_c \qquad (10.4)$$

$$CaMg(CO_3)_2 \rightleftharpoons Ca^{++} + Mg^{++} + 2CO_3^{-2} \qquad K_d \qquad (10.5)$$

Further complications are introduced by the formation of complexes and ion pairs, by the ionic strength of the groundwater, and by interactions with other components in solution. A recent critical evaluation of the above reactions (Plummer & Busenberg 1982) seems to provide the definitive word on the equilibrium constants and their temperature dependence.

Equilibrium calculations of great complexity and sophistication are possible with interactive computer programs. Many of these have been written for carbonate equilibria and some such as WATSPEC (Wigley 1977) are fairly elaborate. Other general purpose speciation programs may be used for carbonate systems, especially WATEQ and WATEQF (Plummer, Jones &

Truesdell 1978). For a comparison of some of the many codes that have been written see Nordstrom *et al.* (1979).

Establishment of a reference rate equation

The long history of the dissolution kinetics of calcite is summarized and evaluated in the review paper of Plummer *et al.* (1979). We discuss only three aspects of this history. First are the experiments of Berner and Morse (1974), which provided the first really detailed experimental data on dissolution over the entire range of undersaturations expected in natural systems. Second is the rate equation proposed by Plummer, Wigley and Parkhurst (1978), which goes far in identifying the controlling mechanisms in the dissolution process. Third are some new experiments from this laboratory, which provide better control over transport processes than has hitherto been available and a check on the range of applicability of the Plummer, Wigley and Parkhurst equation.

Berner and Morse dissolved powdered calcite in seawater and in a synthetic brine they called "pseudoseawater" over a wide pH range up to equilibrium. These data were generated in a pH-stat, a device that maintains constant pH by automatically titrating the reacting solution with HCl. What they found is shown schematically in Figure 10.1. The rate scale is expressed in milligrams

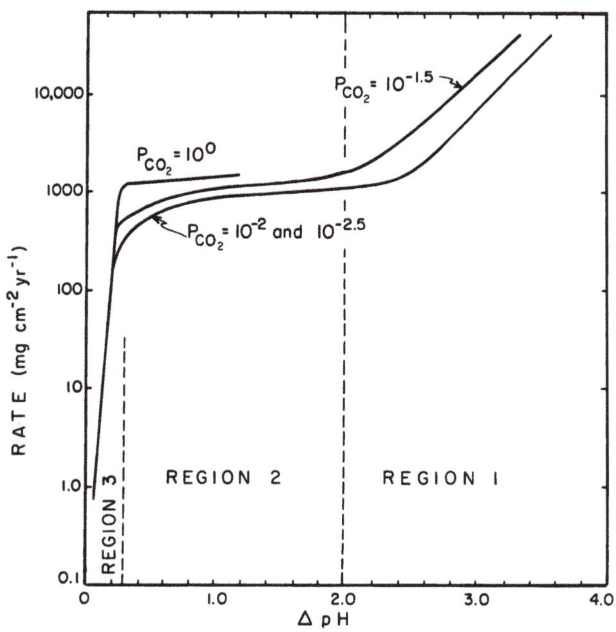

Figure 10.1 The rate of solution of calcite as a function of undersaturation showing three kinetic regimes. Curves plotted from the data of Berner and Morse (1974).

of $CaCO_3$ per square centimeter of surface per year and the experimental data span nearly five orders of magnitude. The ΔpH scale describes the departure of the system from equilibrium; $\Delta pH = 1/2\ SI_c$, the commonly used saturation index for calcite. There are three clearly defined kinetics regimes: Region 1 where the rate increases proportionally to hydrogen ion concentration, Region 2 where rate is nearly independent of pH but has a dependence on CO_2 pressure, and Region 3 where the rate drops precipitously as the solutions approach equilibrium.

Plummer and Wigley (1976) repeated the dissolution of calcite using crushed iceland spar in freshwater solutions at 25° C and at 1 atm CO_2. They chose to express their results as a standard second-order rate equation

$$\frac{dCa}{dt} = k_c \frac{A}{V}(Ca_s - Ca)^2 \tag{10.6}$$

where Ca = molal concentration of Ca^{++}, the subscript s represents saturation, A and V are surface area and solution volume, and k_c is a reaction rate constant.

The above equation was used by White (1977) to demonstrate a kinetic threshold or trigger in the evolution of a solutionally widened fracture into a cave passage. It has also served as the base for Palmer's (1984) kinetics modeling of karst processes.

Most recent and most comprehensive of the calcite rate determinations are the measurements and calculations of Plummer, Wigley and Parkhurst (1978). They reacted crushed iceland spar with freshwater solutions at various temperatures and CO_2 partial pressures using both pH-stat and free drift types of experiments. They concluded that there were three forward rate processes: direct reaction with free protons, direct reaction with unionized carbonic acid, and direct dissolution in water.

$$CaCO_3 + H^+ \rightleftharpoons Ca^{++} + HCO_3^- \tag{10.7}$$

$$CaCO_3 + H_2CO_3 \rightleftharpoons Ca^{++} + 2HCO_3^- \tag{10.8}$$

$$CaCO_3 + H_2O \rightleftharpoons Ca^{++} + HCO_3^- + OH^- \tag{10.9}$$

These were combined with a statement of back reaction rate into a single comprehensive rate equation

$$R = k_1 a_{H^+} + k_2 a_{H_2CO_3*} + k_3 a_{H_2O} - k_4 a_{Ca^{++}} a_{HCO_3^-} \tag{10.10}$$

where the a terms are activities of the ions as given. The three forward reaction rates could be fitted as functions of temperature.

$$\log k_1 = 0.198 - \frac{444}{T} \tag{10.11}$$

$$\log k_2 = 2.84 - \frac{2177}{T} \tag{10.12}$$

$$\log k_3 = -5.86 - \frac{317}{T} \tag{10.13}$$

The back reaction rate constant is more complicated but could be derived from the forward rate constants and was later shown to also describe the precipitation of calcite from supersaturated solutions (Reddy *et al.* 1981).

$$k_4 = \frac{K_2}{K_c}\left[k_1' + \frac{1}{a_{H^+(s)}}(k_2 a_{H_2CO_3(s)} + k_3 a_{H_2O(s)})\right] \tag{10.14}$$

The rate constant k_1' is the forward reaction rate constant for H^+ at the surface and may be 10–20 times larger than k_1 which is a mass transfer coefficient for H^+. The other activities are also expressed as their surface values.

The first term in the equation is dominant at low pH and corresponds to Berner and Morse's Region 1. The second term is dominant at intermediate pH and roughly corresponds to Region 2. The back reaction becomes dominant near saturation (at saturation the back reaction rate equals the sum of the three forward reaction rates) and should correspond to Region 3.

A set of dissolution experiments in this laboratory was designed to look for mass transport control in calcite dissolution. A disc of calcite cut from a large single crystal was spun at various angular velocities in a solution kept saturated with CO_2 with a bubbling device. The uptake of calcium in solution and the pH were monitored as a function of time, temperature and angular velocity of the disc (Herman 1982, Herman & White, in prep). Figure 10.2 shows a

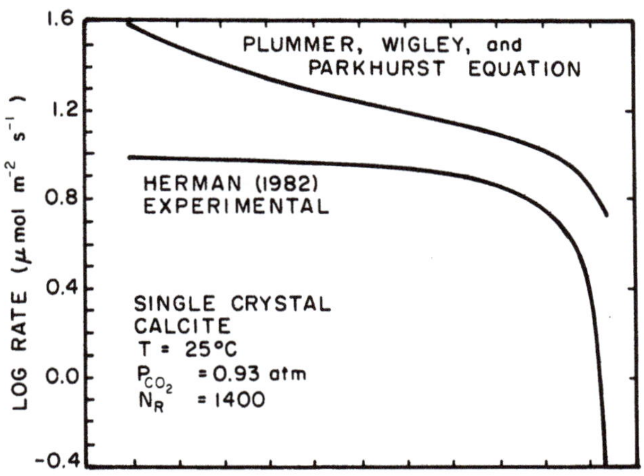

Figure 10.2 Rate of calcite dissolution as a function of saturation index, SI_c. The theoretical rate curve is a plot of Equation 10.10 using Plummer, Wigley and Parkhurst's (1978) values for the rate constants. The experimental curve was measured in the laboratory (Herman 1982) on a single crystal of calcite, rotating at known velocity, in a solution saturated with CO_2.

comparison of these experimental results with the predictions of the Plummer–Wigley–Parkhurst equation. The agreement in the mid-range of saturation index is within a factor of two. We ascribe this to uncertainties in the measurement of surface area of the crushed calcite samples. Agreement at high undersaturation (the edge of Region 1) and the agreement near saturation is less good.

At high undersaturation, the Plummer–Wigley–Parkhurst equation predicts an increase in rate reflecting the increasing importance of R_1. Further, their theoretical curve is supported at still higher undersaturations by many other workers' measurements of the dissolution of calcite and limestone in strong acids. There seems no reason to doubt that direct proton attack is the dissolution mechanism and that the diffusion of protons through the boundary layer is the rate-controlling factor. Why then do our results show a constant dissolution rate from the beginning of the experiments to a saturation index of about -2? The high undersaturation end of the Plummer–Wigley–Parkhurst equation was based on pH-stat data, a technique that uses fully ionized acids as a source of protons. Our experiments used only carbonic acid. Our hypothesis is that reactions are so rapid under high undersaturations that H_2CO_3, the source of protons, is depleted on the crystal surface. The rate-limiting step is then the hydration of aqueous CO_2 to form H_2CO_3:

$$[CO_2]_{Aq} = K_H P_{CO_2} \qquad K_H = 10^{-1.50}(25^\circ\,C) \tag{10.15}$$

$$[H_2CO_3] = K[CO_2]_{Aq} \qquad K = 10^{-2.78}(25^\circ\,C) \tag{10.16}$$

At $P_{CO_2} = 0.93$ atm, $[CO_2]_{Aq} = 0.029$ m whereas $[H_2CO_3] = 4.9 \times 10^{-5}$ m; the H_2CO_3 concentration is only 0.17% of the aqueous CO_2. The hydration rate is very sluggish, $k_+ = 0.03\ s^{-1}$ (Kern 1960). It is thus argued that the concentration of H_2CO_3 will be the critical parameter in surface karst processes, many of which operate far out of equilibrium. An alternative hypothesis is that diffusion of species through the boundary layer to the dissolving surface is rate-controlling. Figure 10.3 shows the dependence of reaction rate on Reynolds number calculated from the angular velocity of the spinning disc. At very high undersaturations, there is about a factor of two increase in rate when turbulence sets in. As the solutions become more saturated, the dependence of solution rate on Reynolds number gradually disappears.

Near saturation, our experimental rate curve falls off much more rapidly than does the Plummer–Wigley–Parkhurst equation. Our result is in closer agreement with Berner and Morse's experimental results. This has considerable significance to the application of these equations to problems of carbonate groundwaters and to the initial phases of cave development, both of which involve waters that are near to saturation. The low solution rates at finite undersaturation allows the "kinetic trigger" mechanism proposed (White 1977) to account for the development of conduit systems in karst aquifers. The Plummer–Wigley–Parkhurst equation would predict a more rapid approach to equilibrium and thus make it unlikely that undersaturated waters would be found at depths in carbonate aquifers. Dreybrodt (1981a) had considerable success in applying the Plummer–Wigley–Parkhurst equation to an interpretation of cavern genesis. However, he found it necessary to draw on a mixing corrosion mechanism to explain the initial enlargement of small joint openings into proto-cave passages (Dreybrodt 1981b). Because there are many examples of covered karst where a mixing corrosion mechanism does not seem likely, the discrepancy between calculated and experimental curves shown in Figure 10.2 may provide an explanation. The reasons for the discrepancies will be dealt with elsewhere (Herman & White, in prep.).

It is instructive to dissect the Plummer–Wigley–Parkhurst equation into its component parts to get an idea of the magnitudes of the forward and back reaction rates. Figure 10.4 shows that the effect of varying CO_2 pressure on

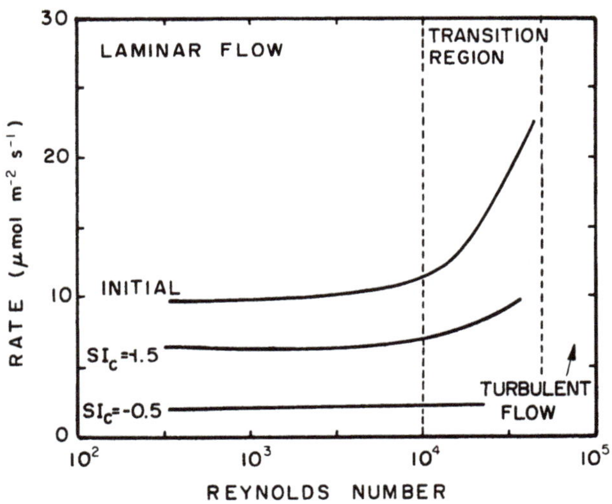

Figure 10.3 Solution rate of calcite (Herman 1982) as a function of Reynolds number for various levels of saturation.

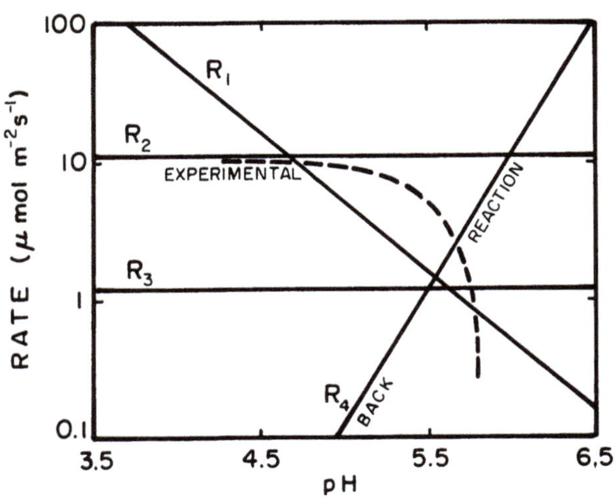

Figure 10.4 Plots of the four components of the Plummer–Wigley–Parkhurst rate equation. A set of experimentally determined rate data (Herman 1982) are shown for comparison.

rate is not all that great. The upper limit of R_2 is illustrated by water saturated with pure CO_2. The lower limit is given by R_3, the rate of calcite dissolution in pure water. The varying CO_2 partial pressures have a substantial effect on the equilibrium solubility of carbonate rocks, but relatively little effect on the rate of dissolution.

The close agreement between the expression for R_2 and the experimental data suggests that the rate equation can best be applied to surface karst landform processes without the R_1 expression. The forward rate far from equilibrium is then given entirely by the temperature and the CO_2 pressure.

Karst denudation

Over the years, a good deal of effort has gone into something called "karst denudation." The concept arose originally from observations of igneous rock boulders perched on top of limestone pedestals in the glaciated karst of northern England. The height of the pedestal represented the general lowering of the limestone surface since the erratics were deposited at the close of Wisconsinan time. Corbel (1959), and many others adopted the concept of karst denudation as a measure of the rate of karst development in various climatic regimes. The denudation rate was estimated from the precipitation and the measured hardness of streams draining the karst area by his famous formula

$$X = \frac{4ET}{100} \tag{10.17}$$

where X is the karst denudation in mm ka^{-1}, E is the precipitation in dm and T is the mean hardness in mg l^{-1}. Although widely quoted and widely used, Corbel's formula is both over-simplified and over-simplistic. It is over-simplified in the sense that the density of limestone was assumed to be 2.5, no account was taken of whether the runoff from which the hardness was measured arose entirely in karst terrain, and no account was taken of evapotranspiration losses. Many later workers seized on these points and revised formulae were proposed by Williams (1963) and Pulina (1972) among others. Most later workers realized that runoff was a better measure of water throughput, and their equations are written in terms of the transported $CaCO_3$ per unit area of drainage basin on carbonate rock determined from discharge of the basin master stream.

The Corbel equation is over-simplistic, however, in assuming that some sort of grand mean can characterize entire regions or climatic regimes without taking account of the tremendous variability of the measured hardness over the seasons of the year, of variability between adjacent drainage basins brought on by differences in topography, rock type, and plant cover and variations due to altitude. Marian Pulina in Poland, in particular, has examined these factors by calculating denudation for individual small drainage basins. His equation is

$$D_m = 12.6\frac{(T - Ta)Q}{P} \tag{10.18}$$

where D_m is denudation in mm ka^{-1} (equivalent to $m^3\ km^{-2}\ yr^{-1}$), T is $CaCO_3$ hardness in mg l^{-1}, Ta is hardness carried into the karst from non-karstic parts of the basin, Q is mean discharge, and P is basin area.

Most of the results on karst denudation through 1975 are summarized by Smith and Atkinson (1976) who make critical comparisons of denudation rates in different climatic regimes. Most of the results reported by them are based on seasonal averages, often with too few data. Smith and Atkinson do the best they can with what they admit is a data set of very heterogeneous quality. They conclude that arctic karsts can be easily distinguished from temperate and tropical karsts but that temperate and tropical karsts cannot be easily distinguished from each other. Runoff (or precipitation) is found to be the most important variable.

The most recent improvement to denudation calculations is Ogden's (1982) introduction of a "mean carbonate leaching," which is based on time series data collected from individual carbonate basins.

The question addressed here is: is the karst denudation rate a useful measure of the overall rate of dissolution of a carbonate rock landmass? Figure 10.5 shows some denudation rate data plotted against precipitation. In spite of the idiosyncracies of the individual determinations and the uncertainties in deducting evapotranspiration losses, the correlation between denudation rate and water throughput is rather good. It is gratifying that the data recently available

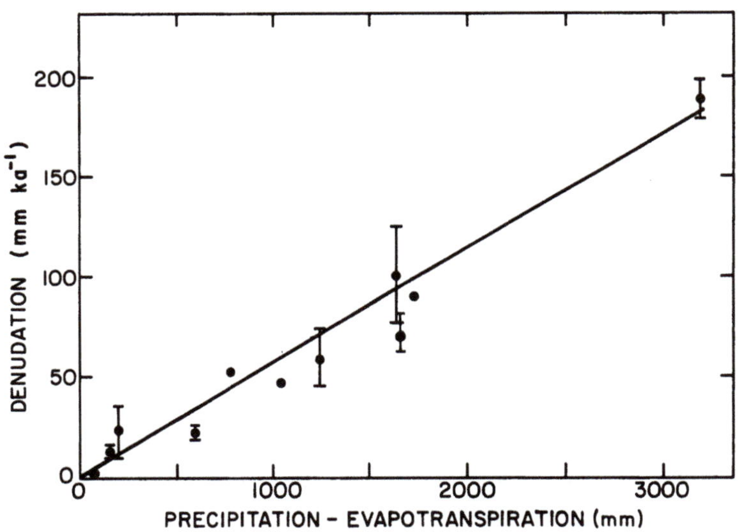

Figure 10.5 Some reported denudation rates as a function of precipitation. Data points reading from left to right have been extracted from Smith (1962) Canadian Arctic; Glazek and Markowicz-Lohinowicz (1973) Poland (2 points); Ogden (1982) West Virginia; Williams (1963) Ireland; Markowicz *et al.* (1972) Bulgaria; Glazek and Markowicz-Lohinowicz (1973) Tatra Mountains, Poland; Williams and Dowling (1979) New Zealand; Gunn (1981a) New Zealand; Miller (1981) Belize; Sweeting (1979) Gunong Mulu, Malaysia.

for the Gunong Mulu karst, perhaps the most active karst on the planet, fits the extrapolation rather well. Figure 10.5 also makes clear that precipitation rate is the most important factor in karstification; all other climatic and hydrogeologic factors are buried within the scatter of the data.

The data points shown on Figure 10.5 can be fitted to a simple linear regression $D = 0.049\,(P - E) + 6.3$ with $r^2 = 0.92$. As might have been expected, this regression line falls between the lines calculated by Smith and Atkinson (1976) for a different data set in which they separated measurements from arctic, temperate, and tropical climates. There seems to be no justification for a power function such as that proposed by Lang (1977). Lang's denudation curve, based largely on the data of Balazs (1973), has been strongly criticized by Gunn (1981b). As will be seen shortly, there are theoretical reasons for a linear relationship between denudation and precipitation.

A more precise relationship between denudation rate and precipitation or runoff can be obtained by simple mass balance considerations.

The karst denudation rate is, in effect, a statement of average mass loss from a carbonate drainage basin in excess of the transported clastic load. Small drainage basins are convenient units for examination. Figure 10.6 shows schematically the annual hydrograph for such a basin and also a typical hardness variation curve. The annual runoff can be calculated from the hydrograph by

$$R_f = \frac{1}{A}\frac{K}{t_R}\int_0^t Q(t)\,\mathrm{d}t \qquad (10.19)$$

where R_f is runoff expressed in mm yr^{-1}, A is basin area in km^2, K is a conversion factor $= 10^{-3}$ for the units given, t_R is the period of record in years, $Q(t)$ is instantaneous discharge in m^3 s^{-1}, and the time unit for integration is in seconds. An exact expression for dunudation is then given by the convolution of the runoff hydrograph and the hardness curve

$$D = \frac{1}{N_L A}\frac{K'}{\varrho}\frac{1}{t_R}\int_0^t Q(t)H(t)\,\mathrm{d}t \qquad (10.20)$$

where D = denudation rate in mm ka^{-1}, N_L is the fraction of the basin underlain by carbonate rocks, $K' = 10^{-12}$ for the units given, ϱ = density of carbonate rock in g cm^{-3}, and $H(t)$ is instantaneous Ca + Mg hardness in g m^{-3} $CaCO_3$.

Many times, of course, continuous records, particularly for hardness, are not available. The denudation rate equation can be approximated by a summation over the individual observations

$$D = \frac{1}{N_L A}\frac{K'}{\varrho}\frac{\Delta t}{t_R}\frac{1}{n}\sum_{i=1}^{n} Q_i H_i \qquad (10.21)$$

which is essentially equivalent to Drake and Ford's (1973) and Ogden's (1982) approach. This equation assumes individual measurements of discharge and hardness. However, it should be remembered that an open karst aquifer responds rapidly to individual storm events, and it is necessary that the observations be as closely spaced as possible and that they be evenly spaced over the period of observation.

Two assumptions (at least) remain in this reformulation of the karst denudation equation. One is the use of a simple area fraction, N_L, to normalize the karstic portion of the drainage basin. If there is cavern development or groundwater flow beneath the non-karstic rocks, this factor would need to be adjusted in some fashion. The second is that the basin is conservative; the master stream draining the basin is gauged in a place where all outlets from the underground system are included. Further, there must be no other underground drainage routes carrying water across drainage divides.

If the period of observation is sufficiently long that changes in storage average to zero, the total runoff from the basin should just balance the input from precipitation, and there should be a good correlation between denudation rate and precipitation rate after correction for evaporation and transpiration losses.

$$R_f = P - E - \Delta S \tag{10.22}$$

If the infiltrating portion of the precipitation comes into equilibrium with the carbonate bedrock during its passage through the soil as diffuse infiltration, into dolines as internal runoff, and into the karst aquifer as sinking surface streams, it is possible to calculate a theoretical denudation rate. The

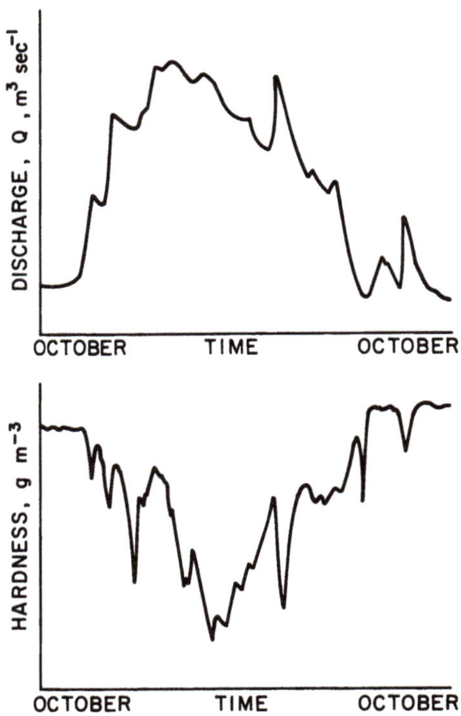

Figure 10.6 Schematic hydrograph and related hardness curve of a karst river basin.

theoretical denudation rate based on chemical equilibrium will represent an upper bound since it assumes that the waters discharging from the karst drainage basin are saturated at the local temperature and at the pressure of CO_2 provided by the local environment.

The dissolution of limestone is written as the net reaction among the dominant species

$$CaCO_3 + CO_2(gas) + H_2O \rightleftharpoons Ca^{++} + 2HCO_3^- \tag{10.23}$$

The equilibrium constant for this reaction can be written

$$\frac{[HCO_3]^{-2}[Ca^{++}]}{P_{CO_2}} = \frac{K_c K_1 K_{CO_2}}{K_2} \tag{10.24}$$

where the equilibrium constant is given as a product of the individual equilibrium constants defined in Equations 10.1–10.5 since these are the ones usually tabulated.

Assuming that no other cations or anions are present, electrical neutrality demands that $2Ca^{++} = HCO_3^-$. Ignoring non-ideality, electroneutrality can be used to solve Equation 10.24 for $[Ca^{++}]$ with the result

$$[Ca^{++}] = \frac{1}{\sqrt[3]{4}}\left(\frac{K_c K_1 K_{CO_2}}{K_2}\right)^{1/3} P_{CO_2}{}^{1/3} \tag{10.25}$$

The hardness of saturated water, H_s, is directly proportional to the calcium ion concentration

$$H_s = 10^5[Ca^{++}] \tag{10.26}$$

The denudation rate given by Equation 10.20 in terms of instantaneous runoff can be rewritten in terms of total annual precipitation and the saturated hardness as

$$D_{max} = 10^{-3}\frac{H_s}{\varrho}(P - E) \tag{10.27}$$

Combining Equations 10.25–27 gives us the theoretical expression for denudation in terms of the chemical variables.

$$D_{max} = \frac{100}{\varrho\sqrt[3]{4}}\left(\frac{K_c K_1 K_{CO_2}}{K_2}\right)^{1/3} P_{CO_2}{}^{1/3}(P - E) \tag{10.28}$$

where D_{max} = denudation rate in mm ka^{-1} for the system at equilibrium, P_{CO_2} is CO_2 partial pressure in atmospheres, P is precipitation in mm yr^{-1} and E is evapotranspiration loss in mm yr^{-1}.

Note that Equation 10.28 has no adjustable parameters. All terms in the equation can be calculated. It combines in a single statement the three climatic variables of temperature, CO_2 pressure, and precipitation. Examination of the theoretical denudation equation provides an explanation of many of the empirical observations that have accumulated.

Denudation is predicted to vary linearly with precipitation as is implied by Figure 10.5 and by the regression lines calculated by Smith and Atkinson (1976).

Denudation rate varies with the cube root of the CO_2 partial pressure. Carbon dioxide pressures in the environment vary by about a factor of 100, from (roughly) $10^{-1.5}$ in certain CO_2–rich soils to $10^{-3.5}$ on bare bedrock exposed to the atmosphere (Fig. 10.7). Because of the cube root dependence, a factor

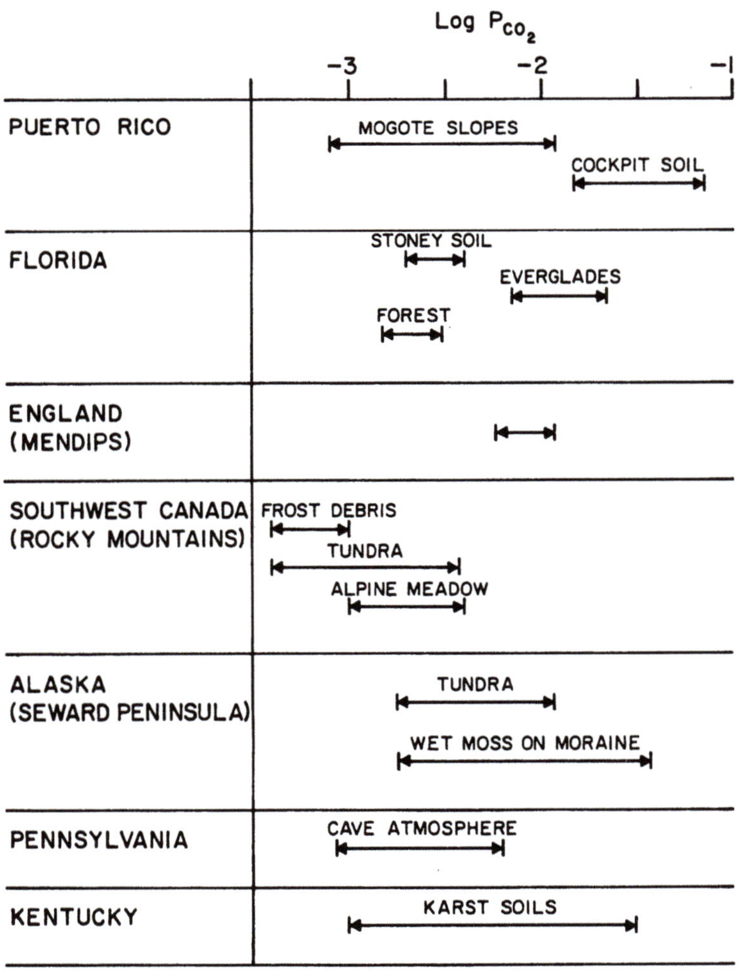

Figure 10.7 Ranges of CO_2 pressures measured in soil atmosphere in various climatic regimes. Data from Kentucky from Miotke (1975). from Pennsylvania from Troester and White (in press). All other data from Miotke (1974).

of 100 in P_{CO_2} provides only about a factor of 5 in the denudation rate. The low denudation rates observed in arctic and alpine climates then arise both from the low precipitation rates and also from the predominance of bare limestone exposures. Also explained is Smith and Atkinson's observation that temperate climates and tropical climates have rather similar denudation rates when the effect of precipitation rate is deducted. Where there are thick soils in both temperate and tropical karst environments, there is a difference of only about a factor of 10 in CO_2 partial pressure and thus about a factor of 2 in denudation rate. This small difference is easily lost in the noise of competing factors in the denudation rate–precipitation rate plots.

The dependence of the denudation rate on temperature is buried in the temperature dependence of the equilibrium constants that appear in Equation 10.28. These are known quite precisely (see Plummer & Busenberg 1982, for analytical expressions). Corbel's hypothesis can now be evaluated quantitatively. While he was certainly correct in asserting that the solubility of CO_2 is higher in cold climates, it is the combined temperature dependence of four equilibrium constants and not just of K_{CO_2} that describes the temperature dependence of the denudation rate. When numerical values are substituted, it is found that equilibrium denudation rate increases by only about 30% between 25° and 5°C. Temperature is thus the least important of the three climatic variables in Equation 10.27.

Figure 10.8 expresses some of these results graphically. Denudation rates are calculated for temperatures of 5°, 10° and 25°C as representative of arctic–alpine, temperate, and tropical climates, and P_{CO_2} of $10^{-3.5}$, $10^{-2.5}$, and $10^{-1.5}$ as representative of bare limestone, normal soils, and organic-rich soils, respectively. Superimposed on the theoretical curves are the observed denudation line from Figure 10.5 and the three regression lines of Smith and Atkinson (1976). It is apparent that there is general agreement between the predications of Equation 10.28 and the observed denudation rates although there is much room for refinement of the details.

This theoretical development assumes that the water and carbonate rocks are in equilibrium. In reality, most karst waters are undersaturated. The more open and well developed the karst system, the faster the throughput, and the farther the water is from equilibrium when it leaves the drainage basin. Taking specific account of the carbonate reaction kinetics is a clearly defined next step in improving the denudation model.

Differential solution and karst landforms

It is difficult to construct chemical–hydrodynamical models for surface landforms. Surface forms (dolines, cutters, cones, towers and cockpits) are often soil-mantled, and it is difficult to estimate flow rates and velocities. Leaving aside questions of detailed geometry, however, it is apparent that much of the difference between arctic, temperate and tropical landforms is a matter of relief. Karst landforms in high alpine regions and in the arctic tend to be solu-

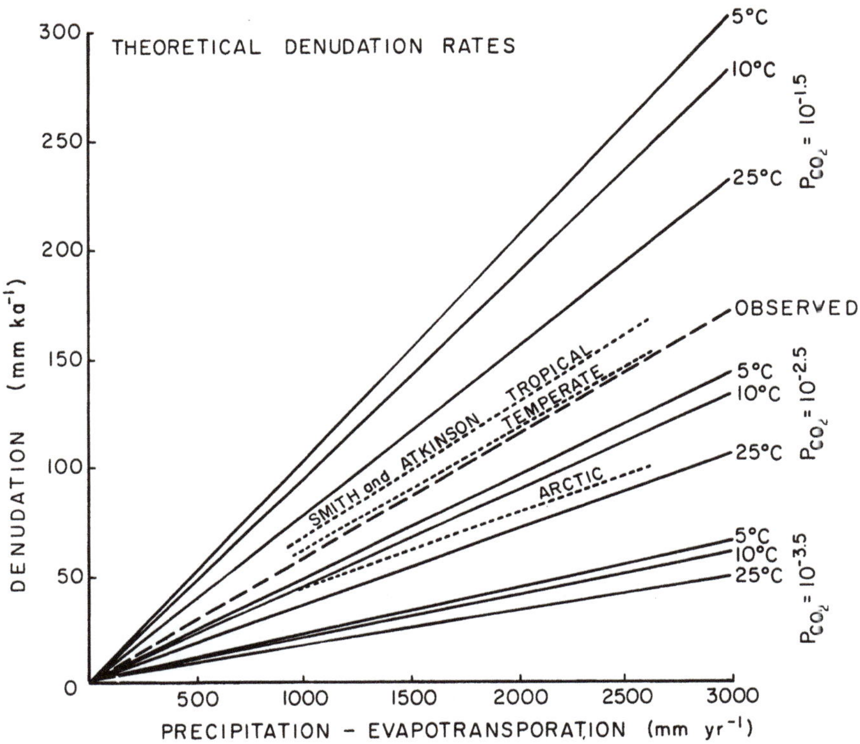

Figure 10.8 Solutions of the theoretical denudation equation for various temperatures and CO_2 pressures. Observed denudation rates from Figure 10 5 and from Smith and Atkinson (1976) are shown.

tion grooves and solution-widened joints (kluftkarren) often with relief of a meter or less. Looking at the distribution of doline depths in the Appalachian Highlands (White & White 1979) or at the depths of cutters exposed in quarries and roadcuts, the relief of temperate climate landforms is in the range of a few meters to a few tens of meters. The relief of cockpits or of cone and tower karst in tropical regions is often in the range of hundreds of meters and in extreme cases, such as the pinnacle karst of New Guinea, is more than 1000 m.

Some of the differential solution is simply the result of flow paths in the karst. Once internal drainage has been established, more and more of the precipitation is funneled into points of internal runoff. Dolines grow larger and larger until the boundaries meet to form a polygonal karst with all internal drainage and no interfluve area at all.

Some of the differential solution is a result of equilibrium chemistry. Depressions tend to collect thicker soils and organic debris with resulting higher CO_2 production and thus more rapid rate of dissolution.

Some differential solution must result from kinetics because the rate at which water moves underground in the karst is usually too fast for it to have come to equilibrium with the bedrock at the soil base. This is a delicate point for there exist almost no data on the chemistry of the water at the base of the soil horizon.

Some of these possibilities are illustrated schematically in Figure 10.9. Suppose there is a thick soil overlying a limestone bedrock that contains widely spaced fractures. Infiltration water moves through the soil to the bedrock where its downward course is blocked by the impermeable limestone. The water must flow in long, shallow, flow paths to the nearest fracture, which acts as a drain to the subsurface. In the inter-fracture regions of the bedrock surface, water has a long residence time in contact with the limestone and might be expected to reach saturation at the P_{CO_2} provided by the soil. If it does, the rate of lowering of the land surface is described by the equilibrium denudation model. There should be little differential solution; only a gradual lowering of the limestone surface. The fractures, which will enlarge to become kluftkarren and which may further enlarge to solutional troughs (cutters) some meters in depth, act as drains and create a local drawdown for the water levels in the soil during periods of precipitation. If the flow is sufficiently rapid, the water does not reach equilibrium with the carbonate rock, and solution continues to depth within the limestone mass. Further, the rapid infiltration of water through the fracture captures infiltrating soil water from some distance on both sides as indicated by the arrows. The dissolving power of this larger area of infiltration is concentrated on the limestone walls of the fracture and leads to differential enlargement and deepening of the solution crevice along the fracture at the expense of surface lowering. A flow model based on this schematic one seems most appropriate for many temperate climate karsts where temperatures are constant over the karst surface and CO_2 pressures are also not likely to vary much from one part of the soil mass to another.

Now consider tropical karsts. The sketch in the lower part of Figure 10.9 is intended to suggest differential solution along fractures with some concurrent lowering of the land surface. If dissolution along the fractures takes the form of deep and narrow gorges, the intervening limestone blocks will evolve into cone and tower karst. If dissolution is concentrated on fracture intersections with concurrent removal from the walls, the system will evolve into a cockpit karst with the intervening blocks sculptured down into the characteristic pyramidal hills. If we examine the three climatic variables in tropical karst regions, which exhibit characteristic cone and tower or cockpit landforms we find:

(a) Temperatures are higher; perhaps 25° C compared with 10° C for temperate karsts.
(b) Carbon dioxide pressures may be higher in local regions, but they are also much more variable.
(c) Precipitation rates are usually much higher
(d) There is less seasonal variability; CO_2 levels and precipitation tend to be more uniformly distributed throughout the year.

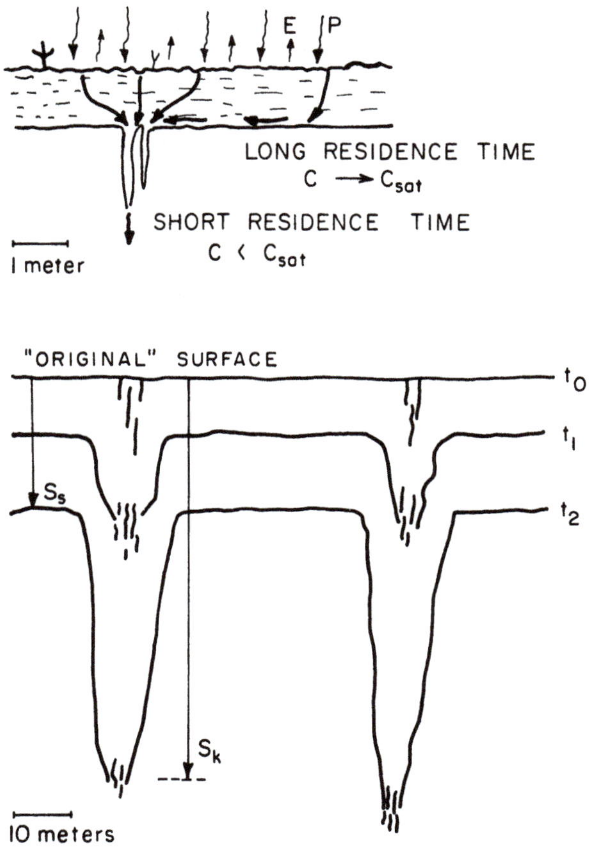

Figure 10.9 Sketch showing differential dissolution of limestone surface in a soil-mantled temperate karst (above) and a rugged massive limestone tropical karst (below).

The problem is not of accounting for the overall average denudation rates, which both measurements and the theoretical model agree are not greatly different from temperate climate denudation rates. The problem is of accounting for the differential solution that has produced the rugged topography that makes up tropical karst.

Tropical storms are characteristically flashy, with very intense precipitation over short periods of time, fast runoff and rapid loss of water into the internal drainways of the karst. Under these conditions, dissolution is controlled mainly by the forward reaction rates since the system is far from equilibrium. Going back to the Plummer, Wigley and Parkhurst rate equation (Eq. 10.10), we can drop the free proton term since we are dealing only with carbonate equilibria, and if we want to examine rates far from equilibrium, we can also neglect the back reaction term. Recalling that $a_{H_2CO_3*} = K_{CO_2}P_{CO_2}$ and that $a_{H_2C} = 1$ in ordinary groundwaters, the rate equation simplifies into

$$R = k_2 K_{CO_2} P_{CO_2} + k_3 \tag{10.29}$$

The temperature dependence of the rate is hidden in the temperature dependence of the rate constants (Eq. 10.11–13) and in the temperature dependence of K_{CO_2}. The rate constants increase with increasing temperature whereas the solubility of CO_2 decreases with increasing temperature. When numerical values are substituted, there is about a factor of 2 increase in rate between 10° and 25° C. Because the temperature of a particular tropical karst region would be expected to be more or less constant, the temperature dependence of the rate has the effect of shortening the time scale for the development of the landscape but does not affect the differential solution.

Equation 10.29 shows that the initial rate of solution, when a CO_2-containing solution comes into contact with limestone, is directly proportional to the CO_2 partial pressure. Figure 10.9 suggests a landscape evolving through time, with a lowering of the inter-fracture surface by an amount S_s while at the same time, more rapid solution along the fractures lowers the bottom of the trench or gorge by an amount S_k. The relief at any instant in time is $S_s - S_k$. The relative rate of lowering of the surface is S_k/S_s. Assuming that conditions have remained constant from t_0 to t_2, i.e. the overall time of evolution of the landscape, the surface lowering is just $S = R(t_2 - t_0)$. Thus $S_k/S_s = R_k/R_s$. To a reasonable approximation, the ratio of rates varies as the ratio of the CO_2 partial pressures.

High precipitation rates, flashy storms and rapid drainage into the subsurface combine to strip soils from exposed surfaces and steep slopes, making the tropical karst topography one of extreme contrasts. The sides and top of cones and towers are often completely denuded of soil whereas rich organic debris collects in the bottoms of trenches between the towers. As suggested by the data of Figure 10.7, there could easily be $P_{CO_2} = 10^{-3.5}$ on the tops and slopes of the inter-fracture limestone blocks, and $P_{CO_2} = 10^{-1.5}$ in the soil and organic debris in the trench between. A factor of 100 in CO_2 pressure implies a factor of 100 in relative rate of solution, and a factor of 100 in the relative lowering of the surface (ignoring all of the hydrodynamic implications). Thus during a period of time in which the original surface is lowered a few meters, differential solution could lower the trenches by several hundred meters. While these back-of-the-envelope calculations ignore nearly all of the detail and say nothing about the final geometry of the landform, they do suggest that the observed relief in tropical karst can be accounted for.

Karst landforms in dolomite terrain

Every student who ever dripped acid onto an outcrop knows that dolomite dissolves more slowly than calcite. The equilibrium solubility of dolomite, however, is not greatly different from that of calcite. The solubility product constant, K_d, (see Eq. 10.5) has been subject to much debate, but the accepted value is near 10^{-17}, which on a mole-for-mole comparison is $10^{-8.5}$ compared with $10^{-8.47}$ for calcite. The marked differences between karst landforms developed on limestone and karst landforms developed on dolomite cannot be accounted for by equilibrium solubility.

The dissolution kinetics of dolomite have been measured experimentally with a precision beyond that of the old acid test only in the past few years. Busenberg and Plummer (1982) followed the dissolution of dolomite cleavage fragments by weight loss over the P_{CO_2} range of 0 to 1 atm and over the temperature range of 1.5 to 65° C. Their results were reduced to the rate equation

$$R = k_1 a_{H^+}{}^n + k_2 a_{H_2CO_3^*}{}^n + k_3 a_{H_2O}{}^n - k_4 a_{HCO_3^-} \qquad (10.30)$$

based on the same arguments used to construct Equation 10.10. The exponent $n = \frac{1}{2}$ for temperatures up to 45° C. The dependence of rate on hydrogen ion activity takes on different values of n at higher temperatures.

Herman (1982) and Herman and White (in prep.) also investigated the dissolution kinetics of dolomite with a spinning disc technique at $P_{CO_2} = 0.93$ atm. Some typical results are shown in Figure 10.10 in comparison with the Busenberg and Plummer (1982) equation. Comparison with Figure 10.2, the dissolution kinetics of calcite, is of interest. The initial rates are not, in fact, very different. The initial dissolution rate of calcite in CO_2-saturated water is $10\,\mu$mol m^{-2} s^{-1} compared with $4\,\mu$mol m^{-2} s^{-1} for dolomite. However, dolomite dissolution does not show a constant-rate intermediate region (Region 2 in the Berner–Morse notation). As the solution becomes more saturated, the dissolution rate of dolomite falls continuously by about an order of magnitude between the initial solution and $SI_D = -2$. The agreement

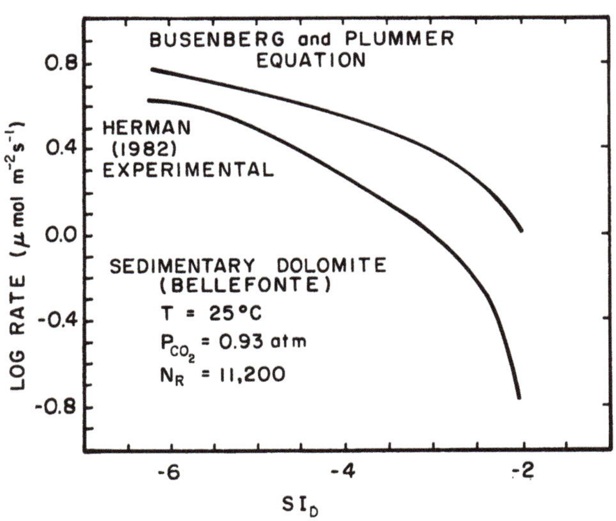

Figure 10.10 Dissolution rates for dolomite comparing the predictions of the Busenberg and Plummer equation with Herman's (1982) experimental curve.

between the controlled hydrodynamics laboratory experiments and the Busenberg–Plummer rate equation is rather good over the entire pH range.

Between SI_D or $SI_C = -2$ and equilibrium, however, the behaviors of dolomite and calcite are dramatically different. The solution rate of calcite remains fairly constant across most of this range and begins to decrease rapidly only quite near equilibrium (see Fig. 10.1 or Fig. 10.2). The dissolution rate of dolomite decreases continuously and at $SI_D = -2$ becomes too slow to measure on a laboratory time scale. Dolomite does eventually come to equilibrium with groundwater because calculations from water in wells (e.g. Langmuir 1971) show that many of these waters are saturated with respect to dolomite. Crude extrapolation of the rate curves to equilibrium suggests that several years would be required for the laboratory solutions to reach equilibrium. Quite clearly, some new dissolution-controlling mechanism takes effect when the dolomite concentration builds up to the level indicated by $SI_D = -2$.

The characteristic feature of dolomite karst is that the landforms are usually subdued compared to limestone karsts. Dolines are shallower, and caves tend to be smaller and fewer. The dissolution kinetics explain the field observations, at least in a qualitative way. The very slow rate at which dolomite waters come to equilibrium means that precipitation can infiltrate through the soil, into fractures in the dolomite, and can travel a long way through the fracture system before coming to equilibrium. Differential dissolution is suppressed. However, because the concentrations at saturation are not very different for calcite and dolomite, the denudation rates, as determined from water chemistry measurements at springs discharging from the drainage basin, are rather similar. The bulk of the mass of dissolved limestone comes from the surface. Dolomite dissolution occurs throughout the fracture system, which may explain why fractured dolomites are often highly productive aquifers.

The carbonate valleys of the folded Appalachians illustrate the absence of strong differentional denudation. The Ordovician carbonate rocks contain cavernous limestone units near the top of the section and mainly dolomite in the lower parts of the section. Because of anticlinal folding, the central portions of the anticlinal valleys are underlain by dolomite whereas the limestones outcrop in bands along the valley sides. There is very little topographic distinction between valley upland underlain by limestone and valley upland underlain by dolomite. The portion of the valley underlain by the Cambrian Gatesburg dolomite does stand out in relief as a low ridge. However, the Gatesburg is a sandy dolomite and forms thick sandy soils, which are sparse in organic matter and support only a thin scrub. The CO_2 pressures in water in the Gatesburg dolomite are about an order of magnitude lower than CO_2 pressures of waters in the Ordovician dolomites, and this factor may be responsible for the difference in relief.

Acknowledgements

I am grateful to Dr. Janet S. Herman for many discussions of carbonate dissolution kinetics and to Dr. Elizabeth L. White for her help and encouragement with this manuscript.

References

Balazs, D. 1973. Comparative investigation of karst waters in the Pacific. *Proc. 6th Int. Speleol Congr.* Olomouc, Czechoslovakia. **4**, 23–32.

Berner, R. A. and J. W. Morse 1974. Dissolution kinetics of calcium carbonate in sea water. IV. Theory of calcite dissolution. *Am. J. Sci.* **274**, 108–34.

Bögli, A. 1980. *Karst hydrology and physical speleology.* Berlin: Springer.

Busenberg, E. and L. N. Plummer 1982. The kinetics of dissolution of dolomite in CO_2-H_2O systems at 1.5 to 65°C and 0 to 1 atm P_{CO_2}. *Am. J. Sci.* **282**, 45–78.

Corbel, J. 1959. Erosion en terrain calcaire. *Ann. Geograph.* **68**, 97–120.

Drake, J. J. and D. C. Ford 1973. The dissolved solids regime and hydrology of two mountain rivers. *Proc. 6th Int. Speleol Congr.* **4**, 53–6. Olomouc, Czechoslovakia.

Dreybrodt, W. 1981a. Kinetics of the dissolution of calcite and its application to karstification. *Chem. Geol.* **31**, 245–69.

Dreybrodt, W. 1981b. Mixing corrosion in $CaCO_3-H_2O$ systems and its role in the karstification of limestone areas. *Chem. Geol.* **32**, 221–36.

Ford, T. D. and C. H. D. Cullingford 1976. *The science of speleology.* London: Academic Press.

Glazek, J. and M. Markowicz-Lohinowicz 1973. Remarks to the use of quantitative methods to karst denudation velocity. *Proc. 6th Int. Speleol Congr.* **3**, 225–30.

Gunn, J. 1981a. Limestone solution rates and processes in the Waitomo District, New Zealand. *Earth Surf. Proc. Landforms* **6**, 427–55.

Gunn, J. 1981b. Prediction of limestone solution rates from rainfall and runoff data: some comments. *Earth Surf. Proc. Landforms* **6**, 595–7.

Herman, J. S. 1982. *The dissolution kinetics of calcite, dolomite and dolomitic rocks in the CO_2-water system.* PhD dissertation. The Pennsylvania State University.

Herman, J. S. and W. B. White in prep. *The dissolution of calcite in the* $CaCO_3-CO_2-H_2O$ *system: mass transfer contributions to the kinetics.*

Herman, J. S. and W. B White in prep. *Dissolution kinetics of dolomite and dolomitic rocks under conditions of controlled mass transport.*

Kern, D. M. 1960. The hydration of carbon dioxide. *J. Chem. Ed.* **37**, 14–23.

Lang, S. 1977. Relationship between world-wide karstic denudation (corrosion) and precipitation. *Proc. 7th Int. Speleol Congr.* 282–3. Sheffield, England.

Langmuir, D. 1971. The geochemistry of some carbonate groundwaters in central Pennsylvania. *Geochim. Cosmochim. Acta* **35**, 1023–45.

Markowicz, M., V. Popov and M. Pulina 1972. Comments on karst denudation in Bulgaria. *Geog. Polonica* **23**, 111–39.

Miller, T. 1981. *Hydrochemistry, hydrology, and morphology of the Caves Branch Karst, Belize,* PhD dissertation. McMaster University.

Miotke, F. D. 1974. Carbon dioxide and the soil atmosphere. *Abh. Karst Höhl.* **9**, 1–49.

Miotke, F. D. 1975. Der Karst im zentralen Kentucky bei Mammoth Cave. *Jahr. Geogr. Gesell. Han.* **1973**.

Nordstrom, D. K., L. N. Plummer, T. M. L. Wigley, T. J. Wolery, J. W. Ball, E. A. Jenne, R. L. Bassett, D. A Crerar, T. M. Florence, B. Fritz, M. Hoffman, G. R. Holdren, Jr., G. M. Lafon, S. V. Mattigod, R. E. McDuff, F. Morel, M. M. Reddy, G. Sposito, and J. Thrailkill 1979. A comparison of computerized chemical models for equilibrium calculations in aqueous systems. In *Chemical modeling in aqueous systems*, Symposium series 93, E. A. Jenne (ed.), 856–92. Washington: American Chemical Society.

Ogden, A. E. 1982. Karst denudation rates for selected spring basins in West Virginia. *Nat. Speleol Soc. Bull.* **44**, 6–10.

Palmer, A. N. 1984. Geomorphic interpretation of karst features. In *Groundwater as a geomorphic agent,* R. G. LaFleur (ed.), 173–209. Boston: Allen & Unwin.

Plummer, L. N. and T. M. L. Wigley 1976. The dissolution of calcite in CO_2-saturated solutions at 25° C and 1 atmosphere total pressure. *Geochim. Cosmochim. Acta* **40**, 191–202.

Plummer, L. N., T. M. L. Wigley and D. L. Parkhurst 1978. The kinetics of calcite dissolution in CO_2-water systems at 5° to 60° C and 0.0 to 1.0 atm CO_2 *Am. J. Sci.* **278**, 179–216.

Plummer, L. N., B. F. Jones and A. H. Truesdell 1978. *WATEQF – a Fortran IV version of WATEQ, a computer program for calculating chemical equilibrium of natural waters.* US Geol Surv. Water Resour. Invest. 76–13.

Plummer, L. N., D. L. Parkhurst and T. M. L. Wigley 1979. Critical review of the kinetics of calcite dissolution and precipitation. In *Chemical modeling in aqueous systems,* Symposium series 93, E. A. Jenne (ed.), 537–73. Washington: American Chemical Society.

Plummer, L. N. and E. Busenberg 1982. The solubilities of calcite, aragonite, and vaterite in CO_2-H_2O solutions between 0 and 90° C and an evaluation of the aqueous model for the system $CaCO_3—CO_2—H_2O$. *Geochim. Cosmochim. Acta* **46**, 1011–40.

Pulina, M. 1971. Observations on the chemical denudation of some karst areas of Europe and Asia. *Stud. Geomorph. Carpatho-Balcanica* **5**, 79–92.

Pulina, M. 1972. A comment on present-day chemical denudation in Poland. *Geogr. Polonica* **23**, 45–62.

Reddy, M. M., L. N. Plummer and E. Busenberg 1981. Crystal growth of calcite from calcium bicarbonate solutions at constant P_{CO_2} and 25° C: a test of a calcite dissolution model. *Geochim. Cosmochim. Acta* **45**, 1281–9.

Smith, D. I. 1962. *The solution of limestone in an Arctic environment.* Inst. B. Geogs. Spec. Publn. **4**, 187–200.

Smith, D. I. and T. C. Atkinson 1976. Process, landforms, and climate in limestone regions. In *Geomorphology and climate,* E. Derbyshire (ed.), 367–409. London: Wiley.

Sweeting, M. M. 1979. Weathering and solution of the Melinau limestones in the Gunong Mulu National Park, Sarawak, Malaysia. *Ann. Soc. Geol. Belgium* **102**, 53–7.

Troester, J. W. and W. B. White in press. Seasonal fluctuations in the carbon dioxide partial pressure in a cave atmosphere. *Water Resour. Res.*

White, W. B. 1977. Role of solution kinetics in the development of karst aquifers. *Int. Assoc. Hydrogeols Mem.* **12**, 503–17.

White, E. L. and W. B. White 1979. Quantitative morphology of landforms in carbonate rock basins in the Appalachian Highlands. *Geol Soc. Am. Bull.* **90**, 385–96.

Wigley, T. M. L. 1977. *WATSPEC: a computer program for determining the equilibrium speciation of aqueous solutions. Br. Geomorph. Res. Grp Tech. Bull.* **20**, 1–46.

Williams, P. W. 1963. An initial estimate of the speed of limestone solution in County Clare. *Irish Geogr.* **4**, 432–41.

Williams, P. W. and R. K. Dowling 1979. Solution of marble in the karst of the Pikikiruna Range, Northwest Nelson, New Zealand. *Earth Surf. Proc.* **4**, 15–36.

11
Theoretical considerations on simulation of karstic aquifers

James J. Cullen IV and Robert G. LaFleur

Introduction

Simulation of groundwater flow through a karstic aquifer requires interdisciplinary application of principles of hydrology and geomorphology, quantified by selected hydrologic numerical methods. In this chapter, relationships contributed by each discipline are reviewed first, followed by examination of appropriate flow laws and a description of one technique found useful in modeling karstic aquifer behavior.

Other than for heuristic reasons, justification for creating karstic aquifer simulation models is provided by Yevjevich (1981): "Although water resources of karstified carbonate terranes are more expensive to investigate, less safe to develop and more costly to tap, sooner or later ongoing population pressures will extend from nearly completely developed, less expensive resources to those of karstified areas."

If one considers the large proportion of karstified lands that exist in the east-central and southeastern United States, and also recalls the drought conditions that recently occurred in these areas (see, e.g. Alder *et al.* 1981), it is easy to believe that the "pressures" of which Yevjevich speaks may be felt in this decade.

Karst hydrogeologic and geomorphologic influences on motion of subsurface water

Principles that control water movement through porous media are well understood and are discussed in elementary hydrology texts (Davis & DeWiest 1966, Freeze & Cherry 1979, Todd 1980). However, the mechanics of water flow through fractured, conduit-charged karstic aquifers is more diverse and complex. One can easily see why differing geometries of the two aquifer types require separate concepts when it becomes necessary to derive governing flow equations.

In non-indurated granular sediments, governing equations for fluid flow are derived by combining Darcy's law (Fig. 11.1) with a continuity equation (Fig. 11.2). Darcy's law is an empirically derived energy balance equation that quantifies the amount of fluid moving through a porous medium as a function of certain physical characteristics of the medium and permeant, and as a function of the hydraulic head (potential energy or energy per unit weight of the permeant) gradient. Although the equation was derived for phreatic (saturated) flow, it may be equally applicable to flow in the vadose (unsaturated) zone by recognizing that the hydraulic conductivity tensor (**K**) is an hysteretic function of the soil water potential.

Consider an elemental volume of sediment in which the grains are assumed to remain stationary for unsaturated flow, but for saturated flow, are permitted to enter or leave the elemental volume. The constraint is required because in saturated flow, changes in the amount of water stored within the elemental volume are attributed not only to changes in the density of the fluid but also to changes in the arrangement of the grains. Water moving into or out of the elemental volume is accounted for (conventionally on a volume basis for unsaturated flow and on a mass basis for saturated flow), and any net difference

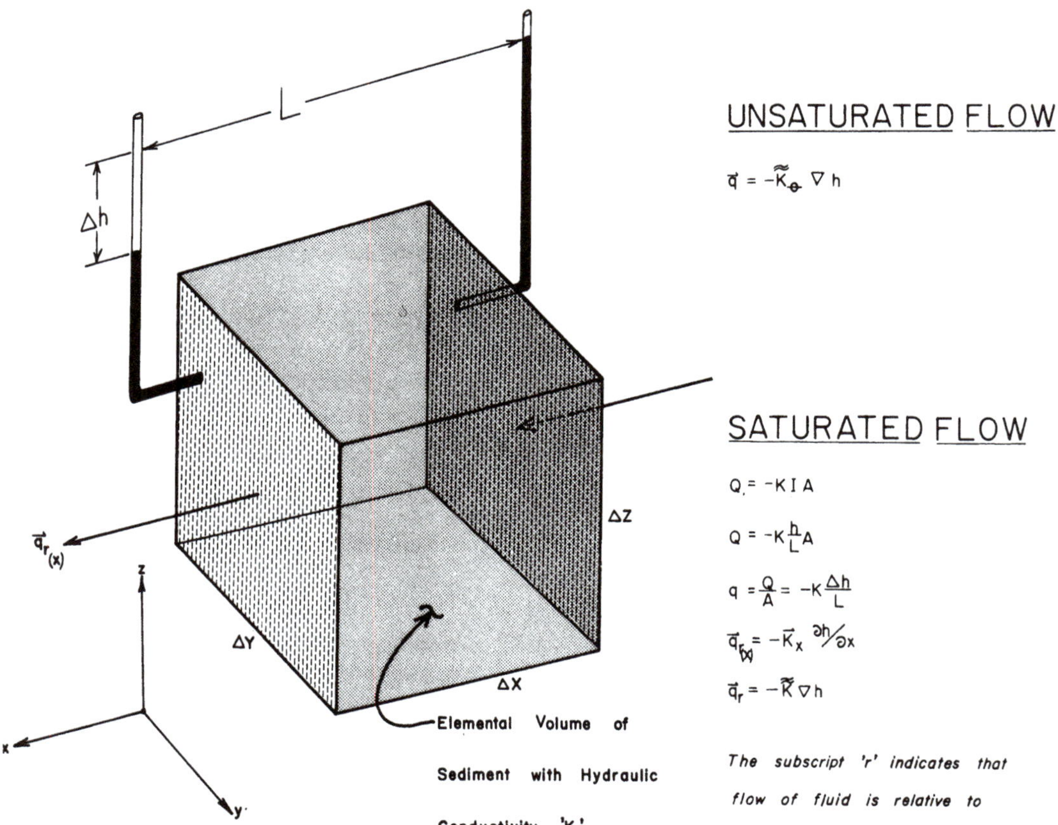

Figure 11.1 Darcy's Law.

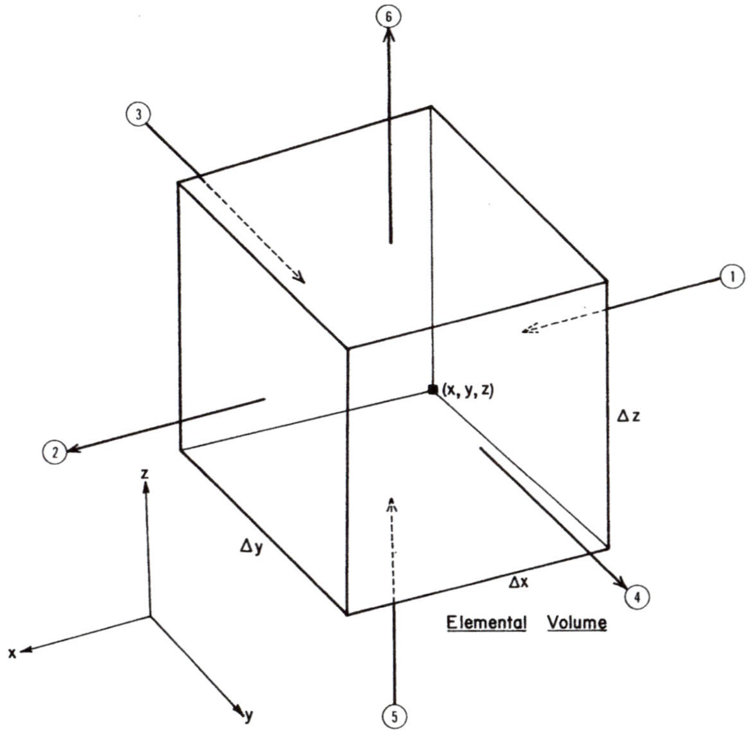

Figure 11.2 Making use of the principle of conservation of mass to derive equations of continuity for the unsaturated and saturated flow of water through a porous medium, assuming that the sediment grains remain stationary with respect to the coordinate axes.

UNSATURATED FLOW-VOLUME BASIS

① $\vec{V}_x|_x \;\Delta Y \Delta Z \Delta t$

② $(\vec{V}_x|_x + \partial \vec{V}_x/\partial x)\,\Delta Y \Delta Z \Delta t$

③ $\vec{V}_y|_y \;\Delta X \Delta Z \Delta t$

④ $(\vec{V}_y|_y + \partial \vec{V}_y/\partial y)\,\Delta X \Delta Z \Delta t$

⑤ $\vec{V}_z|_z \;\Delta X \Delta Y \Delta t$

⑥ $(\vec{V}_z|_z + \partial \vec{V}_z/\partial z)\,\Delta X \Delta Y \Delta t$

STORAGE- $\partial \theta/\partial t \;\Delta X \Delta Y \Delta Z \Delta t$

$$\boxed{-\nabla \cdot \vec{V} = \partial \theta/\partial t}$$

SATURATED FLOW–MASS BASIS

① $(\rho \vec{q}_x)|_x \;\Delta Y \Delta Z$

② $-[(\rho \vec{q}_x)|_x + \partial(\vec{q}_x)/\partial x]\;\Delta Y \Delta Z$

③ $(\rho \vec{q}_y)|_y \;\Delta X \Delta Z$

④ $-[(\rho \vec{q}_y)|_y + \partial(\vec{q}_y)/\partial y]\;\Delta X \Delta Z$

⑤ $(\rho \vec{q}_z)|_z \;\Delta X \Delta Y$

⑥ $-[(\rho \vec{q}_z)|_z + \partial(\vec{q}_z)/\partial z]\;\Delta X \Delta Y$

STORAGE- $n\partial \rho'/\partial t$

$$\boxed{-\nabla \cdot (\rho \vec{q}) = n\partial \rho/\partial t}$$

is attributed to changes in storage. Storage changes due to the grain rearrangement mechanism are commonly three to four orders of magnitude larger than those produced by changes in fluid density. The continuity equation derived in Figure 11.2 for saturated flow is a restatement of the principle of conservation of mass for water entering or leaving the elemental volume. A similar expression can be written for sediment grains entering or leaving the volume, with any net difference being attributed to changes in the porosity of the medium. Compare this with the case of unsaturated flow where changes in water storage are simply accounted for by changes in soil moisture content.

Factors that influence flow of fluid through non-indurated sediments are summarized in Table 11.1. The table has been constructed in a manner to emphasize that on any scale of examination other than a very local one, spatial variability of these factors may be of major importance. Many of the factors are common to both vadose and phreatic flow. However, soil water potential factors are important only in vadose flow whereas factors influencing specific storage are important only in phreatic flow.

The geometries of karstic aquifers are considerably more complex than their granular counterparts. As a result, the convenient distinction between saturated and unsaturated zones of sedimentary aquifers no longer applies in karstic aquifers. A brief review of karst geomorphology will illustrate why this is true.

Table 11.1 Factors that influence the flow of water through a non-indurated sediment.

Saturated flow	Unsaturated flow
governing flow equation	governing flow equation
$\nabla \cdot \mathbf{K} \nabla h = S_s \partial h/\partial t$	$\nabla \cdot \mathbf{K}_{(\psi)} \nabla h = \partial\theta/\partial t$
spatial variance of:	spatial variance of:
grain size distribution	grain size distribution
physical makeup of the grains	physical makeup of the grains
physical arrangement of the grains	physical arrangement of the grains
chemical makeup of the grains	chemical makeup of the grains
chemical makeup of the permeant	chemical makeup of the permeant
density of the permeant	density of the permeant
viscosity of the permeant	viscosity of the permeant
hydraulic head gradient	hydraulic head gradient
compressibility of water	soil water potential factors:
compressibility of the porous medium	temperature
porosity	pressure
	osmotic potential
	matric potential

In karst, water commonly moves across bedding-plane partings, through joints, through solutional openings such as caves and caverns, and through the rock matrix. Although a popular notion contends that most karst is joint controlled, this is not true. Karst geomorphologists recognize numerous classifications of cave and cavern systems such as joint controlled, maze, branchwork, anastomosing and spongework (see Jennings 1971).

Joint-controlled caves are formed by bedrock dissolution along pre-existing joints. Typically, these joints occur in conjugate sets that strike in two or more directions. For example, in Virginia, Ward Cove Cave (Fig. 11.3) exhibits four

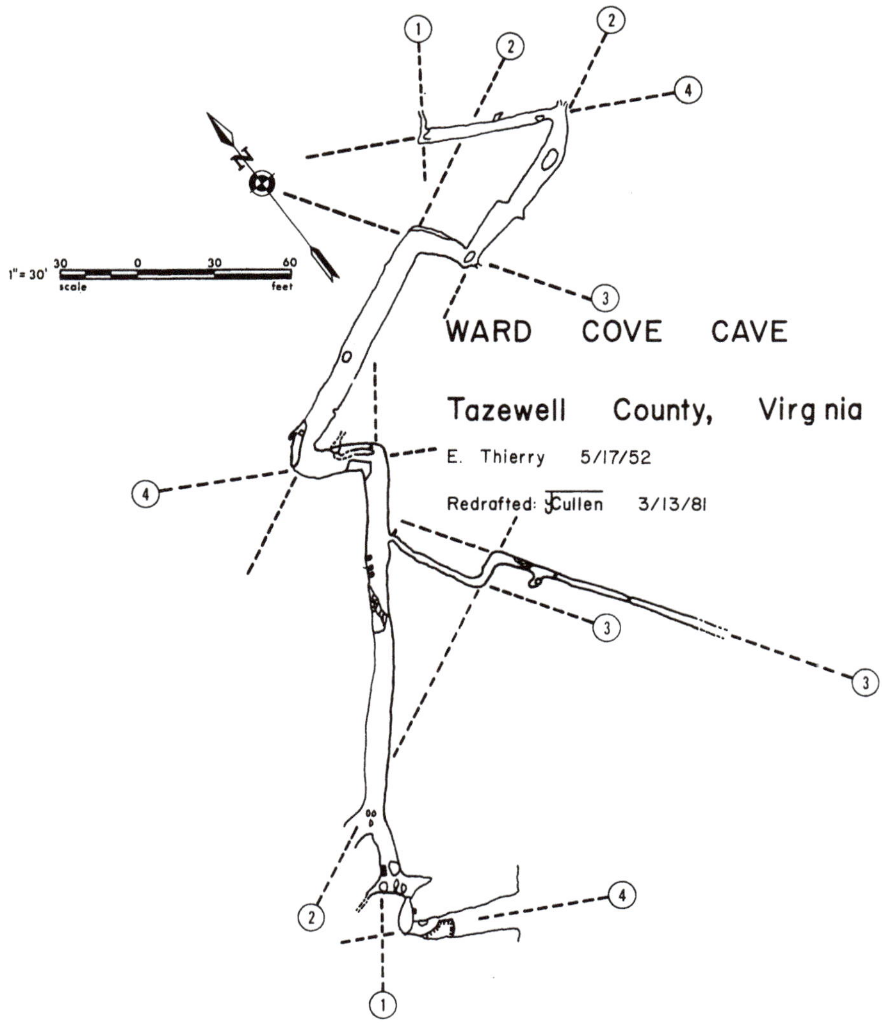

Figure 11.3 An example of a joint-controlled cave (Douglas 1964).

principal fracture directions (indicated by the circled numerals). Knox Cave (Fig. 11.4) in Albany County, New York, is another example of a joint-controlled cave. The two major passages visible in the photograph have developed along two parallel joints that trend N19° E; these are the principal joints controlling speleogenesis. A second set of joints at 90° to the principal set influences speleogenesis in a more subtle way: the face of the bedrock spur (just to the right of center in the photograph) was produced by dissolution along the second set of joints (about N69° W). The portion of the cave that has been explored occupies an area of about 300 m by 40 m suggesting strong anisotropy. These observations may agree with the reader's impression of karstic terranes.

Caves that are maze-like in appearance may form by one of at least two mechanisms. Some are simply joint-controlled caves that form where bedrock

Figure 11.4 The two major passages visible in this photograph have developed as a result of dissolution along parallel joints, Knox Cave, Albany County, New York.

is highly fractured. However, Palmer (1975) observed that other maze caves form by an entirely different mechanism. In this case, a fractured limestone is overlain by an insoluble but fractured caprock. The caprock fractures serve as a distributary system for percolating meteoric waters, uniformly directing dissolution of the limestone. Clark's Cave, Bath County, Virginia (Fig. 11.5) is a good example of a maze cave.

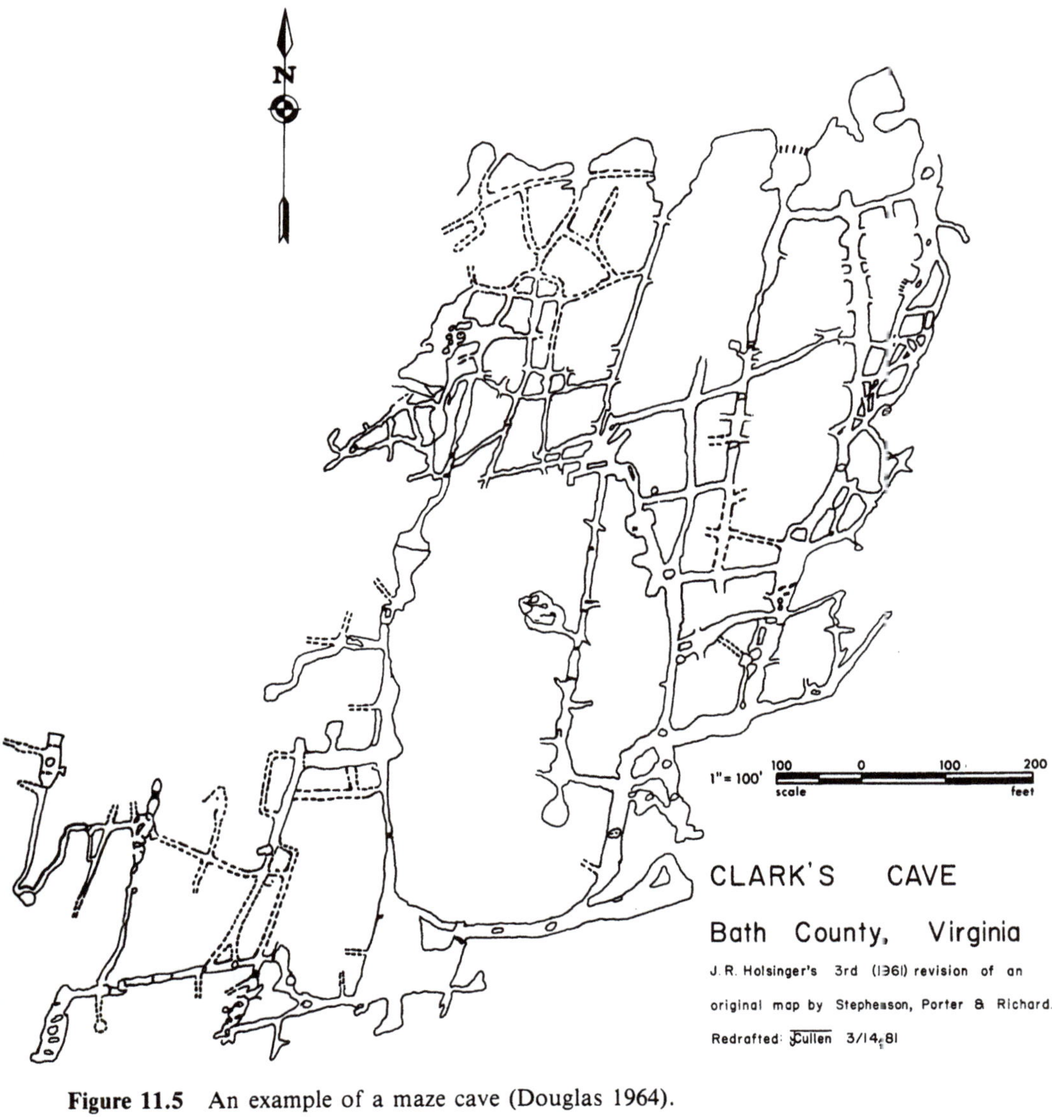

Figure 11.5 An example of a maze cave (Douglas 1964).

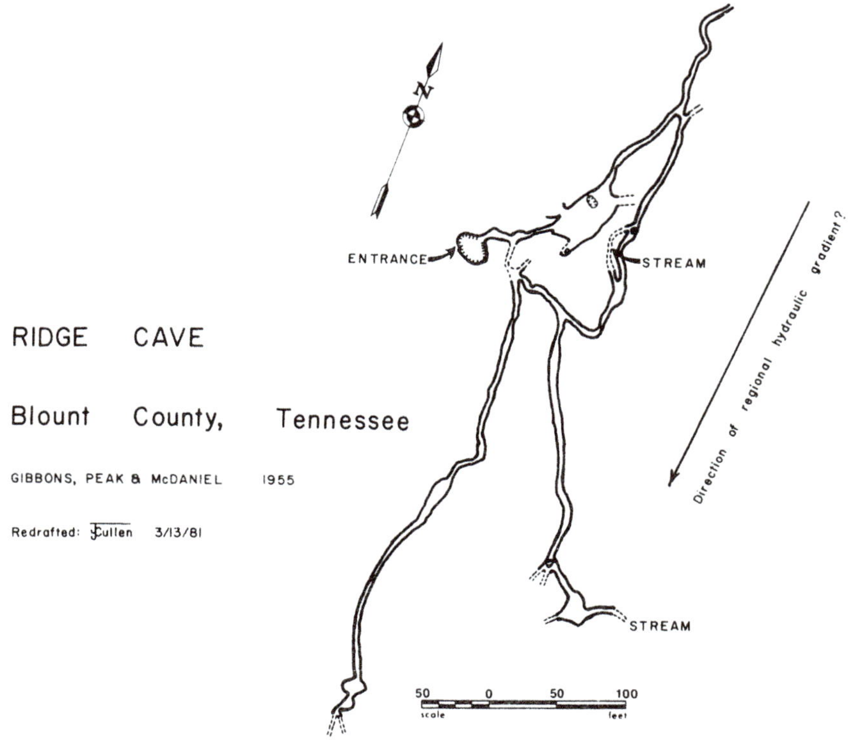

Figure 11.6 An example of a branchwork cave (Barr 1961).

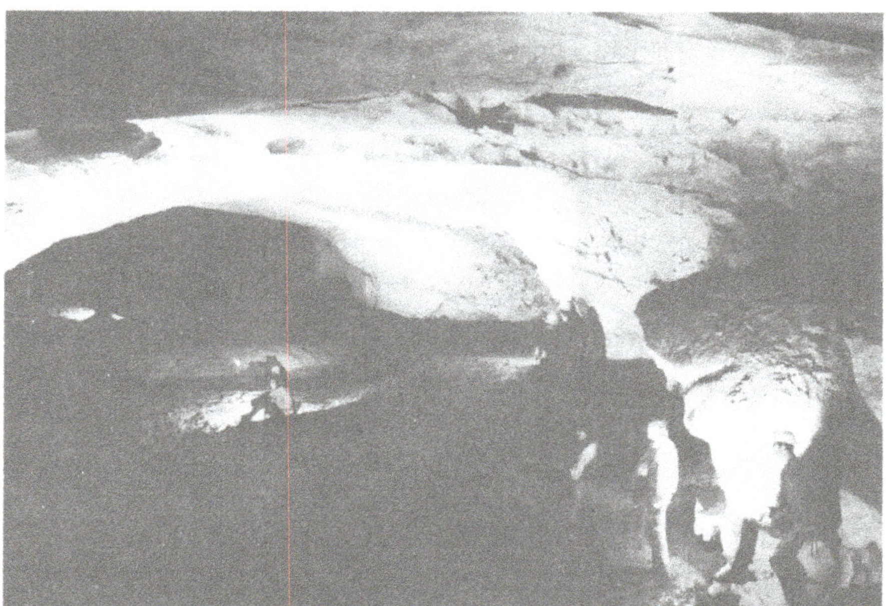

Figure 11.7 A typical passage in a branchwork cave. Note the meandering stream channel that has been cut into the clay-filled floor, Greenville Saltpeter Cave, Monroe County, West Virginia.

Branchwork caves (Fig. 11.6) may exhibit either dendritic or joint-oriented passages. Although dendritic branchwork passages appear to have formed under influence of a regional paleohydraulic gradient, the dip of the controlling fracture or parting is the primary feature responsible for directing joint-oriented branchwork passages. Where branchwork passages develop along joints or partings, they do not form closed loops as do their joint-controlled and maze counterparts. Greenville Saltpeter Cave, Monroe County, West Virginia (Fig. 11.7), exhibits a branchwork morphology. Note that the limestone is visibly unfractured. In recent times, a stream has meandered across the clay-filled floor of the passage, cutting meander niches in the bases of the passage walls. The occurrence of clay fills in cave systems should be kept in mind: clay fills have hydrologic significance to be discussed later.

Anastomosing caves (Fig. 11.8) appear to form in a manner similar to

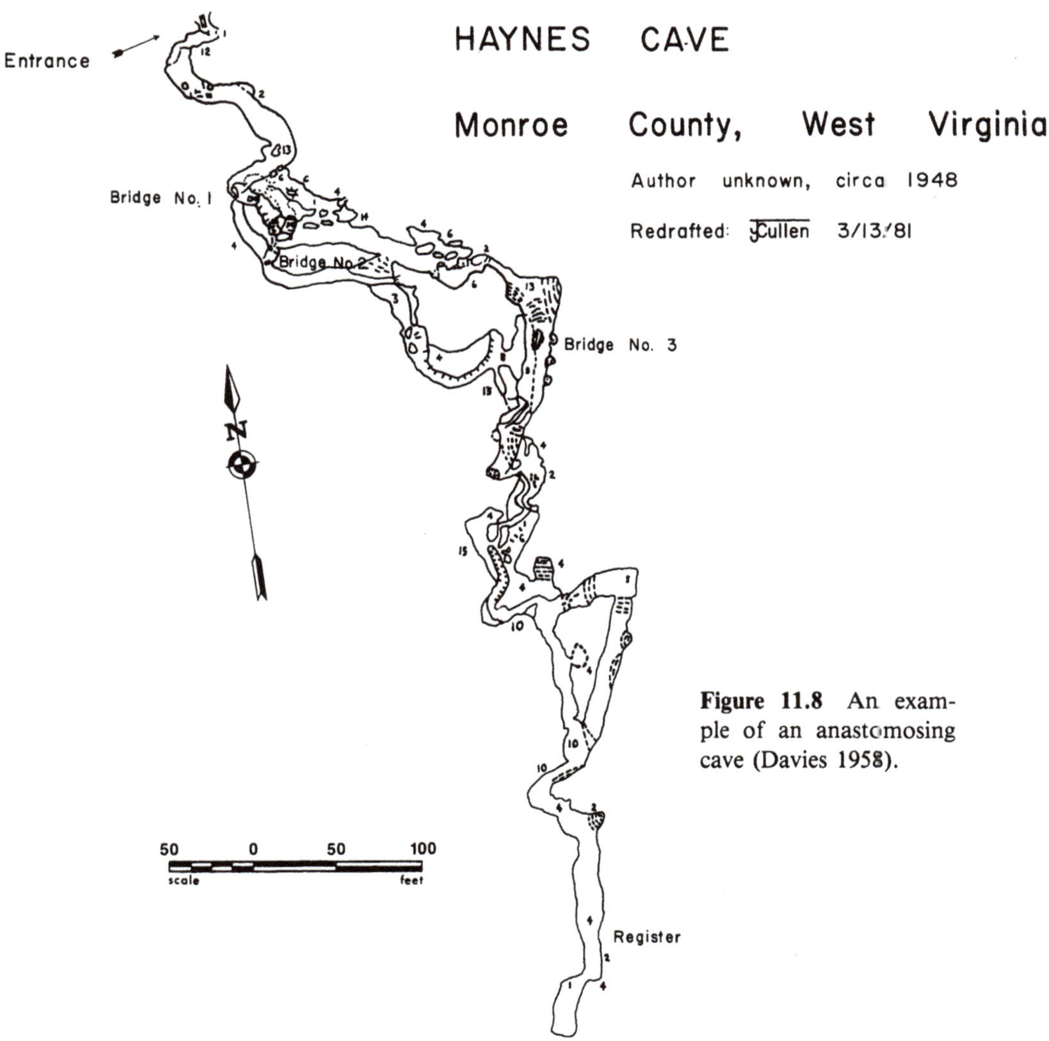

Figure 11.8 An example of an anastomosing cave (Davies 1958).

branchwork caves, but under steeper hydraulic gradients induced by floodwater conditions. Typically, their passages meander and braid, and they commonly develop on multiple levels. These various levels appear to have been selected in response to favorable lithologic partings. Eventually, the levels become connected when floors of upper levels collapse into lower passages, perhaps as a result of loss of bouyant support of a receding saturated condition. Passage appearances are similar to those found in branchwork caves. Phreatic tubes are common. Indeed, at Howe Caverns, Schoharie County, New York, a section of the main cave passage is one large phreatic tube with numerous smaller tubes intertwining in the passage side walls.

The passages of a spongework cave (Fig. 11.9) resemble the interstices of a sea sponge. There is little, if any, preferred orientation. Spongework caves are believed to form under stagnant water-table conditions or through a flushing action within the zone of seasonal fluctuation of the water table. For reasons of equilibrium geochemistry, the second hypothesis is preferred. A third,

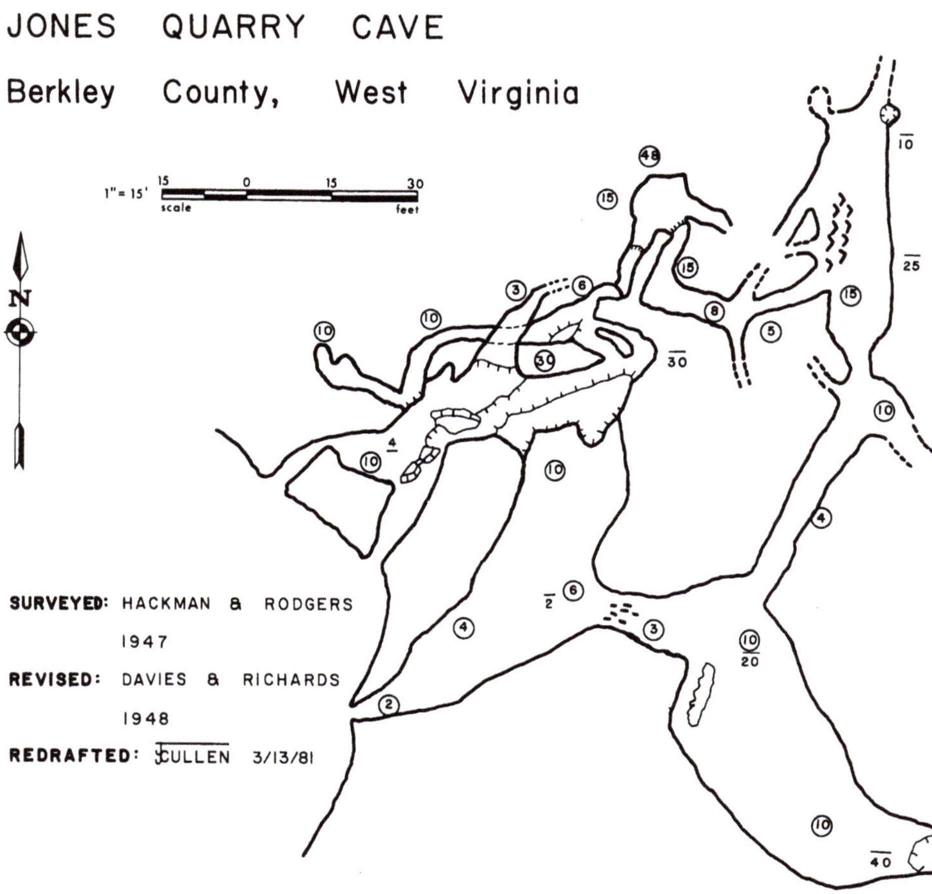

Figure 11.9 An example of a spongework cave (Davies 1958).

though less accepted hypothesis, is that some spongework caves are, in reality, maze or joint-controlled caves that have undergone an extreme degree of dissolution. Blowing Cave (Fig 11.10), developed in a highly permeable limestone in southern Georgia, exhibits spongework erosion. Note the pocket-like erosion pits on the walls. Pieces of the limestone that have undergone incasion reveal the same pitted erosion pattern, suggesting the cave formed by dissolution of individual vugs that grew and coalesced.

Because cave morphologies are extremely varied, any attempt to make use of existing aquifer simulation programs to model karstic aquifers should encounter numerous difficulties. The first problem involves defining positions of the phreatic and piezometric surfaces within karstic aquifers. This is more than a problem of semantics, for many existing hydrologic models require that the user be able to define, with some degree of precision, the positions of the phreatic and piezometric surfaces.

By way of comparison, recall that aquifers developed in non-indurated sediments are classified as being confined (artesian), semi-confined (leaky), or unconfined (water table). Unconfined aquifers are recharged either by direct infiltration or by predominantly horizontal flow from areas of higher hydraulic head within the same aquifer. The boundary between the saturated zone and the zone of soil moisture, called the water table or phreatic surface, is at an absolute pressure of 1 atm. Confined aquifers are recharged primarily by lateral flow and exist at non-atmospheric pressures; these pressures are

Figure 11.10 Random dissolution and enlargement of vugs has produced this spongework chamber, Blowing Cave, Decatur County, Georgia.

caused by overlying confining layers of impermeable strata that restrict areas of recharge. These recharge areas are higher than, and are generally at some distance from, the area of interest. Water levels in piezometers or wells penetrating a confined aquifer may rise to an elevation above the bottom of the confining layers. This elevation, when extended as a surface over the confined aquifer, is called the piezometric surface. Semi-confined aquifers are transitional between confined and unconfined conditions. The separating strata are semi-permeable, permitting vertical movement of groundwater into, or out of, the underlying leaky confined, overlying leaky confined or unconfined aquifers.

In karstic aquifers, distinctions between confined, semi-confined and unconfined conditions are not always clear. Karstic aquifers do not form in some medium that overlies bedrock; rather, they develop in the bedrock itself. The bedrock may or may not be heavily jointed, may have undergone any degree of lithification, may or may not have major bedding-plane partings, and simply because it has undergone a high degree of surface karstification, may or may not have major solutional channels, caves and/or caverns developed in it. Also, what is a very minor solutional feature in a geomorphologic sense may be a very important feature in a hydrologic sense.

Figure 11.11 shows schematically some of the complications that can occur when one attempts to attach classifications to segments of a karstic aquifer. Water-table conditions exist under pump A. A well developed at this location should exhibit behavioral patterns characteristic of wells developed in fractured media. Water-table conditions also exist under pump B. Indeed, in a three-dimensional sense the wells developed under pumps A and B actually penetrate the same aquifer even though the well developed under pump B could conceivably produce much greater quantities of water than the one developed under pump A. The well developed under pump C penetrates a section of aquifer vertically confined under a clay fill. Depending on the conductivity of the fractured bedrock, the well could behave in any number of ways, ranging from no flow to large flows. In fact, if the vertical fractures were fairly open, the pump might even draw air. Unquestionably, the well developed under pump D would produce relatively large quantities of water, but only for a short period of time. Karstic aquifers in dense limestones have low storage coefficients, even taking secondary porous features into account. Because the perforated section of the well casing extends below a relatively impermeable bedrock and because the water source for the well is recharged under an elevated hydraulic head at some distance from the well, should this aquifer be classified as confined (even though pump tests could be interpreted otherwise)? If this *gedankenexperiment* serves only one purpose, it should point out that traditional definitions and existing models are not always applicable in the modeling of karstic aquifers.

There are other major complicating factors. Karstic aquifers are subject to rapid flooding and draining. As cave passages become flooded, higher, normally dry fractures and passages may begin to conduct water, rerouting

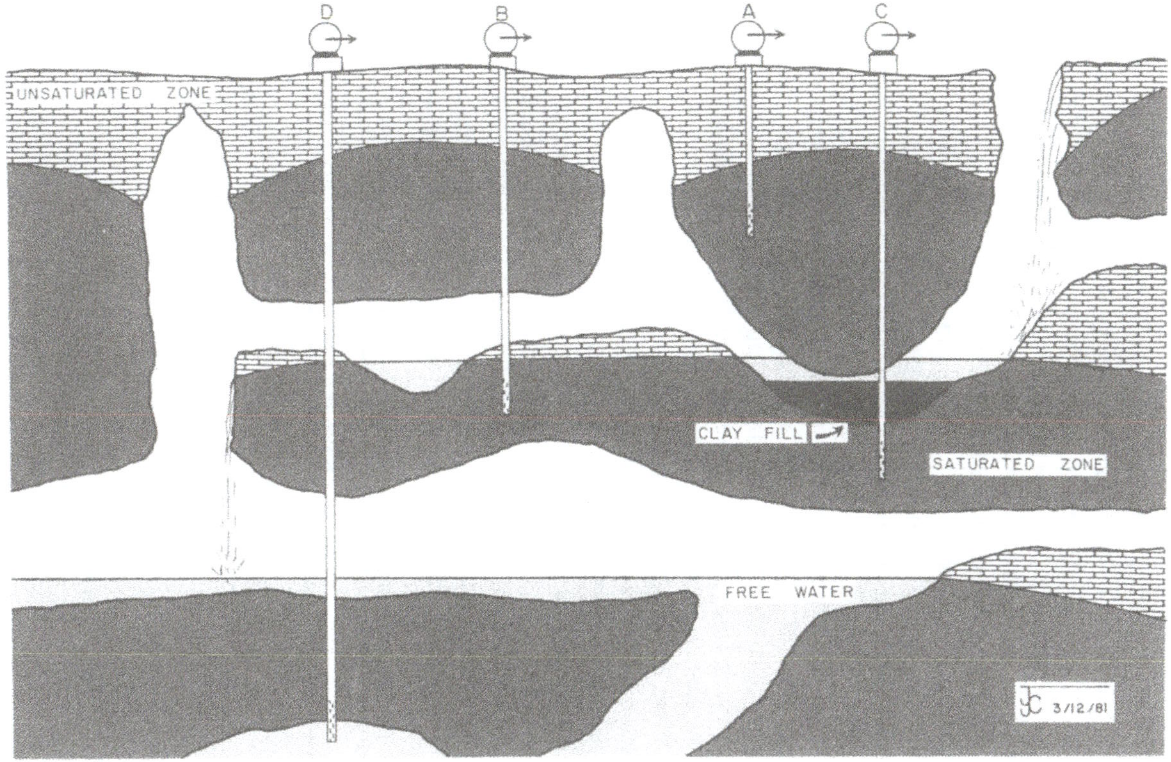

Figure 11.11 Geomorphologic configurations that complicate delineation of the position of a water table or phreatic surface in a karstic aquifer.

storm drainage and altering the entire hydrologic system. Springs that were dry for extended periods of time may begin to flow. Scouring actions of floodwaters may alter the location and character of sediment fills, and after the peak discharge has passed, draining may occur within a few hours or take several weeks. In localized sections of some karstic aquifers, Torrecellian effects will influence flow patterns. These are effects produced when elevated or reduced pressures are exerted against a water surface, most pronounced in places where air becomes trapped in ceiling pockets. Karstic aquifers may contain ebbing and flowing springs as a result of the existence of inverted syphons. For a detailed discussion of landforms that influence karst hydrologic behavior, the reader is referred to Bögli (1980).

Finally, there is the problem of appreciating the total aquifer geometry. Generalizations about aquifer properties (transmissivity, storativity, stratigraphy) often used in programs that model non-indurated aquifers cannot be applied to karstic aquifers. Certain portions of caverns or hydrologically important landforms may be unexplorable. Minor geomorphologic features may have major hydrologic significance. Phreatic cave passages of great importance to the hydrologist are difficult, if not impossible, to explore and chart. If such features cannot be examined first hand, a large part of any computer simulation soon becomes an exercise in black-boxing.

Despite these difficulties, karstic aquifers have been modeled successfully with conventional techniques. To understand how and to suspect when conventional techniques might go awry, we must turn to continuum theory.

Continuum theory focuses attention on the fact that the nature of a problem depends on the scale at which the problem is examined. In modeling karstic aquifers, the scale of the model becomes all important in determining which factors play a dominant role in affecting the model. For example on a megascopic or regional scale (Table 11.2), individual cave systems or zones of

Table 11.2 A modeling approach generated by the continuum concept.

Scale	Karst type	Proposed model
megascopic or regional	joint-controlled maze	conventional
	anastomosing branchwork	anisotropic
	spongework	conventional isotropic
local	joint-controlled maze	fracture models?
	anastomosing branchwork	pipe flow models?
	spongework	conventional isotropic

high fracture density are modeled best by using conventional techniques that simulate these areas as regions of increased hydraulic conductivity, which behave either isotropically, in the case of spongework type karst, or anisotropically for the other four cases discussed previously. On a more localized scale, details of fracture widths, spacings and orientations, cave-passage sizes and locations, and other landforms that influence hydrologic behavior have to be taken into account. On a very large scale, karst has been modeled as a porous medium; on a local scale, it certainly could not be.

Later in this chapter, a local-scale karstic aquifer simulation program will be examined. To evaluate the worthiness of such a model, we should first examine some basic concepts of numerical analysis as used in hydrologic modeling.

Simulating aquifers with a digital computer

Consider the hypothetical aquifer shown in Figure 11.12. Some of the hydrologic characteristics of this aquifer may be known, but a more detailed evaluation is desired. To make such an evaluation through the use of a computer simulation program, one proceeds in the following way.

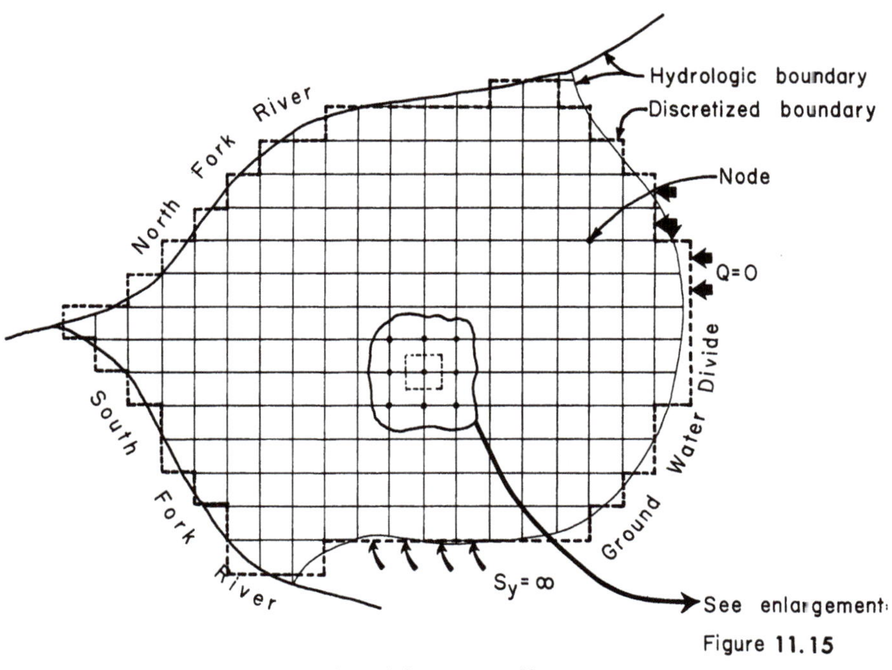

Figure 11.12 One method of discretizing an aquifer.

First, the boundaries of the study area must be selected, and critically so, because the accuracy or convergence rate of the model can be detrimentally affected by a nebulous choice of boundary conditions. Hydrologic features that exert some known control on water movement are the best boundaries. For example, flow perpendicular to a groundwater divide, and thus transmissivity normal to the divide, may be set equal to zero; this is called a constant (zero) flux boundary. Rivers, which are not severely drawn down by nearby pumping wells, have large storage and act as recharge or constant head boundaries for aquifers that are in direct hydraulic connection.

After the extent of the aquifer is defined, the second step is to differentiate those parameters that are reasonably well known from those that are to be examined by digital simulation. In other words, the modeler must decide what it is that he wants the computer to tell him.

By drawing a distinction between well known and less well known parameters, the modeler has chosen one of two categories of aquifer simulation programs (Fig. 11.13). If transmissivity, storativity, and infiltration rates are to be considered as well known variables, the modeler has chosen what has traditionally been a "forward" model. In such a case, hydraulic head becomes the dependent variable. If, on the other hand, the modeler feels confident

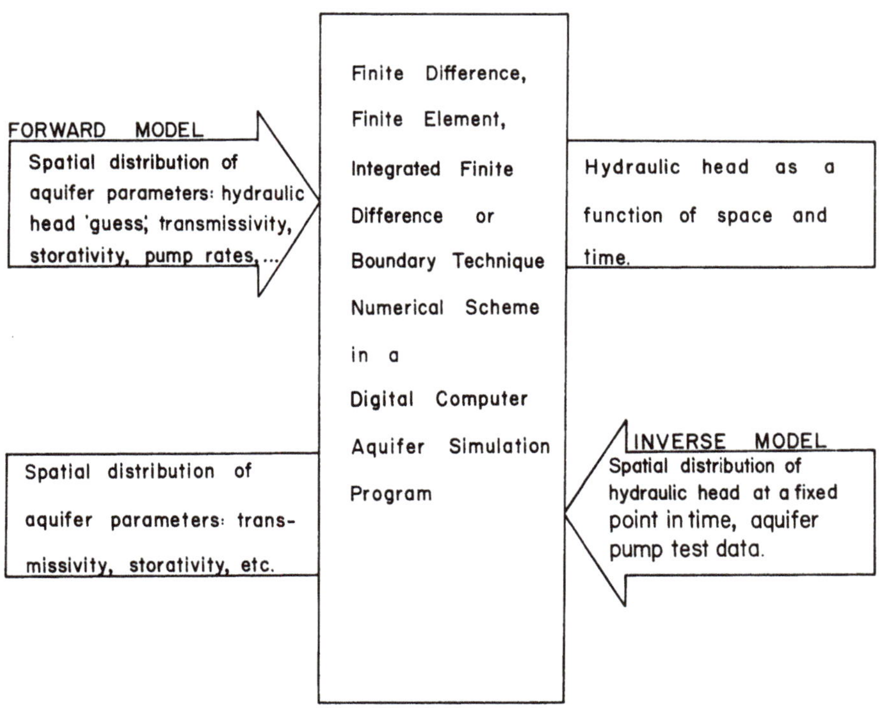

Figure 11.13 A comparison of input and output parameters for forward and inverse aquifer simulation programs.

about existing data on spatial distribution of hydraulic head, and he wishes to make use of this information to learn about aquifer transmissivity and storativity, he would run what has been termed an "inverse" model. The remainder of this chapter will deal strictly with forward models.

Broadly speaking, forward models can be analyzed by any one of at least four categories of numerical techniques. These include finite difference techniques, finite element techniques, integrated finite difference techniques (which are hybrid techniques employing the better parts of the finite difference and finite element methods), and boundary techniques. Within each of these categories, variations abound. Table 11.3 lists different numerical techniques that can be used to solve simultaneous equations developed by a finite difference scheme. The karst model proposed herein makes use of an iterative, alternating direction, line successive overrelaxation procedure. Differences in the various schemes will affect the number of iterations required to achieve a certain degree of precision, will influence conditions under which convergence will take place, or will affect the amount of computer storage required to perform the simulation. Cooley (1974), working in the saturated zone, and

Table 11.3 Numerical techniques that can be used to analyze equations generated by a finite difference scheme.

Direct solution techniques	Iterative solution techniques
Gaussian elimination	point Jacobi method
pivoting strategies	$x_i = (1/a_{ii})\left(-\sum_{j=1, j\neq i}^{N} a_{ij}x_j + b_i\right); \; i = 1, 2, 3, \ldots N$
LU-decomposition	
compact schemes	
Doolittle's method	Gauss–Seidel method
Crout's method	$x_i^{(m+1)} = (1/a_{ii})\left(-\sum_{j=1}^{i-1} a_{ij}x_j^{(m+1)} - \sum_{j=i+1}^{N} a_{ij}x_j^{(m)} + b_i\right); \; i = 1, 2, 3, \ldots N$
Choleski's method	
out-of-core solvers	
frontal method	
	point successive overrelaxation (SOR)
	$x_i^{(m+1)} = (1-\omega)x_i^{(m)} + (\omega/a_{ii})\left(-\sum_{j=1}^{i-1} a_{ij}x_j^{(m+1)} - \sum_{j=i+1}^{N} a_{ij}x_j^{(m)} + b_i\right); \; i = 1, 2, 3, \ldots N$
	line successive overrelaxation (LSOR)
	block successive overrelaxation
	line successive overrelaxation (SLORC) – with correction
	power method
	marathon method
	iterative alternating direction implicit procedure (ADI, ADIPIT)

Haverkamp *et al.* (1977) and Haverkamp and Vauclin (1979), working in the unsaturated zone, independently reached the comforting conclusion that when allowed to operate on the same data set, most of these numerical methods will produce essentially the same results.

Returning to Figure 11.12, the next operation involves dividing the aquifer into a number of pieces. The nature of this subdivision process depends on the analytical scheme being employed. Finite element models frequently make use of polygonal shapes whereas finite difference models use rectangles. In the present example and in the karst aquifer simulation models to follow, finite difference techniques after the methods of Prickett and Lonnquist (1971) are employed. Although Figure 11.12 shows the aquifer as having been divided into squares, mostly for mathematical convenience, rectangles are often used to concentrate computer analysis around points of high interest. Such a procedure tends to minimize the number of iterative calculations required in areas of low interest, decreasing program cost. The points of intersection of the grid lines are called nodes. A detailed look at one of these nodes is the next step.

Inherent in subdividing the aquifer is the assumption that its behavior can be represented by the interactive behaviors of its pieces. In an analogous manner, the assumption is now made that the properties of the square region surrounding a node (Fig. 11.14) are everywhere identical and that they may be represented by the properties that exist at that node. Of course, increasing the number of nodes increases the validity of this assumption, albeit at an increase in the cost of computer run time. For a forward model, the assumption is made that the transmissivity, storativity, leakage rates and infiltration rates of the aquifer at all points within the dashed line shown in Figure 11.14 are constant and are equal to the values found at the central node. The values of these properties may differ from those found at adjacent nodes. Also, one must consider that this square area actually represents a volume. For confined aquifers, the thickness of the block is equal to the thickness of the aquifer. For unconfined aquifers, the thickness of the block is equal to the saturated thickness of the aquifer. In karstic aquifers, the thickness of the block is more difficult to delineate because the aquifer may not be bounded by impermeable strata. Ways of dealing with this problem will be considered shortly.

Water flowing through the control volume can be analyzed in terms of two-dimensional components. An overall mass balance equation is written so that any net water flowing into or out of the control volume is accounted for by changes in storage. An additional source–sink term is added to account for other factors that influence the overall mass balance, such as evapotranspiration, leakage and recharge. Similar material balance equations are written for each node in the model. Because all nodes are interrelated, either to one another or to the boundary conditions, a large system of simultaneous mass balance equations is produced.

For an aquifer operating under steady state conditions, the inter-node flow values are constant. However, if the aquifer is disturbed by some external

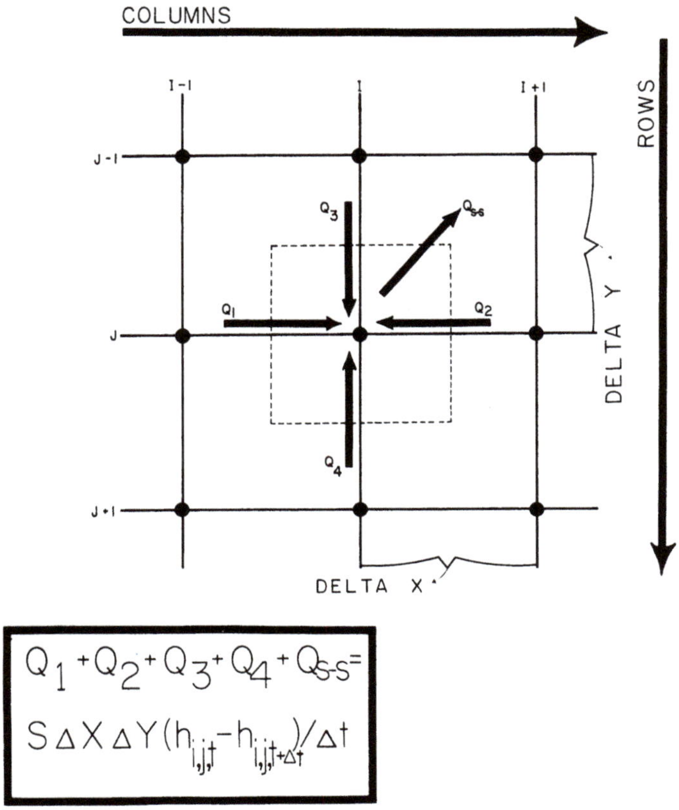

Figure 11.14 One means of constructing a mass balance equation about a node.

stress, such as by changing the flow rate of a well pump, the magnitude of the inter-node flow lines may change with time. In such situations, the purpose of a forward aquifer simulation program becomes one of calculating values of hydraulic head at each node at discrete intervals of time. The gradient of the head represents the driving force responsible for producing the inter-node flows. The task is accomplished as follows:

(a) A flow law is chosen that must stipulate and interrelate those factors that influence the inter-node flow values. Traditionally, Darcy's law has been used. However, when certain types of karstic aquifers are examined on a local scale, other laws are more appropriate.
(b) The flow law is combined with a mass balance equation to generate a governing flow equation.
(c) Because governing flow equations are usually written in the form of partial differential equations, they must be discretized before they can be

manipulated by a digital computer. Figure 11.15 demonstrates one means by which this can be accomplished.

(d) Finally, the aquifer simulation program solves the set of discretized equations, one per node, either simultaneously or in a specified sequence depending on the numerical scheme being employed, for the dependent variable. When Darcy's law is used, the solution is performed in terms of such independent variables as transmissivity and storativity. When non-Darcian flow laws are used, the independent variables can include storage factors, friction factors and shape factors. Computer output from a forward model is usually in the form of tables of hydraulic head values as a function of space and time.

Figure 11.16 shows an alternative manner of displaying the output from a forward model. Qualitatively, this output style is far easier to interpret than are pages of numbers. A ten-by-ten node grid has been superimposed over an imaginary aquifer in which the east–west transmissivities have been set to a greater value than the north-south ones. At the beginning of the run, the shape of the piezometric surface is flat. The three distinct dark areas that appear on the grid surround three pumping wells. The thickness of the inter-node flow lines indicates the relative magnitude of groundwater flow between the nodes: the thicker the line, the greater the flow. Arrows are used to indicate the direction of flow; generally, they point towards the pumping wells. Three

Governing equation for 2-dimensional flow of ground water in an heterogeneous, anisotropic, porous medium:

$$\frac{\partial}{\partial x}\left(T_x \frac{\partial h}{\partial x}\right) + \frac{\partial}{\partial y}\left(T_y \frac{\partial h}{\partial y}\right) = S\frac{\partial h}{\partial t}$$

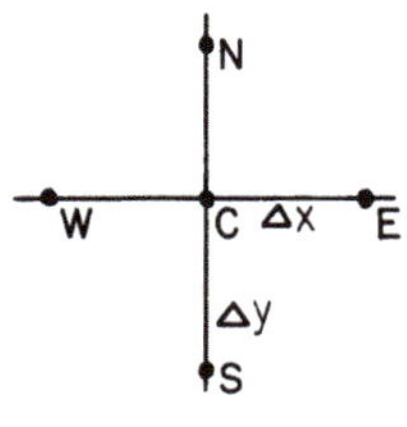

Five point star arrangement of nodes

coordinate axes aligned with principle transmissivity directions

$q_{NC} - q_{EC} - q_{SC} - q_{WC}$ = net volumetric accumulation rate =

$$T_y\frac{(h_N-h_C)}{\Delta y}\Delta x + T_y\frac{(h_S+h_C)}{\Delta y}\Delta x - T_x\frac{(h_E+h_C)}{\Delta x}\Delta y - T_x\frac{(h_W+h_C)}{\Delta x}\Delta y =$$

$$\frac{S(h_C^{(k+l)} - h_C^{(k)})\Delta x \Delta y}{\Delta t}$$

$$\boxed{T_y(h_N-2h_C+h_S)/\Delta y^2 + T_x(h_E-2h_C+h_W)/\Delta x^2 = S(h_C^{(k+l)} - h_C^{(k)})/\Delta t}$$

Figure 11.15 Discretizing a governing flow equation.

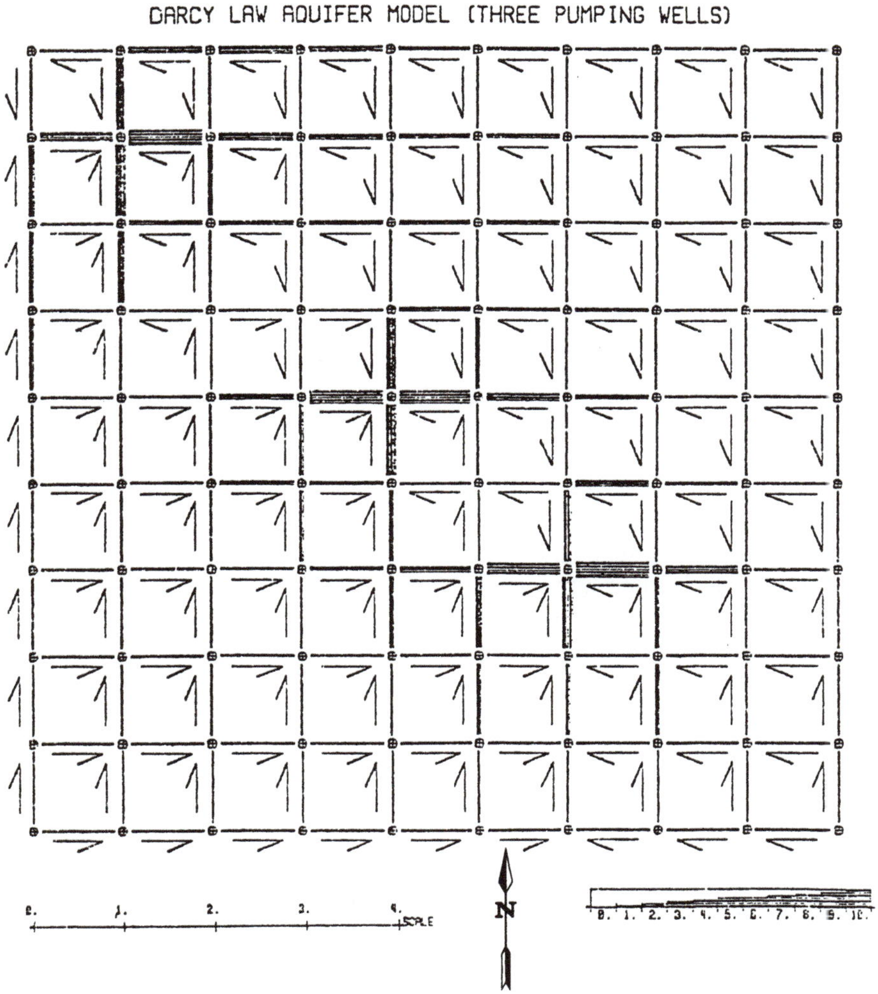

Figure 11.16 An example of an aquifer map that has been constructed by a digital computer.

oval-shaped cones of depression are visible. The oval shape is a result of the differences between the x- and y-direction transmissivities. Some interference between the lower two wells can be seen, and the pattern around the upper well shows the effect of assigning zero transmissivities perpendicular to the model boundary.

Figure 11.16 was drawn solely by computer in about six seconds using an IMLAC Corporation vector scan CRT. It probably represents the first application of interactive computer graphics to the field of groundwater hydrology. Hard copies were produced with a vertical and horizontal resolution of 200 pixels per inch on a Versatec (Xerox Corporation) electrostatic printer. Also, the drawdown process could be animated. A sequence of 20 frames required about 4 minutes of actual run time on a very heavily used PRIME 500 time-shared mainframe.

Modeling karstic aquifers

Having introduced the idea that karstic aquifers have been (and probably should be) simulated on a regional scale with existing models, it follows that examination on a local scale will require the development of new models. A short thought experiment will serve to illustrate the multiplicity of governing flow equations that will be required for local-scale examination.

Two types of flow predominate in a karstic vadose zone. Surface streams may insurge via caves, caverns or sinkholes. Open-channel flow equations could be used to model such flows. Meteoric water percolates through a karstic vadose zone primarily along fractures. Factors that influence fracture flow in the vadose zone are virtually unaddressed in the literature.

When meteoric waters reach the phreatic surface, equations governing groundwater flow come into play. Cave streams will act as point recharge sources, i.e. variable flux boundaries, active at the point where they sump and not where they insurge. One must avoid the temptation to believe that surface water becomes groundwater as soon as it enters the ground. Many cave streams travel for kilometers within the vadose zone. Their waters only become groundwater when they sump at the water table. (If, on the other hand, the water resurges at a spring or cave entrance, it may never actually have become groundwater.) Also, the effects of seasonal flow variations can alter the locations of sumps, can dry up some and can create others. Vertical seepage through the vadose zone should be treated as an heterogeneous source of diffuse recharge.

Within the phreatic zone, anastomosing and branchwork caves may be modeled best with pipe flow laws, whereas fracture flow laws could be employed to simulate groundwater flow through maze and joint-controlled karst. Spongework karst could be modeled with a traditional isotropic approach, as could matrix flow, although under all but very abnormal conditions, matrix flow would be insignificant when compared to secondary porous flow.

Karstic aquifers exhibit laminar, transitional and turbulent flow conditions. Different governing equations are required to describe each of these flow regimes.

Finally, the fact that some hydrologically significant conduits would either go undiscovered or would be inaccessible during charting requires that the modeler be very open minded during the calibration (inverse) phase of model construction. Unexpected groundwater vectors or zones of high transmissivity will inevitably appear. Indeed, karstic aquifer models could prove to be one of the more esoteric tools of exploring for new cave passages.

Joint-controlled flow

Figure 11.17 shows a modeling scheme used to simulate joint-controlled flow. Two conjugate sets of joints are superimposed over a finite difference grid. The following constraints are required to apply this scheme to a karstic aquifer simulation program:

(a) All fractures are assumed to be vertical. Whereas assumptions of two-dimensional flow can prove valid in aquifers developed in porous media (as long as the aquifer is thin compared to its lateral extent), karstic aquifers almost certainly have significant flow components in three dimensions. Thus, this first assumption may seriously limit the validity of the model.

(b) The fractures are assumed to extend vertically downward to the full depth of the aquifer; by "full depth" we mean that at a certain elevation either the fractures no longer contribute significantly to groundwater flow or they are healed (i.e. their width changes from some constant value to zero).

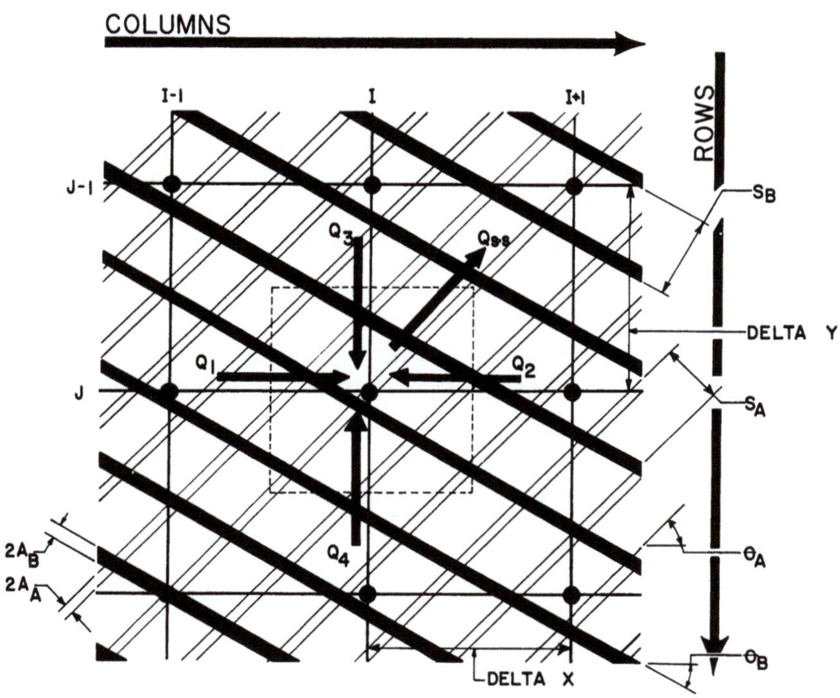

Figure 11.17 Superimposing joints on a finite difference grid.

(c) Joint orientations, spacings and widths are defined as shown in the figure.
(d) Mixing effects at joint intersections are omitted.
(e) Friction factors for joint walls (Fig. 11.18) are determined empirically, in a manner analogous to that used by engineers for pipe flow friction, after the methods of Castillo *et al.* (1972). Note that for a relatively smooth wall, the transition from laminar to turbulent flow takes place at a Reynolds number of about 2200; this is the same criterion used for pipe flow. Transition flow is ignored, primarily because of a lack of any suitable technique for dealing with it; the assumption is made that flow jumps from a laminar to a turbulent state at a discrete Reynolds number. As the wall roughness increases, a point is reached at which the laminar/turbulent transition takes place at decreasing Reynolds numbers, finally approaching a value of 400 when the friction factor goes to one. Castillo extended this work by developing a series of four flow equations (two each for laminar and turbulent flow at high and low Reynolds numbers) that would simulate flow through a single fracture. These equations (Table 11.4), which can be shown to be similar to Darcy's law in form, though non-linear in the case of turbulent flow, are applied to the model shown in Figure 11.17.

The concept of a state matrix (Fig. 11.19) was developed so that fracture zones and cavern systems can be entered into the karstic aquifer simulation program. Those inter-node flow lines designated as state 1 are to be modeled using Darcy's law; state 2 assumes the use of one of the four fracture flow equations given in Table 11.4; state 3 is reserved for pipe flow equations. The model is also designed to evaluate inter-node Reynolds numbers; these values are required so that the program is able to make a proper selection of governing flow equations. Reynolds number calculations are also used to warn the modeler of instances when the assumptions inherent in Darcy's law are being violated.

Table 11.4 Empirical equations that describe flow of groundwater through fractures and porous media (after Castillo *et al.* 1972). k/D_h is a relative roughness factor; k^* represents intrinsic permeability.

Conditions	Wall roughness		Fluid constant	Medium constant	Friction factor	Area term	Hydraulic gradient
laminar	low	Q =	$g\varrho/\mu$	1	1	$(2a)^3D/12$	dh/dl
	high	Q =	$g\varrho/\mu$	1	$1/(1 + 8.8k/D_h)$	$(2a)^3D/12$	dh/dl
turbulent	low	Q =	$4g^{\frac{1}{2}}$	1	$\log(3.7D_h/k)$	$(2a)^{1.5}D$	$(dh/dl)^{\frac{1}{2}}$
	high	Q =	$4g^{\frac{1}{2}}$	1	$\log(1.9D_h/k)$	$(2a)^{1.5}D$	$(dh/dl)^{\frac{1}{2}}$
Darcian	none	Q =	$g\varrho/\mu$	k^*	1	A	dh/dl

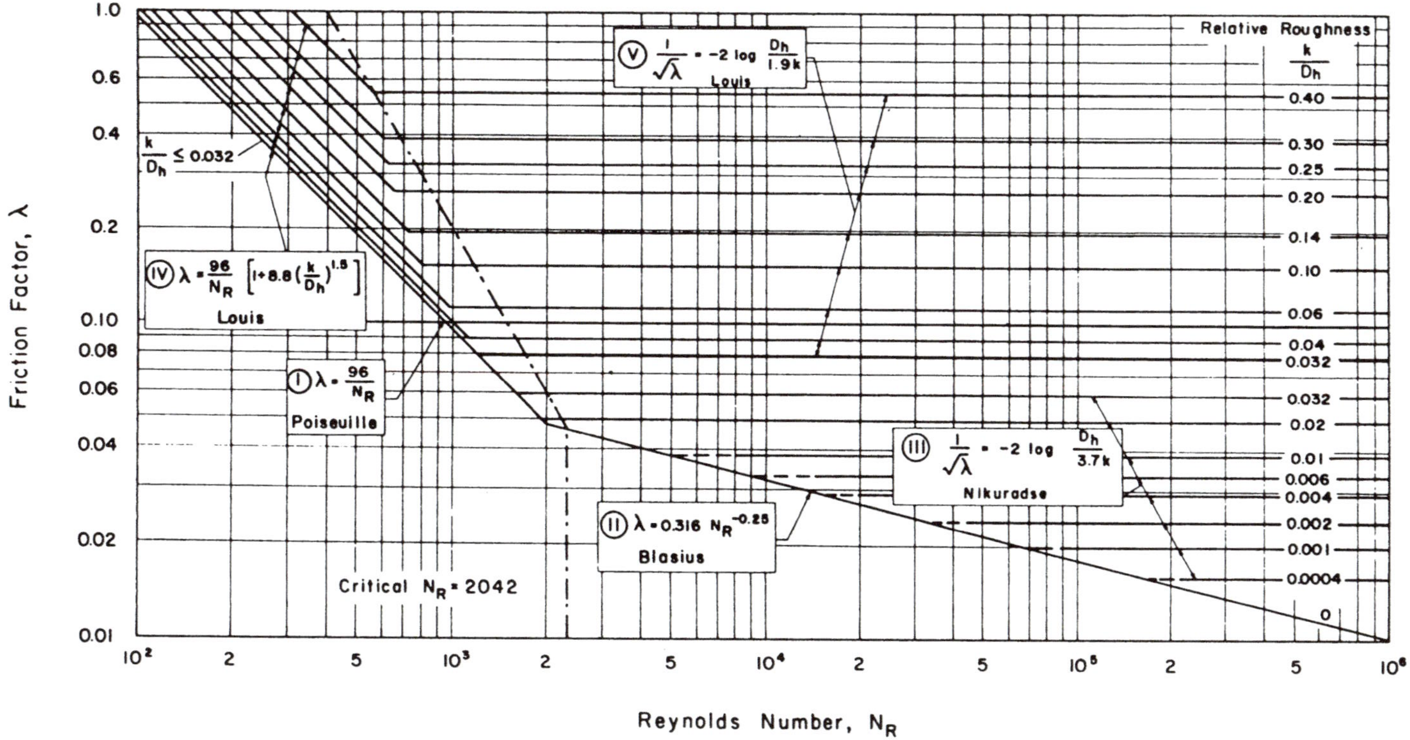

Figure 11.18 Joint wall friction factors (from Castillo *et al.* 1972).

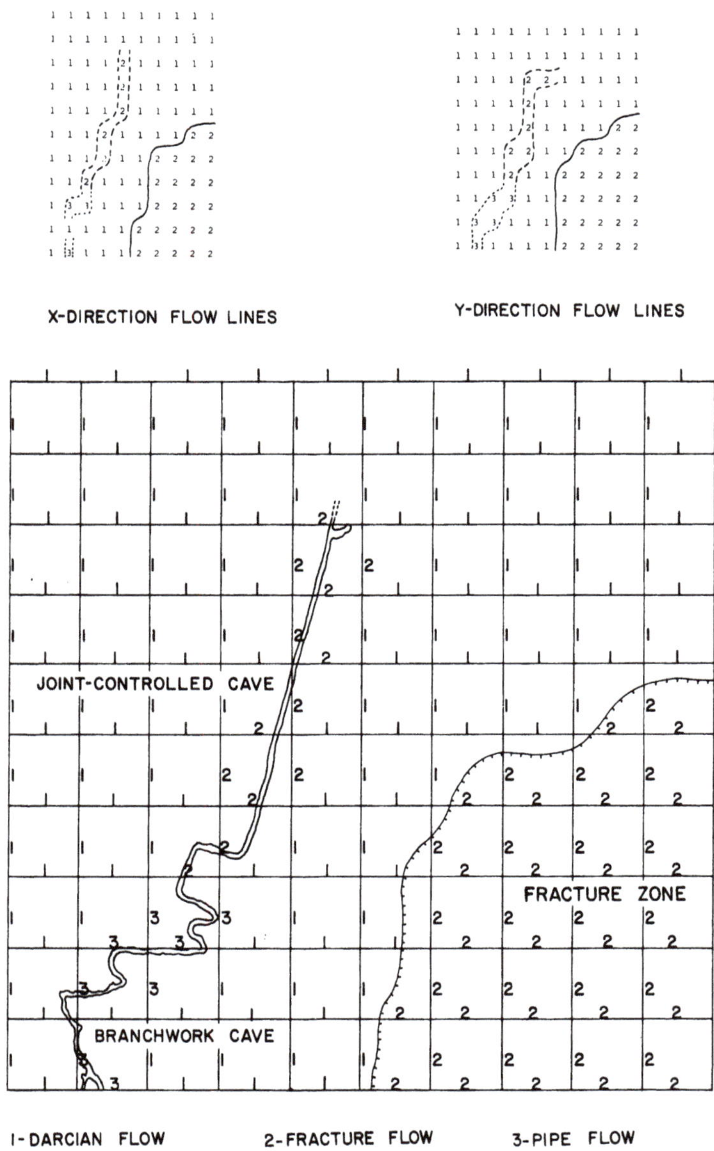

Figure 11.19 An example of a state matrix.

The state matrices shown in Figure 11.20 indicate that this hypothetical aquifer is to be treated as if it behaved in a Darcian manner with the exception of one horizontal inter-node flow line located immediately to the east of a pumping well that is to be modeled with fracture flow governing equations. Such state matrices can be indicative of a fairly tight, homogeneous limestone massif that contains a small but highly transmissive east–west fracture zone. Because the limestone is relatively impermeable compared to the fractures, the cone of depression becomes relegated to a small area surrounding the pump, except in the area of the fractures where the cone extends outward. In effect, a fracture acts as an extension of the well bore.

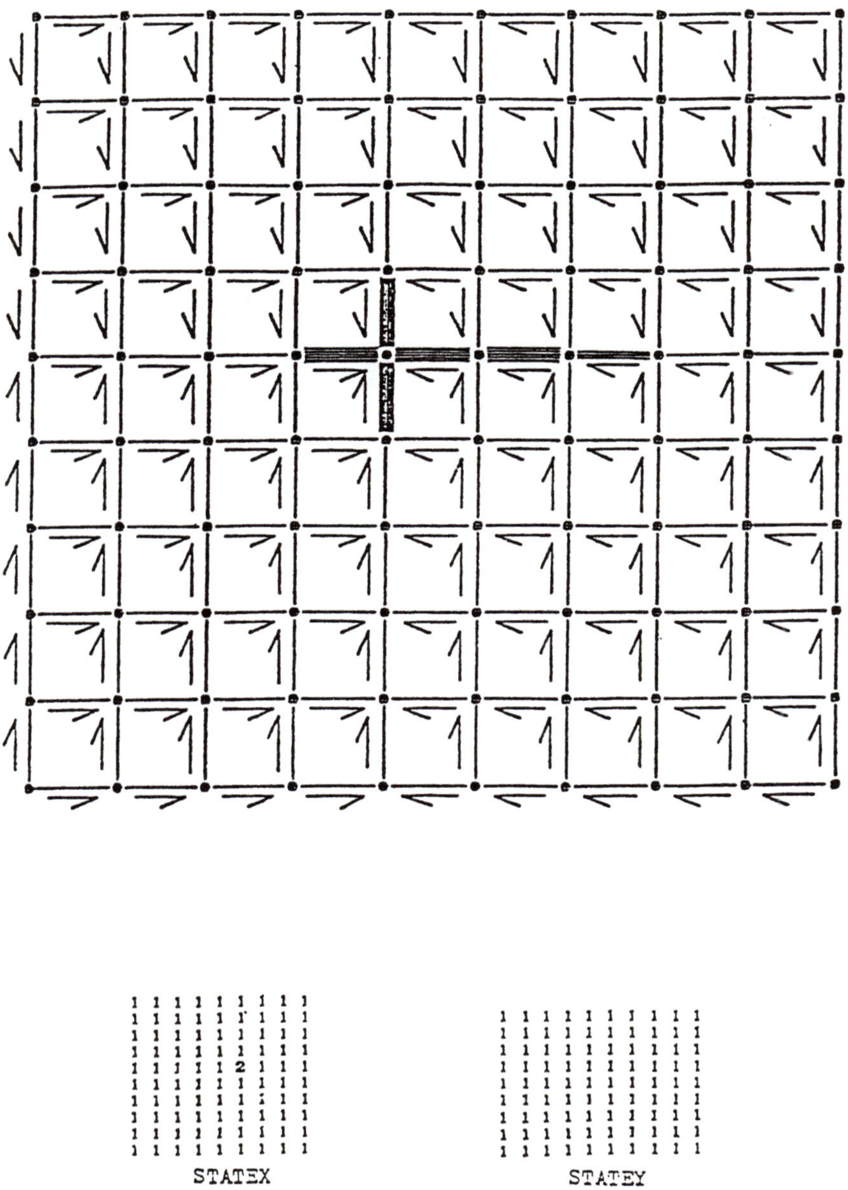

Figure 11.20 An example of a hypothetical Darcian aquifer that contains one small area of east–west fractures immediately to the east of a pumping well.

In Figure 11.21, the western half of the aquifer is to be modeled by a Darcian approach whereas the eastern half is to be treated as being heavily jointed. Very few inter-node flow lines have developed in the eastern region indicating that flows are relatively small. This may result from a relatively flat piezometric surface produced by the highly transmissive joints but more likely, it is an artifact caused by the restriction on mixing at joint intersections. In

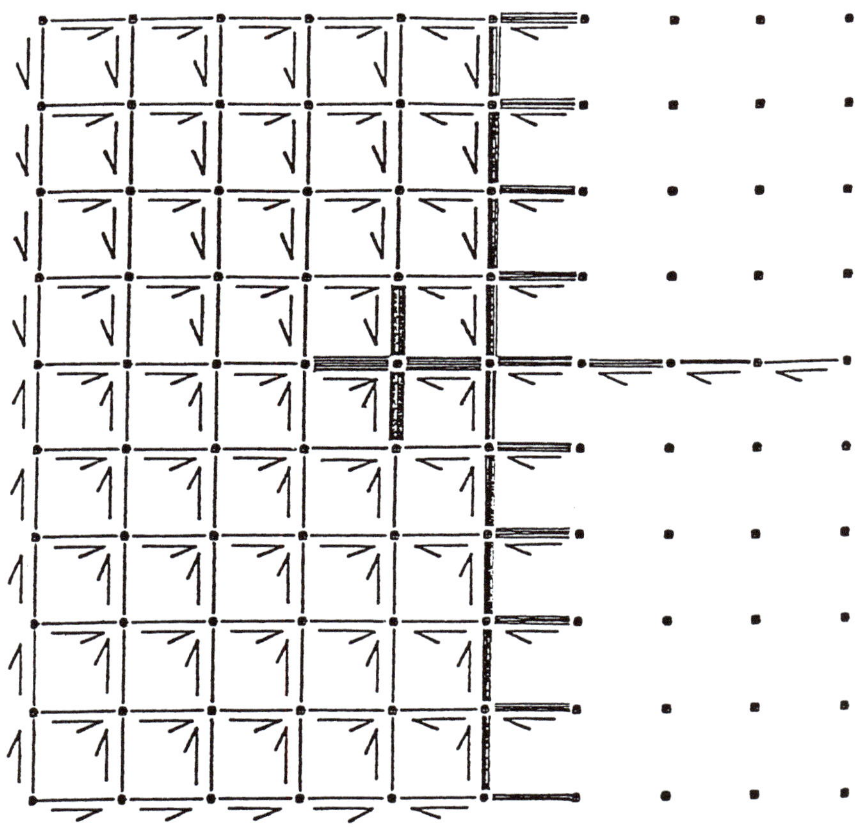

1	1	1	1	1	2	2	2	2
1	1	1	1	1	2	2	2	2
1	1	1	1	1	2	2	2	2
1	1	1	1	1	2	2	2	2
1	1	1	1	1	2	2	2	2
1	1	1	1	1	2	2	2	2
1	1	1	1	1	2	2	2	2
1	1	1	1	1	2	2	2	2
1	1	1	1	1	2	2	2	2
1	1	1	1	1	2	2	2	2

STATEX

1	1	1	1	1	2	2	2	2	2
1	1	1	1	1	2	2	2	2	2
1	1	1	1	1	2	2	2	2	2
1	1	1	1	1	2	2	2	2	2
1	1	1	1	1	2	2	2	2	2
1	1	1	1	1	2	2	2	2	2
1	1	1	1	1	2	2	2	2	2
1	1	1	1	1	2	2	2	2	2
1	1	1	1	1	2	2	2	2	2

STATEY

Figure 11.21 An example of a hypothetical Darcian aquifer in which the eastern portion is replaced with a major area of highly transmissive joints.

this example, the fractures were oriented parallel to the model's principal transmissivity directions. This eliminates cross-product terms in the transmissivity tensor. Such a restriction, coupled with the joint-intersection mixing restriction, may explain the observed flow pattern. A large head difference appears at the boundary between the jointed rock and the limestone massif, reflecting contrasting transmissivities.

Other runs made on this model indicate that it responds nicely to variations in joint orientation. Curiously, nearly identical flow patterns are generated by modeling either large, widely spaced joints or small, closely spaced joints so long as the total joint cross-sectional area per unit vector volume remains constant. This implies that negligible head loss occurs due to joint-wall friction and that only laminar flow develops. Under certain, albeit unknown, conditions the model behaves in an unstable manner, perhaps caused by turbulent flow. Conducting a stability analysis on the model may aid in identifying this problem.

Pipe-controlled flow

Pipes are superimposed on a vector volume in a manner similar to that used for joint-controlled flow (Fig. 11.22). Whereas joints are able to extend to the full depth of the aquifer, pipes, due to their discrete radii, cannot. This requires that the pipes be layered. To minimize the amount of computer memory required, the constraint is imposed that multiple layers of pipes have to lie directly above or below one another. This constraint places certain limitations on the types of caves that can be modeled. Single layer branchwork caves are restricted to two major passage directions. This is not all that unrealistic an assumption, especially if the node spacing is varied to accommodate the cave's

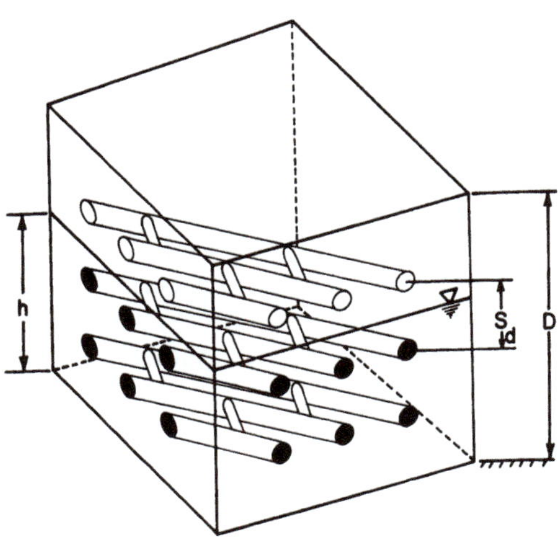

Figure 11.22 Pipes superimposed on a vector volume.

configuration. Multiple levels of anastomosing caves can be handled, but again, passage orientations are restricted to two major directions. In this case, the assumption is far less realistic. The fact that all of the pipes must be horizontal is an acceptable assumption unless the grid spacing becomes excessive (> 500 m or so). All of the pipes within a single swarm have to be of the same diameter. Finally, under unconfined conditions, pipes are assumed to go dry instantaneously if the elevation of the phreatic surface drops below the elevation of the center of the pipe; this eliminates the need to deal with open-channel flow conditions.

Bernoulli's equation, either in its frictionless form or in its complete form, is used as the governing flow equation. Some hypothetical runs are tried. Model instabilities again interfere with interpreting some of the results, but intuitively, the pipe model appears to exhibit behavior similar to that of the joint-controlled model. This is especially true with respect to such variables as pipe diameter, orientation and spacing. Mixing at pipe intersections is not considered, but pipe-fitting friction factors are employed, the assumption being that the pipe-fittings are capped in directions perpendicular to flow. The fittings seem to have no noticeable effect on flow patterns implying, again, that stability exists only under Darcian conditions.

Conclusions

Space limitation prevents presentation of a fully supported karstic aquifer simulation program. For these details, the reader is referred to Cullen (1979). Although existing Darcian models may work for karstic aquifers on a regional scale, local-scale simulation requires development of models that make use of flow laws more apropos to the movement of karstic waters.

Before field calibration and testing of any model is attempted, a period of extensive hypothetical testing to determine the influence of all newly introduced, karst-oriented variables, is indicated. Additional work in fracture-flow hydrology, both in the saturated and unsaturated zones, is necessary, as are investigations in phreatic-zone karst hydrology and geomorphology.

Only after these variables and others yet to be considered have been evaluated and ranked will the models begin to simulate accurately karstic groundwater flow. Models are useful in solving local groundwater flow problems and may reduce the number of situations exemplified in Fig. 11.23. But the impact of modeling is greatest when we are helped to understand the complete behavior of entire karstic aquifer systems.

Acknowledgements

Thanks are expressed to John Thrailkill of the Department of Geology at the University of Kentucky and to William B. White at the Materials Research

Figure 11.23 This domestic water supply system is located in one of the four Higgenbotham caves in Greenbrier County, West Virginia. Water flows from a surface stream into the mouth of the cave where a small portion of the flow is diverted by a steel pipe into a small holding tank. Next, the water is pumped into a pressure tank and is finally fed through a plastic line to the owner's home.

Center, The Pennsylvania State University, the former for initially stimulating interest in the problem and the latter for providing much insight on the nature of karst hydrology. Eugene S. Simpson, John E. Mylroie and Arthur N. Palmer critically reviewed the manuscript and provided thought-provoking discussion that contributed significantly in shaping its final form. James J. Cullen III, Phyllis J. Cullen and H. Allethaire Cullen reviewed the manuscript. Particular acknowledgement is made of financial support provided by the National Speleological Society Research Advisory Committee under the direction of Jack Hess and by the Interactive Computer Graphics Center at Rensselaer Polytechnic Institute.

References

Alder, J., W. J. Cook, S. McGuire, G. C. Lubenow, M. Kasindorf, F. Maier and H. Morris 1981. The browning of America. *Newsweek* Feb. 23, 26–37.

Barr, T. C. 1961. *Caves of Tennessee*. Tenn. Dept. Conserv. Comm. Div. Geol. Bull. 64.

Bögli, A. 1980. *Karst hydrology and physical speleology*. New York: Springer.

Castillo, E., G. M. Karadi and R. J. Krizek 1972. Unconfined flow through jointed rock. *Water Resour. Bull.* **8**, 266–81.

Cooley, R. L. 1974. *Finite element solutions for the equations of ground water flow*. Univ. Nevada Sys. Cen. Water Resour. Res. Desert Res. Inst. Tech. Rep. Ser. H-W Publn 18.

Cullen, J. J., IV 1979. *Digital computer simulation of karst groundwater flow*. MS thesis. Rensselaer Polytechnic Institute.

Davies, W. E. 1958. *Caverns of West Virginia*. W. Vir. Geol Econ. Sur. XIX a.

Davis, S. N. and R. J. M. DeWiest 1966. *Hydrogeology*. New York: Wiley.

Douglas, H. 1964. *Caves of Virginia*. Huntsville, Alabama: Virginia Region of the National Speleological Society.

Freeze, R. A. and J. A. Cherry 1979. *Groundwater*. Englewood Cliffs, NJ: Prentice-Hall

Haverkamp, R. M., M. Vauclin, J. Touma, P. J. Wierenga and G. Vachaud 1977. A comparison of numerical simulation models for one-dimensional infiltration. *Soil Sci. Soc. Am. J.* **41**, 285-94.

Haverkamp, R. M. and M. Vauclin 1979. A note on estimating finite difference interblock hydraulic conductivity values for transient unsaturated flow problems. *Water Resour. Res.* **15**, 181–7.

Jennings, J.N. 1971. *Karst*. Cambridge, Massachusetts: MIT Press.

Palmer, A. N. 1975. *The origin of maze caves. Nat. Speleol Soc. Bull.* **37** (3), 56–76.

Prickett, T. A., and G. C. Lonnquist 1971. *Selected digital computer techniques for ground water resource evaluation*. Ill. St. Water Surv. Bull. 55.

Todd, D. K. 1980. *Groundwater hydrology*. New York: Wiley.

Yevjevich, V. 1981. *Karst water research needs*. Littleton, Col.: Water Resources Publications.

12
Role of groundwater in shaping the Eastern Coastline of the Yucatan Peninsula, Mexico

William Back, Bruce B. Hanshaw and J. Nicholas Van Driel

A scientific and engineering need to understand the physics of seawater encroachment into coastal aquifers has had a significant impact on development of the science of hydrogeology (Back & Freeze 1983). The main physical controls on the position of the saltwater-freshwater interface were independently established by Badon Ghyben (1889) and Herzberg (1901). Although Du Commun published a statement of the principle in 1828 (Carlston 1963), the relationship is referred to as the Ghyben – Herzberg principle. It shows (Fig. 12.1) that under static conditions each foot of freshwater head above sea level depresses the interface 40 ft below sea level according to the relationship:

$$\varrho_s g z_s = \varrho_f g (z_s + z_w)$$

$$z_s = \frac{\varrho_f}{\varrho_s - \varrho_f} z_w$$

where g is gravity constant, z_w is freshwater head above sea level, and z_s is

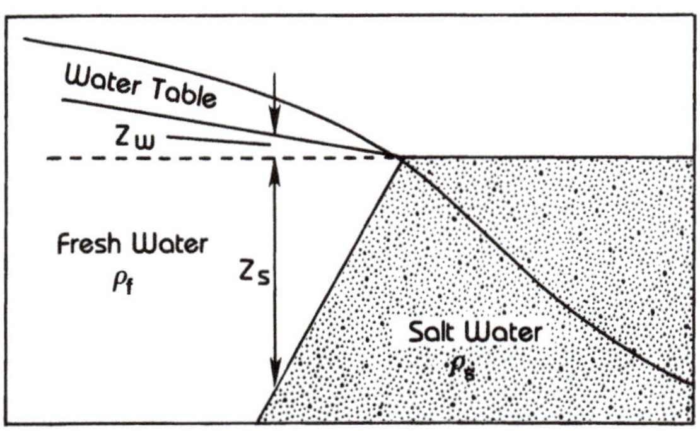

Figure 12.1 Sketch showing position of freshwater–saltwater interface according to the Ghyben–Herzberg principle.

distance of interface below sea level. Where freshwater density ϱ_f, $= 1.0$ and seawater density, ϱ_s, $= 1.025$, $z_s = 40z_w$. Hubbert (1940) expanded this concept by pointing out that the freshwater is not static and that the interface does not intersect the water table at the coastline as assumed by Badon Ghyben and Herzberg. Rather, a dynamic relationship caused by groundwater flow exists in which the position of the interface is controlled by the head distribution (Fig. 12.2). The position of the interface can be determined by flow-net analysis in which the change in freshwater head, Δh, between two adjacent flow lines is the control, and the interface is deeper than it would be if static conditions existed. For this analysis, Hubbert assumed that the interface is a sharp boundary and that the saltwater is static. Cooper (1959) provided a

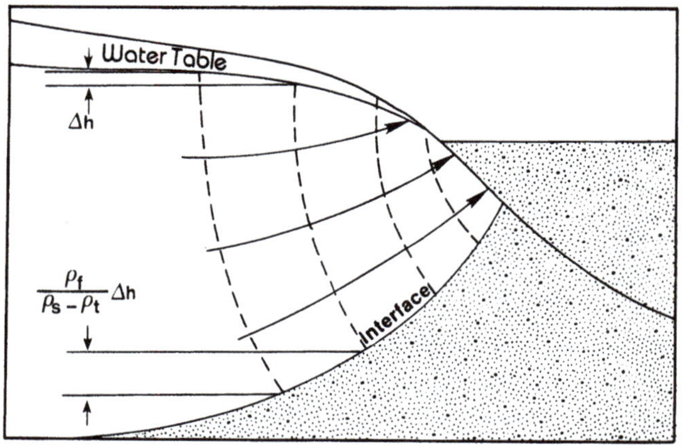

Figure 12.2 Sketch showing position of interface as modified by Hubbert (1940).

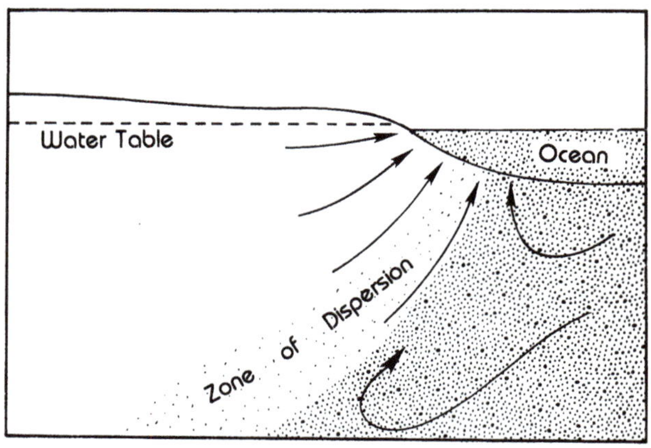

Figure 12.3 Sketch showing cyclic flow of ocean water and discharge of brackish water from zone of dispersion.

theoretical basis to demonstrate that the discharge of freshwater causes a concomitant cyclic flow of seawater from the ocean floor into the aquifer (Fig. 12.3). This seawater mixes with the fresh groundwater to form a zone of dispersion and leads to the discharge of brackish water. Kohout (1960) verified these theoretical considerations with field observations in the Biscayne aquifer in Florida.

This chapter demonstrates the geochemical significance of the mixing zone and formulates a hypothesis for the geomorphic evolution of part of the eastern coastline of the Yucatan, a stretch extending ~ 100 km from near Puerto Morelos southwest to Tulum (Fig. 12.4). Most of the Yucatan Peninsula is composed of flat-lying Tertiary limestone. The limestone along the east coast discussed in this chapter was deposited as a Pleistocene reef ~ 125 000 years ago. It is overlain by beach ridges and eolian deposits. The limestone is highly fractured with four sets of vertical joints; the dominant ones are at N25 – 30°E, essentially parallel to the coast, and N60 – 65°W, perpendicular to the coast (Back *et al.* 1979).

Figure 12.4 Map showing configuration of Yucatan coastline and sites mentioned in text.

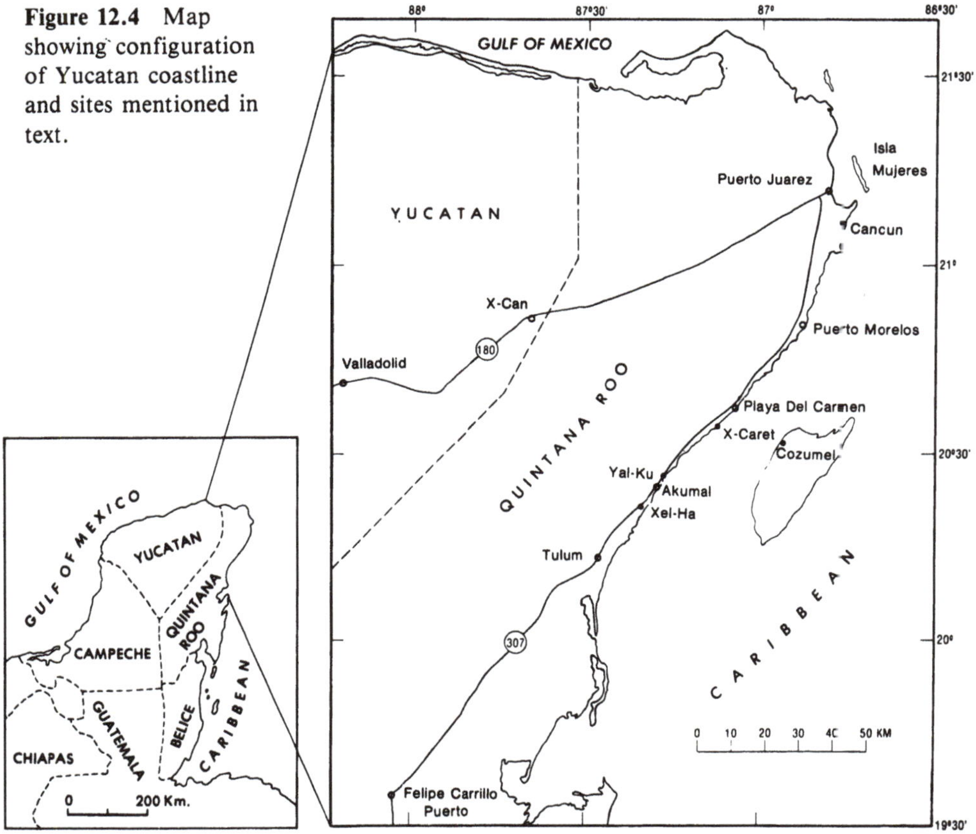

This coastline is characterized by a series of coves, caves and crescent-shaped beaches. We hypothesize these geomorphic features were formed as a result of mixing freshwater and seawater in the aquifer. The mixed groundwater became subsaturated with respect to calcite and dissolved the limestone.

The pertinent geochemical theory has been available since the late 1950s as mentioned briefly in Back and Hanshaw (1965), discussed comprehensively by Runnels (1969) in his classic paper, and applied to carbonate aquifers by Hanshaw and Back (1979). The essence of the theory is that mixing owes its geochemical significance to non-linearity of mineral solubility as a function of variables such as salinity, partial pressure of carbon dioxide, temperature, and ionic activity (effective concentration). An example of non-linearity is given in Fig. 12.5, which shows the relationship between ionic strength (a measure of dissolved-solids content) and activity coefficient, γ (a factor to correct for ion interaction and complexing). The activity coefficient of a single ionic species such as calcium is defined as:

$$\gamma_{Ca} \equiv \alpha_{Ca}/M_{Ca}$$

where M_{Ca}, the molality of calcium, is the analyzed concentration and α_{Ca}, the activity of calcium, is the effective concentration used in thermodynamic calculations. In very dilute solutions, α_{Ca} approaches M_{Ca}, and γ_{Ca} approaches unity (Fig. 12.5). When two solutions of different ionic strength are mixed, the resulting concentration of dissolved solids and of any ion in the mixture (e.g. M_{Ca}) is linear and therefore directly proportional to the ratio of volumes of the two solutions. However, the activity coefficient is a non-linear function of ionic strength, and the curve is typically concave upward (Fig. 12.5). If a linear relationship existed between these two parameters, the activity coefficient, γ_i, of any ion, i, resulting from mixing solution A with solution B to attain a solution with ionic strength given by the line C–C', would have an activity coefficient corresponding to the point C'. However, owing to non-linearity, the true

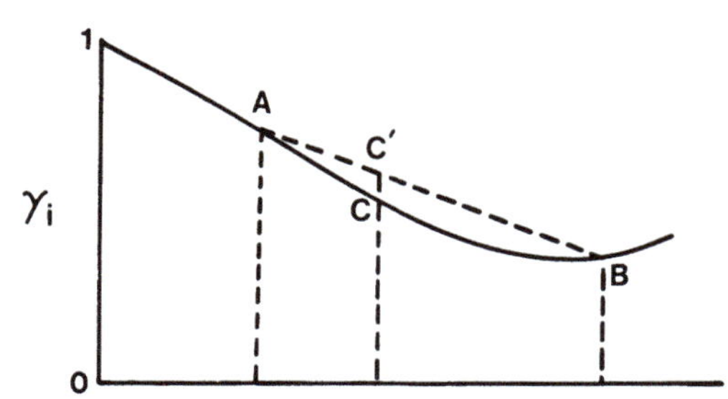

Figure 12.5 Graph showing the upward concavity of activity coefficient as a function of ionic strength.

value is at *C*. Thus, if solutions *A* and *B* had been in equilibrium (i.e. at saturation) with a mineral such as calcite, the resulting solution, C, would be subsaturated and therefore capable of dissolving additional calcite.

In like manner, any other non-linear relationships may result in super- or subsaturation of mineral species depending upon the shape of the function curve. It is also possible under certain conditions for mixing to produce water that is supersaturated with respect to a particular mineral. However, for the present discussion, we are restricting our interest to mixing that produces only subsaturation.

Plummer (1975) published a series of calculations to show the relative effects of various parameters on the saturation index of calcite. (Saturation Index = log Ion Activity Product/Equilibrium Constant; equilibrium, or saturation, is indicated by a value of zero.) One of his curves (Fig. 12.6) shows the change in the saturation index in a range of solutions resulting from mixing normal seawater with a typical fresh groundwater of the Yucatan. This curve shows that even though the freshwater is supersaturated with respect to calcite, the addition of $<10\%$ seawater causes the water to become subsaturated and remains subsaturated until nearly 70% of the mixture is composed of seawater. This wide range of subsaturation is due largely to the high partial pressure of carbon dioxide in the freshwater.

Combining geochemical theory with the cyclic flow of the mixing zone

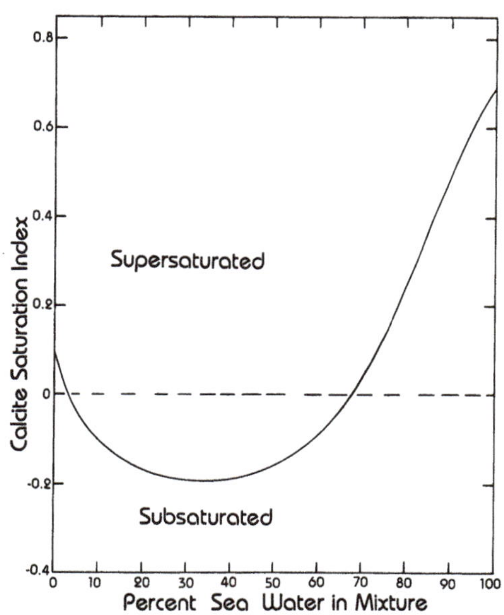

Figure 12.6 Change in saturation index as a function of mixing seawater with a sample of groundwater from the Yucatan (from Plummer 1975).

demonstrates which part of the groundwater regime is subsaturated (Fig. 12.7) and therefore has the greatest potential for porosity development. It has also been proposed that this mixing zone is an area of dolomitization (Hanshaw *et al.* 1971) and chertification (Knauth 1979).

Geologic significance of subsaturation can be emphasized by remembering that every marine limestone now containing freshwater has been subjected to dissolution and diagenesis caused by the mixing zone phenomenon at least once. Most have probably undergone the effects of this process repeatedly. For example, at any stand of sea level, the zone of dispersion will occupy a certain position within the limestone aquifer (Fig. 12.8a). As sea level drops, the zone of dispersion will follow the lowering sea level, thereby subjecting additional carbonate rocks to these processes (Fig. 12.8b). Subsequent rise of sea level will permit the zone of dispersion to migrate back up through the aquifer and cause the limestone to undergo additional diagenesis and differential dissolution (Fig. 12.8c).

A field study (Back *et al.* 1979) was undertaken at Xel Ha Caleta (lagoon) (Fig 12.4) to apply the mixing zone concept to development of this lagoon and other dissolution features observed along the coast of the Yucatan. Examination of aerial photographs and field observations demonstrated that dissolution features along the coast are fracture controlled, and landward extension of the many beaches forms a line parallel to the major northeast joint system. Reconnaissance sampling of water at Xel Ha demonstrated this area to be one of significant groundwater discharge. We believed that the Xel Ha lagoon was probably formed by dissolution of limestone by groundwater discharging along fractures, and that this would be an ideal location for field verification of the theoretical calculations. However, more detailed mapping of the chemical character of water showed that most of both the upper layer of brackish water and the lower body of seawater were supersaturated with

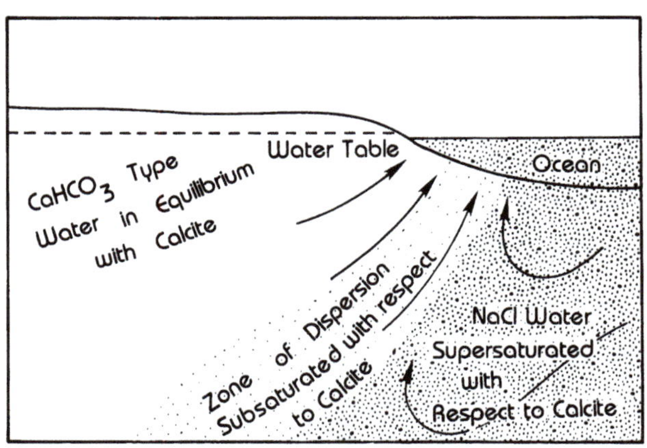

Figure 12.7 Chemical type water and relative calcite saturation in coastal carbonate aquifers.

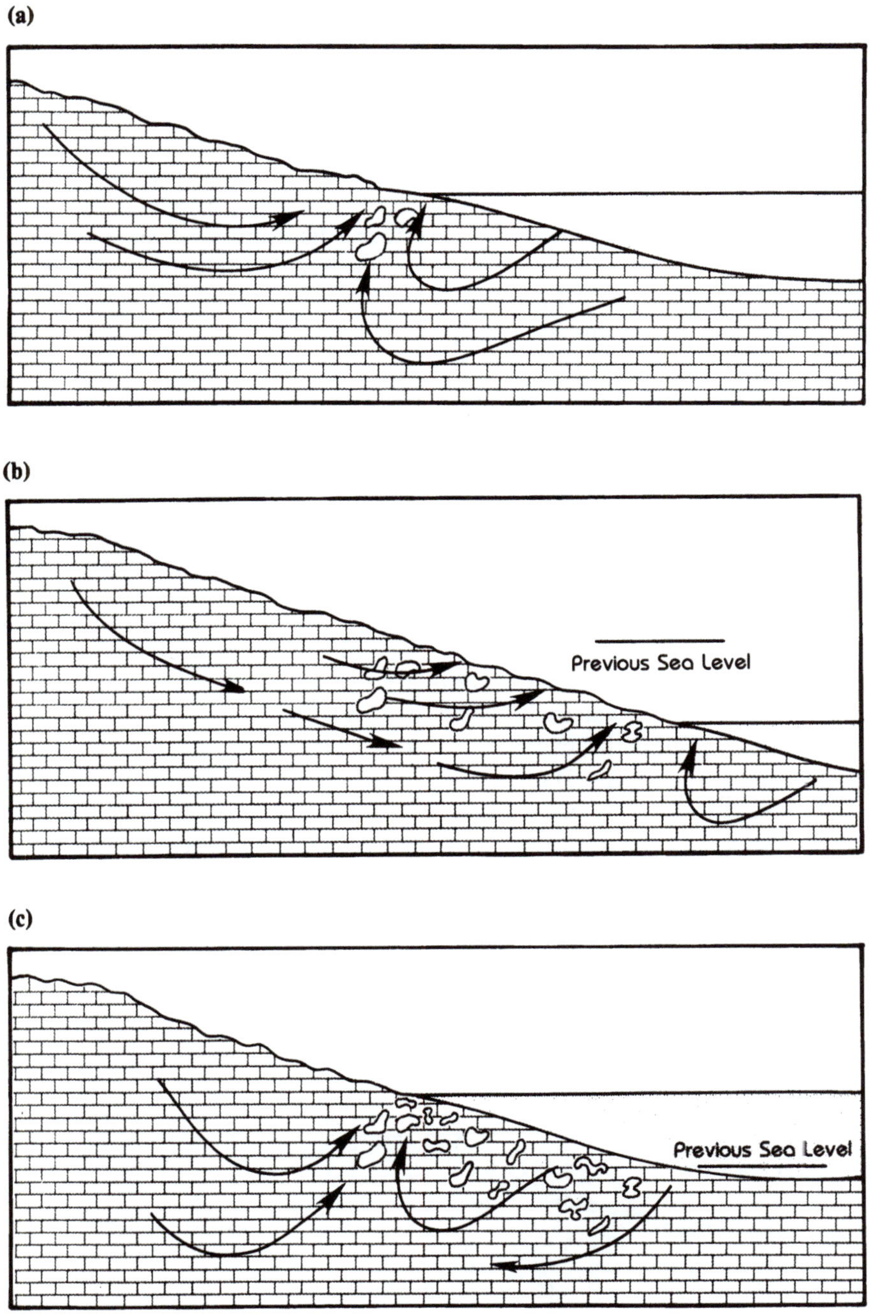

Figure 12.8 Sketch showing increase in dissolution of limestone aquifer in response to sea-level changes. (a) Initial sea-level stage. (b) Zone of dispersion moves through aquifer as sea level lowers. (c) Limestone is subjected to second period of dissolution as sea level rises.

respect to calcite. The only dissolution occurring in the caleta is in the southern "arm" where water is subsaturated with respect to calcite. This is also the area of greatest discharge into the lagoon where water flows through a surface channel heavily forested with mangroves that probably supply carbon dioxide to the water.

Chemical modeling indicates that dissolution is not occurring throughout the lagoon owing to rapid outgassing of carbon dioxide, which causes the water to become supersaturated. Therefore, dissolution of limestone was postulated to be occurring within the aquifer before the groundwater discharges into the lagoon (Fig. 12.9). Using the subsurface dissolution as our conceptual model, calculations were made (Hanshaw & Back 1980b) that demonstrated the vital importance of dissolution in the brackish-water zone. Freshwater, some of which had travelled for more than 100 km within the freshwater zone, had dissolved ~2.5 mmoles of calcite per liter of water, whereas in the zone of mixing nearly half that amount, or 1.2 mmoles of calcite was dissolved in an area only 1 km wide. Even though these calculations were based on field evidence, the conclusions were conjectural because of the inability to observe the dissolution that was calculated to be occurring in the mixing zone.

The opportunity for observation of dissolution was provided in the cave of Xcaret that had been explored by speleologists (Exley 1980). They had observed the mixing zone, i.e. a layer of fresher water ~2 m thick floating on top of the more saline water within the cave. We then undertook another field investigation to map the chemistry of the water in the cave and in other lagoons along the coast (Hanshaw & Back 1980a; Back *et al.* 1981). The calculations and interpretation based on the chemical analyses demonstrate that the water is, indeed, subsaturated with respect to calcite; the differential dissolution is readily observable within the cave. Above the water level, the walls of the cave are quite smooth with no effect of differential

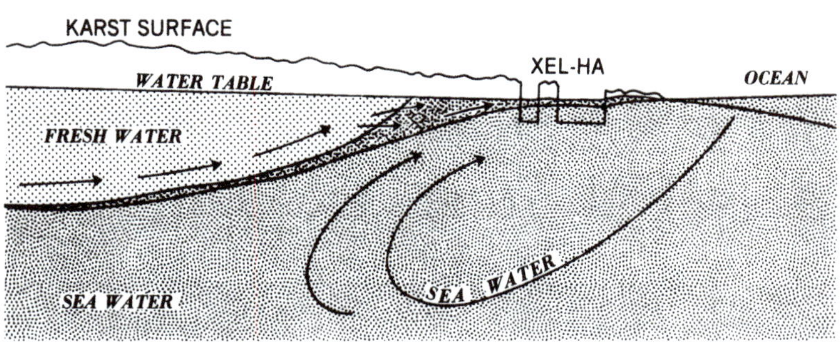

Figure 12.9 Diagrammatic cross section showing discharge of brackish water from zone of dispersion at Xel Ha (from Back *et al.* 1979).

dissolution. Within the mixing zone, however, the differential dissolution is quite dramatic and produces extremely high porosity, which causes much of the limestone to have a general appearance of Swiss cheese (Fig. 12.10).

Based on these theoretical considerations and field observations, we hypothesize that the evolutionary development of the coastline is as follows. After fracturing of the limestone, groundwater that was flowing from the inland part of the Peninsula was channeled along the dominant fractures in its effort to discharge along the coast. This flowing water mixed with the seawater that had encroached into the aquifer. The mixing formed a zone of dispersion from which brackish water discharged. The major flow, and therefore, the major dissolution, occurred along fractures with widest openings; this led to establishing major discharge points along the coast where a dominant fracture intersected it. These discharge points were the location of incipient coves (Fig.

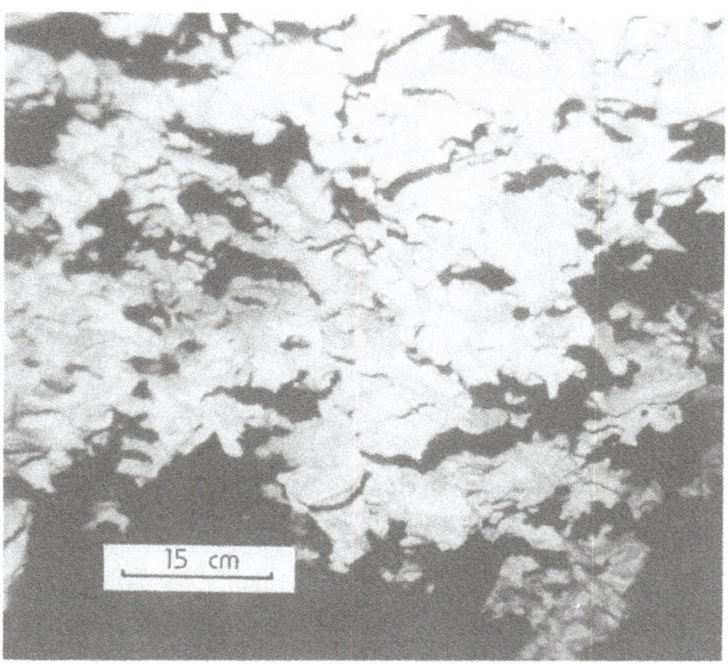

Figure 12.10 Underwater photograph of the "swiss cheese" rock showing development of extreme porosity.

12.11). The area of discharge and groundwater flow pattern became a self-perpetuating mechanism that expanded the area of dissolution to form a branching network of subsurface solution channels that coalesced to form a cave system whose orientation was controlled by the fracture pattern. As the dissolution continued, it removed support for the 1 – 2 m thick slab of limestone, the bottom of which is the cave roof and the top of which is land surface. Eventually, there remained no support for the roof of the cave, and it collapsed, as at Yalku (Fig. 12.12). This process is occurring at the present time. The roof blocks submerge into the zone of dispersion where they are dissolved as the caleta is enlarged. This enlargement opens the lagoon to wave action that permits physical erosion and provides a habitat for marine organisms that cause further erosion by biological activity. These processes continue until the lagoon is an open body of water separated from adjacent lagoons by a headland that is gradually eroded by wave action until the coastline develops a serrated configuration resulting from the coalescence of the many crescent-shaped beaches (Fig. 12.13).

We believe that throughout geologic time this mixing-zone dissolution has played a major role in development of porosity and provides an alternative hypothesis for many of the features that have been referred to as "paleokarst," generally attributed to subaerial erosion as a function of sea-level change. We

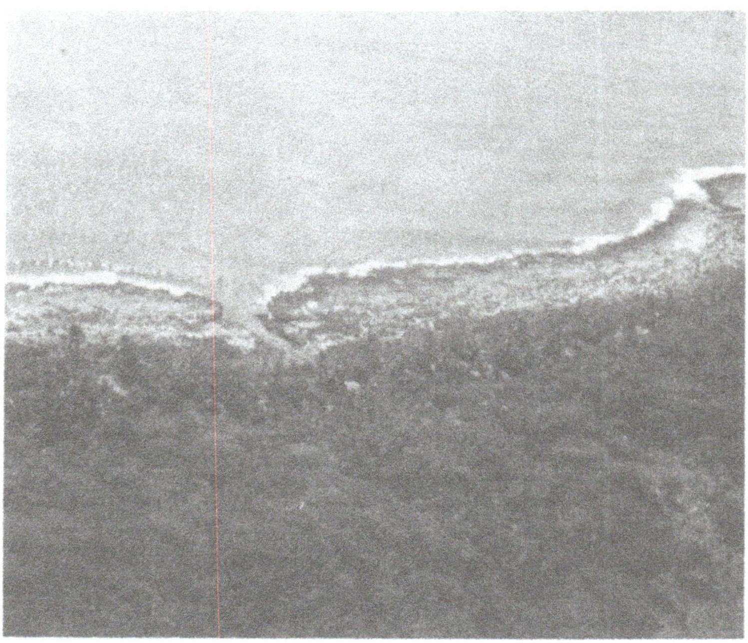

Figure 12.11 Early stage of development of cove by dissolution and discharge of groundwater. Mouth of cove is 20 m wide.

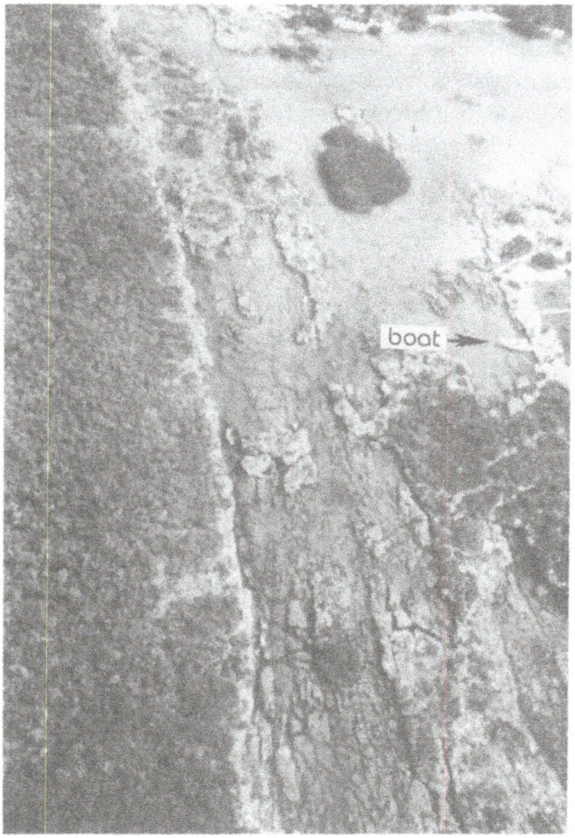

Figure 12.12 Aerial photograph of collapsed roof blocks of the cave at Yalku (10 m boat is shown for scale). Cave development is controlled by the dominant northeast-trending fracture system. Darker roof blocks are submerged. This cave system extends beyond the caleta and is traversable with scuba at least 100 m in the area beyond the top of the picture.

Figure 12.13 Advanced stage of coastline development before the coalescence of the crescent-shaped beaches.

further expect that the dissolution phenomena occur in many other limestone coastal areas and will be recognized as a major geomorphic agent when other investigators study the development of porosity and chemistry of discharging groundwater in coastal caves and lagoons.

Acknowledgements

Many of the ideas expressed here have been developed by numerous discussions in the field with A. E. Weidie, University of New Orleans, who first introduced us to the geology and other fascinations of the Yucatan, and William C. Ward, whose patient and enthusiastic encouragement has sustained our interest. Juan Lesser, formerly of Secretaria de Recursos Hidraulicos, now president of Lesser and Associates, Inc., has long made significant contributions in many ways. We thank Leonard F. Konikow and Waite Osterkamp, United States Geological Survey, for their review and clarification of the manuscript. Martina Johnson prepared the illustrations, assisted in library research and editing. Joanne Taylor, with her usual cheerful attitude, typed the manuscript. The work is a continuation of that supported by National Science Foundation and Consejo Nacional de Ciencia y Tecnoloǵia of Mexico.

References

Back, W. and R. A. Freeze (eds) 1983 *Benchmark papers in geology*, vol 73: chemical hydrogeology. Stroudsburg, Pa.: Hutchinson & Ross.

Back, W. and B. B. Hanshaw 1965. Chemical geohydrology. In *Advances in hydroscience*, vol. 2, V. T. Chow (ed.), 49–109. New York: Academic Press.

Back, W., B. B. Hanshaw, T. E. Pyle, L. N. Plummer and A. E. Weidie 1979. Geochemical significance of groundwater discharge and carbonate solution to the formation of Caleta Xel Ha, Quintana Roo, Mexico. *Water Resour. Res.* **15**(6), 1521–35.

Back, W., B. B. Hanshaw, J. N. Van Driel, W. C. Ward and E. J. Wexler 1981. Chemical characterization of cave, cove, caleta and karst creation in Quintana Roo. *Geol Soc. Am. Abs. Prog.* **13**(7), 400.

Badon Ghyben, W. 1889, Nota in verband met voorgenomen put boring nabij Amsterdam. *Koninkl. Inst. Ing. Tijdschr. 1888–89*, 21.

Carlston, C.W. 1963, An early American statement of the Badon Ghyben–Herzberg principle of static fresh-water–salt-water balance. Am. J. Sci. **261**, 88–91. Also in Back & Freeze (1983).

Cooper, H. H., Jr. 1959. A hypothesis concerning the dynamic balance of fresh water and salt water in a coastal aquifer. *J. Geophys. Res.* **64**(4), 461–7.

Exley, S. 1980. Diving beneath the Mayan city of Xcaret. *Caving Int. Mag.* **8**, 38–40.

Hanshaw, B. B. and W. Back 1979. The major geochemical processes in the evolution of carbonate-aquifer systems. *J. Hydrol.* **43**, 287–312.

Hanshaw B. B. and W. Back 1980a. Chemical reactions in the salt-water mixing zone of carbonate aquifers. *Geol Soc. Am. Abs. Prog.* **12**(7), 441–2.

Hanshaw, B. B. and W. Back 1980b. Chemical mass-wasting of the northern Yucatan Peninsula by groundwater dissolution. *Geology* **8**, 222–4.

Hanshaw, B. B., W. Back and R. G. Deike 1971. A geochemical hypothesis for dolomitization by groundwater. *Econ. Geol.* **66**(5), 710–24.

Herzberg, A. 1901. Die Wasserversorgung einiger Nordseebader. *Gasbeleucht. Wasserversorg. Jahrb.* **44**.

Hubbert, M. K. 1940. The theory of ground water motion. *J. Geol.* **48**, 785–944.

Knauth, L. P. 1979. A model for the origin of chert in limestone. *Geology* **7**, 274–7.

Kohout, F. A. 1960. Cyclic flow of salt water in the Biscayne aquifer of southeastern Florida. *J. Geophys. Res.* **65**(7), 2133–41. Also in Back and Freeze (1983).

Plummer, L. N. 1975. *Mixing of seawater with calcium carbonate groundwater: quantitative studies in the geological sciences*, Geol Soc. Am. Mem. 142, 219–36.

Runnels, D. D. 1969. Diagenesis, chemical sediments, and the mixing of natural waters. *J. Sed. Pet.* **39**(3), 1188–201. Also in Back & Freeze (1983).

13 Karst landform development along the Cumberland Plateau escarpment of Tennessee

Nicholas C. Crawford

Introduction

This study is an investigation of groundwater as a geomorphic agent in karst landform development along retreating escarpments where carbonate rock is overlain by less soluble and less permeable caprock. Of primary importance is the relationship between subterranean stream invasion, conduit cavern development and slope retreat. It is believed that karst processes have played an important and largely unrecognized geomorphic role in the erosion of extensive areas of the eastern United States.

Large areas of the eastern United States are presently capped, or have been capped in the past, by a thick sequence of Pennsylvanian and late Mississippian sandstones, conglomerates and shales. Of particular interest is the Pennsylvanian caprock that overlies, or has overlain, the Mississippian and older carbonates of the Interior Low Plateaus, Appalachian Plateaus, and Valley and Ridge physiographic provinces. It appears that a major factor in the geomorphic history of these provinces has been, and still is to a large degree, a story of caprock removal. Over large areas, the resistant caprock has been completely removed, exposing chemically less resistant limestones while in other areas the caprock continues to protect the erosionally weaker limestones beneath. The most extensive area lies in the southern Appalachian Plateaus Province, the Cumberland Plateau of Tennessee.

Cumberland Plateau

All of middle Tennessee was at one time capped by a thick sequence of Pennsylvanian sandstones, conglomerates and shales. Over large areas, the resistant caprock has been completely removed, exposing chemically less resistant limestones. Only in the Cumberland Plateau area does the caprock continue to protect the underlying limestones from rapid solution (Fig. 13.1). The plateau surface is an area of gentle relief with a general elevation of ~550 m. The Pennsylvanian caprock is underlain by Mississippian carbonates that are much less resistant to solution. The difference in lithology has resulted in an escarpment of ~300 m that rims the plateau.

Although the concept of dynamic equilibrium may not be applicable to all areas, it does appear to fit quite well in the Cumberland Plateau, Highland

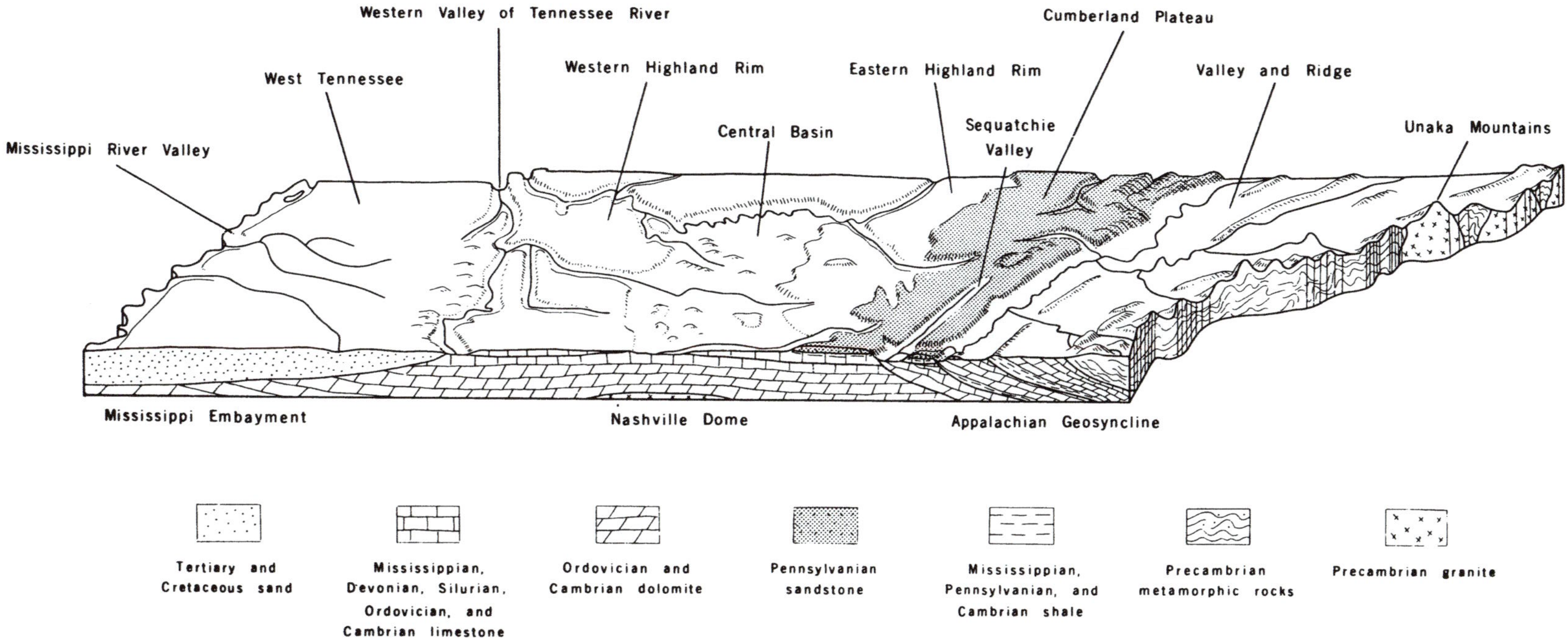

Figure 13.1 Cumberland Plateau of Tennessee (modified from Miller 1974).

Rim Low Plateau and Central Basin areas of Tennessee (Fig. 13.1). The present topographic forms appear to be in adjustment with present-day erosional processes and geologic framework (Hack 1966). According to the dynamic equilibrium concept, the present topography forms by the continuous lowering of the surface by erosion, a process that involves slope retreat on beds of different resistance. The origin of the western escarpment of the Cumberland Plateau, therefore, appears to be associated with the removal of the Pennsylvanian caprock from the Nashville Dome, a structural high along the Cincinnati Arch. The excavation of the Nashville Dome may have begun during the middle Mesozoic era. Once the resistant Pennsylvanian sandstones were removed by erosion from the central part of the structure, exposing the underlying Mississippian limestones, slope retreat by sapping began. This breaching of the once continuous expanse of Pennsylvanian caprock formed an escarpment and initiated its subsequent retreat in all directions away from the dome. Erosion continued both downward and outward in the area of the dome and a plain-like surface developed upon the cherty, erosionally resistant, lower Mississippian rocks that formed the floor of the expanding Central Basin by the late Cretaceous (Miller 1974).

The resistant Mississippian Fort Payne Formation was breached by erosion during the Tertiary and Quaternary periods, thus exposing the underlying Ordovician limestones. The breaching of the Fort Payne caprock resulted in the Highland Rim escarpment, which is presently retreating as the Central Basin expands. Sapping of the underlying limestones is primarily responsible for the steep slope angles along both the Highland Rim and Cumberland Plateau escarpments. The escarpments are retreating down dip toward the southeast, away from the Nashville Dome.

Along the Cumberland Plateau escarpment, one finds impressive examples of the weathering and erosional processes associated with caprock removal. The Cumberland Plateau caprock is being destroyed both vertically and horizontally. Vertical downwasting of the surface of the plateau results from chemical and mechanical weathering and erosion, the weathered material being removed by streams that descend abruptly from the tableland plateau. Along the edge of the plateau where the underlying Mississippian limestones are exposed, the caprock is being eroded by slope retreat. The limestones, highly vulnerable to chemical solution, are eroding chemically at a rate appreciably greater than that of the overlying non-carbonate caprock, resulting in a steep escarpment. This rapid chemical solution of the lower limestones forces the silicious caprock to erode mechanically, often in the form of massive rockslides. Thus by sapping the escarpment retreats, the caprock is removed, and the plateau area is reduced in size.

The eastern escarpment of the Cumberland Plateau is an eroded fault escarpment and, therefore, was not included in this investigation. The western escarpment has a narrow band of karstic topography; sinking streams, karst valleys, large springs and caves are abundant in this region. A sinkhole plain, usually 10–15 km in width, is located at the base of the escarpment (Fig. 13.2).

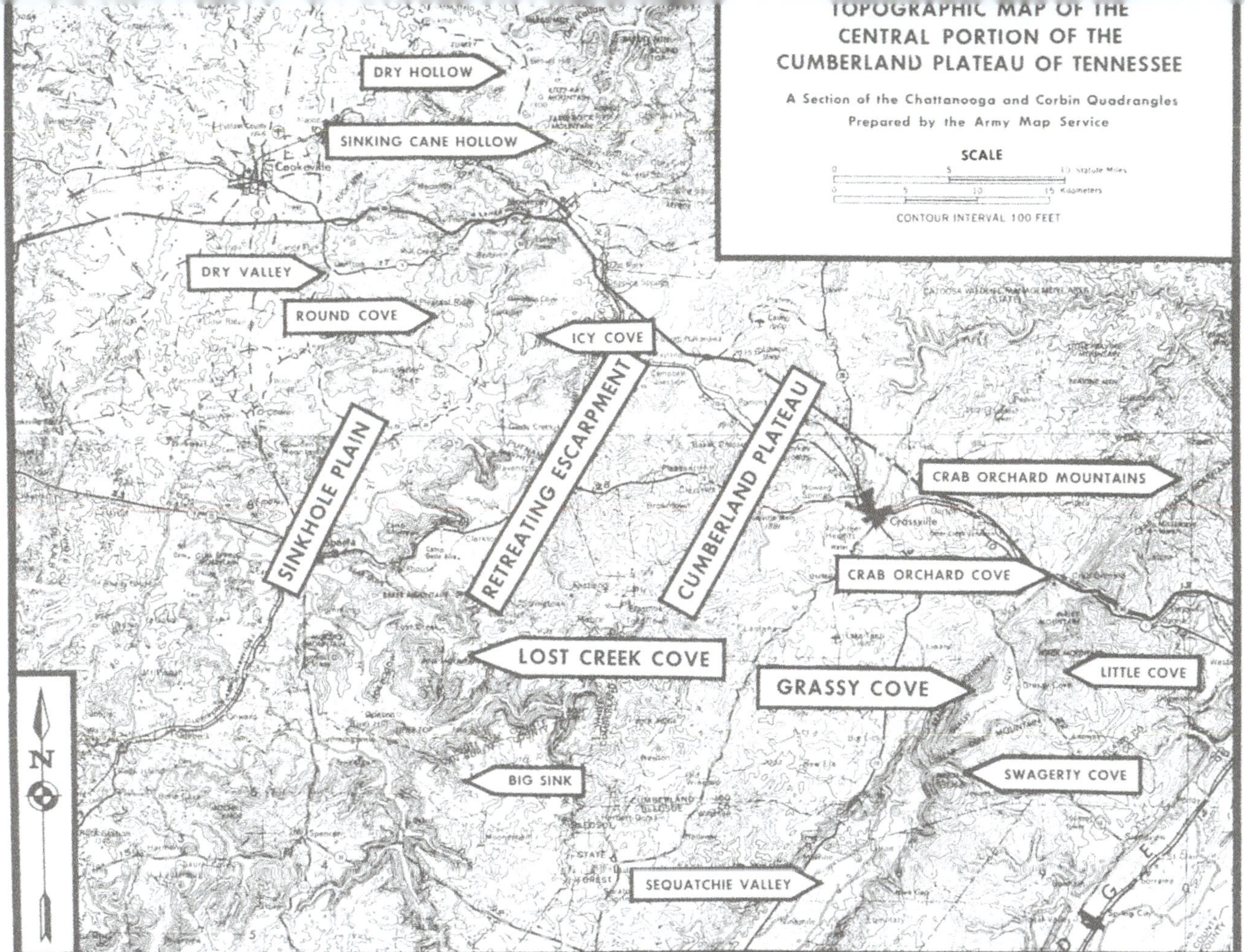

Figure 13.2 Topographic map of the central portion of the Cumberland Plateau of Tennessee showing the sinkhole plain, the retreating escarpment and the caprock plateau. Lost Creek Cove, Grassy Cove and other karst valleys are also indicated.

Of primary importance in this investigation is a hypothesized relationship between subterranean stream invasion, conduit cavern development and slope retreat. Conduit caves have at least one swallet, where a surface-flowing stream sinks, and at least one resurgence, where the stream resurfaces as a spring. The discharge of subsurface streams is usually increased by percolation water, but the caves are primarily conduits through which subsurface streams flow from swallets to resurgences. It is postulated that conduit caves along the Cumberland Plateau escarpment result primarily from subterranean invasion of surface streams flowing off the plateau and that they are directly related to caprock removal by slope retreat.

Subterranean stream invasion, conduit cavern development and slope retreat

A schematic model of the hypothesized relationship between slope retreat and conduit cave systems along the western escarpment of the Cumberland Plateau is presented in Figure 13.3. Subterranean stream invasion occurs along the edge of the retreating escarpment where surface streams leave the sandstone and shale caprock plateau and flow onto the Bangor Limestone. Subterranean stream invasion often occurs at or near the contact between the overlying shales and underlying carbonates as indicated in Figure 13.3. Invasion is initiated as some of the stream begins to flow into and through the joints and bedding planes of the Bangor Limestone, resurfacing on top of the impermeable Hartselle Formation halfway down the escarpment. As the stream water flows through the Bangor, it gradually enlarges the most efficient route through the joints and bedding planes by corrosion, thus creating a cave. As the subsurface conduit is gradually enlarged, more and more of the stream begins to flow through until finally the entire stream is diverted underground. With enlargement of the subsurface conduit, suspended and bed loads begin to travel through the system with the subsurface stream, further enlarging the cave by abrasion.

The Hartselle Formation is very resistant to erosion and virtually impermeable. Primarily a sandstone, it is locally calcareous with lenses of shale near the top and bottom. The Hartselle tends to be a structural elevation control for stream caves that form in the overlying Bangor Limestone. Cave passages form along the bedding plane that separates the Bangor Limestone from the Hartselle, but only if the dip is toward the escarpment. If the dip is away from the escarpment, the cave passage will form above the Bangor – Hartselle contact except at the point where the cave stream breaches the Hartselle. The Hartselle Formation thus acts as a "spillover" layer and, therefore, as an elevation control for cave development in the overlying Bangor Limestone.

Where the strata are horizontal, the cave passage varies back and forth from joint to bedding plane in the Bangor Limestone, and the Hartselle Formation forms the floor of the cave only where it is breached by the cave stream at or near the edge of the escarpment (Fig. 13.3). Because the stream cannot have

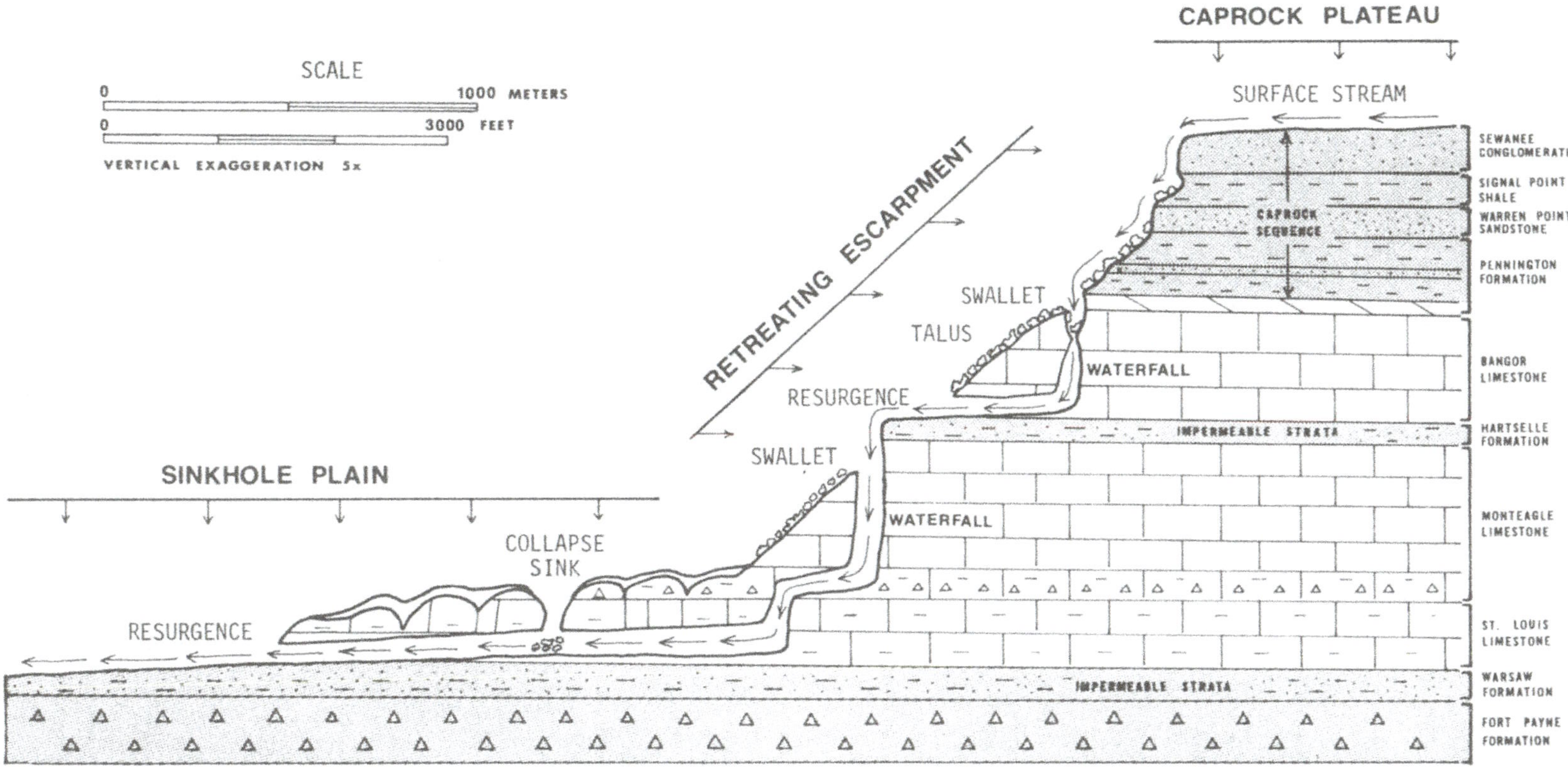

Figure 13.3 Schematic model of subterranean stream invasion and slope retreat along the Cumberland Plateau escarpment of Tennessee.

zero gradient, which would be the case if it flowed directly on top of the horizontal Hartselle Formation, it will establish the lowest possible gradient above the Hartselle in the Bangor Limestone. Although the Hartselle will remain at constant elevation as the escarpment retreats, the control point (where the stream drops off the Hartselle) will be moving upstream. This will cause the stream to lower its bed by downcutting or possibly to erode a lower passage. It is believed that in almost all cases the impermeable Hartselle Formation controls the elevation of the stream caves in the overlying Bangor Limestone.

The erosionally resistant Hartselle Formation usually forms a structural bench about halfway down the escarpment. Here, a resurgence of an invading caprock stream sometimes occurs as depicted in the model (Fig. 13.3). The stream flows for a short distance as a surface stream before dropping off the Hartselle into a sink or vertical shaft, which it has enlarged in the Monteagle Limestone below. Near the base of the Monteagle are resistant and relatively impermeable layers of chert and shale that appear to be the control for the numerous large cave systems in this formation. The resurgences of many of these subsurface streams occur near the base of the Monteagle Limestone. However, in other cases, subsurface streams breach this control and drop into the St. Louis Limestone below, often resurging on top of chert, shale or dolomite layers. The top of the Warsaw Formation appears to be the lowest control for subsurface streams in the overlying St. Louis Limestone. The Warsaw is a sandy limestone and has thin zones of shale throughout. It leaches to a sandstone and often outcrops as a sandy bench along major streams at the foot of the escarpment. However, in some areas, a relatively pure limestone occurs at the top of the Warsaw Formation, and in these areas, the control is usually about 20 m below the top of the Warsaw. Therefore, the resurgence at the base of the escarpment will occur somewhere between the lower Monteagle and upper Warsaw limestones. The resurgence is normally perched upon impermeable strata and not concordant with the surface-flowing streams, which reveal the elevation of the water table at the base of the escarpment. The ability of a subsurface stream to breach the several shale and chert control layers depends upon: (a) stratigraphic variation in the thickness and permeability of the control layers, (b) the size of the subsurface stream, (c) the extent to which surface streams at the base of the escarpment have lowered their channels.

In Figures 13.2 and 13.3 the area depicted as the sinkhole plain begins at the base of the retreating escarpment and extends outwards for several kilometers. It is hypothesized that it is a product of caprock removal by slope retreat and that it is left behind as the escarpment retreats. Rather small dendritic caves characterize the subsurface hydrologic environment of the sinkhole plain; they are fed by percolation and by short ephemeral streams that flow over the terra rossa cover of the plain into swallets after hard rains. The aggressive water of these sinking streams probably plays a major role in the development of the caves. The sinkhole plain is being lowered by vertical downwasting associated with corrosion along the terra rossa–bedrock interface and by sinkhole collapse into the numerous small streams that underlie the plain.

It is believed that the terra rossa which covers the sinkhole plain has formed from the weathering of detritus from the caprock plateau deposited at the base of the escarpment, combined with residual material resulting from the weathering of the lower Monteagle and St. Louis limestones. It appears that deposition of caprock material at the base of the Cumberland Plateau escarpment is an integral part of the process of slope retreat. Extensive areas of Quaternary alluvium and colluvium from the caprock cover the sinkhole plain near the escarpment. Alluvial fans are deposited by small streams, often ephemeral, at the escarpment base. The remains of rockslides where massive sandstone and conglomerate boulders fell, rolled and slid down the escarpment in the past are evident on the sinkhole plain as far as 1 km from the present escarpment. It is postulated that this caprock detritus cover weathers into terra rossa and is lowered as the underlying Monteagle and St. Louis limestones are eroded by solution along the terra rossa–bedrock interface. Less soluble materials, particularly chert, are thus incorporated into the gradually subsiding terra rossa. Vertical downwasting proceeds as the St. Louis Limestone is progressively destroyed by chemical solution, thus lowering the terra rossa onto the Warsaw Formation. A surface drainage system replaces the subsurface system of the sinkhole plain as the terra rossa is lowered onto the upper Warsaw. In this fashion, new sinkhole plain is continuously developing at the base of the retreating escarpment while old sinkhole plain, now several kilometers from the escarpment, is being destroyed by vertical downwasting. Thus, the sinkhole plain, a product of slope retreat and caprock removal, follows the retreating escarpment. This investigation dealt with the role played by karst processes along the retreating escarpment, and an investigation of the origin of the sinkhole plain has only just begun. However, preliminary findings indicate that (a) discernible caprock deposits diminish with distance away from the escarpment and (b) residual chert in the terra rossa increases with distance away from the escarpment.

Ruhe and Olson (1980) and Olson *et al.* (1980) have studied the terra rossa of the karst of southern Indiana and conclude that it is not a residual soil; Quinlan and Ewers (1981) maintain that the terra rossa in the Mammoth Cave region of Kentucky consists principally of detritus from rocks younger and older than those on which it now rests. Both locations are similar to that along the western escarpment of the Cumberland Plateau. Both have retreating escarpments capped by Pennsylvanian caprock overlying Mississippian carbonates with a terra rossa-covered sinkhole plain at the base of the escarpment. In southern Indiana and the Mammoth Cave region of Kentucky, streams flowing off the older rocks exposed along the Cincinnati Arch flow through and under the sinkhole plain and into the escarpment. These streams may be the source of alluvially deposited older rocks found in the terra rossa. Because the streams flow off younger rocks toward the Cincinnati Arch in the

Cumberland Plateau escarpment area, it is doubtful that rocks older than the underlying carbonates are a part of the terra rossa.

Subterranean stream invasion and karst valley development

The strata along the retreating Cumberland Plateau escarpment are rarely horizontal. Where the local dip is toward the escarpment, caprock removal is often accelerated by subterranean stream invasion which often occurs several kilometers in back of the retreating escarpment. Figure 13.4 is a hypothesized schematic model of subterranean stream invasion and slope retreat as it occurs in a structurally high area, such as a slight anticline, near the retreating escarpment. Slope retreat proceeds in all directions away from the site of the initial invasion, thus creating a large karst depression, called a karst valley, that is completely surrounded by caprock (Fig. 13.2).

Research design

To investigate the role of groundwater as a geomorphic agent in karst landform development along the Cumberland Plateau escarpment, two areas were selected for intensive study. The hypothesized model was tested in the Lost Creek Cove area, an area of near-horizontal structure along the western escarpment, and in the Grassy Cove area, an area of folded structure along the Sequatchie Anticline at the head of Sequatchie Valley (Fig. 13.2).

Testing the model in the field involved geologic mapping, cave mapping, determination of joint systems, and extensive dye tracing of subsurface streams. For dye tracing, automatic water samplers were built that pumped samples at predetermined intervals. The samplers were stationed at the most likely risings for the subsurface drainage system being traced. At weekly intervals after injecting Rhodamine WT dye (20% solution) into a sinking stream, the sample bottles were collected from each sampler and analyzed for dye concentration on a fluorometer. This technique not only proved subsurface drainage connections but also permitted the calculation of the peak and mean flow-through times as well.

To quantitatively investigate subsurface erosional processes, cave systems along the Cumberland Plateau escarpment in the Grassy Cove area were monitored for 15 months. The caves were treated as open systems with input of mass and energy occurring at swallets and from diffuse sources, the output occurring at risings. An attempt was made to select large but simple conduit cave systems, those with only one swallet and one rising.

Recording stream gauges were installed at the swallet and at the rising for each cave system. About 270 water samples were taken during the 15 month period at swallets and risings to investigate changes that occur as water flows through the systems. The percolation input was monitored by taking drip samples from within the caves. Data collection and analysis dealt with the solutional and suspended loads of the cave systems.

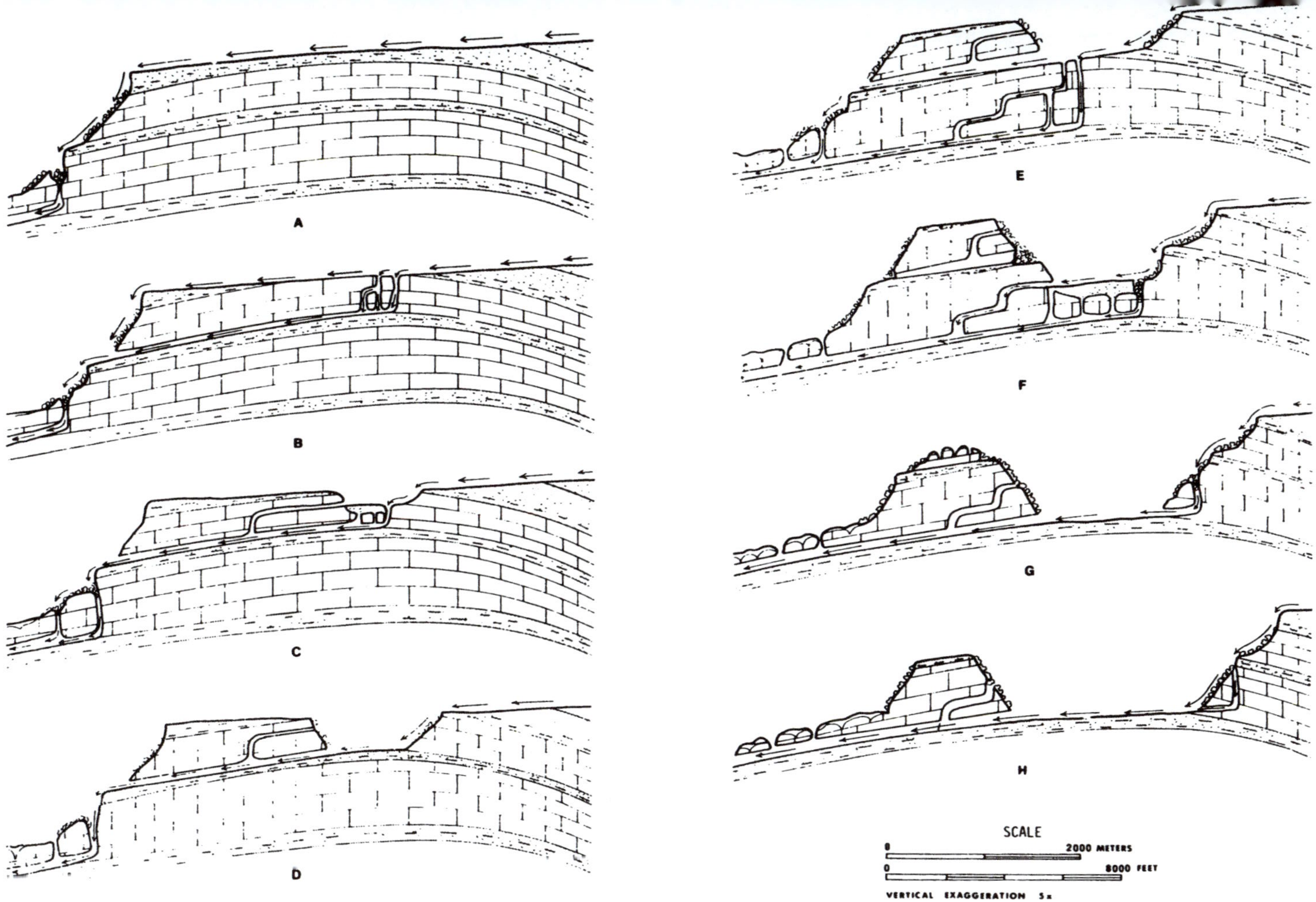

Figure 13.4 Schematic model of subterranean stream invasion and karst valley development in back of the Cumberland Plateau escarpment of Tennessee.

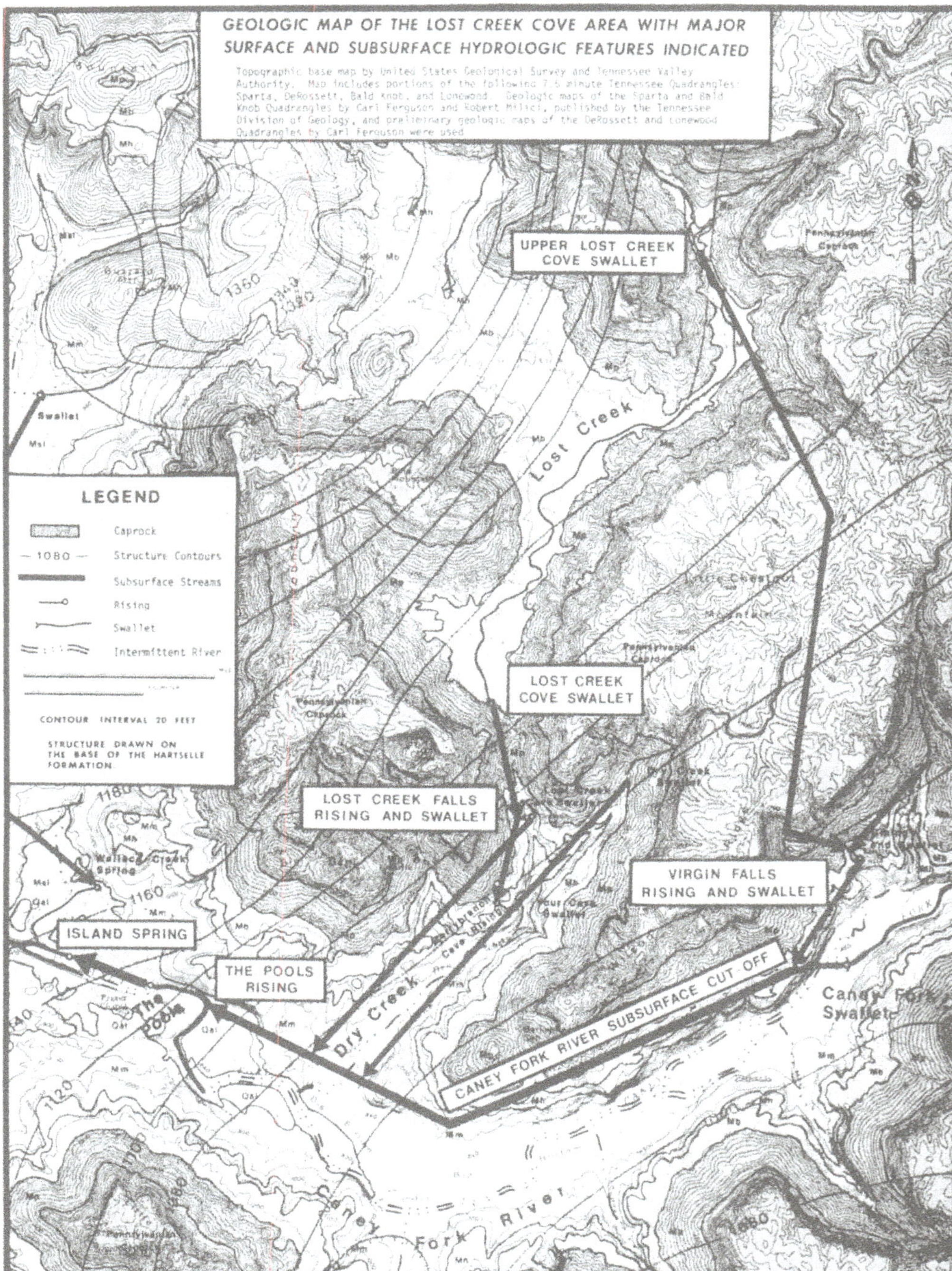

Figure 13.5 Map of the Lost Creek Cove area showing relationships between subterranean stream invasion, topography, geologic structure and stratigraphy, surface and subsurface drainage and conduit cavern development.

Lost Creek Cove area, White County, Tennessee

Subsurface drainage of Lost Creek

Investigation of the Cumberland Plateau escarpment in the Lost Creek Cove area revealed an extremely complex subsurface drainage system (Fig.13.5). Dye traces showed that Lost Creek, which sinks at the southern end of Lost Creek Cove, divides underground with most of its discharge resurging at Upper Dodson Cave where it drops off the Hartselle Formation at Lost Creek Falls (Fig. 13.6) into a large, collapsed sinkhole in the underlying Monteagle Limestone and then flows into Lost Creek Cave. The other branch of the subsurface Lost Creek resurges from Merrybranch Cave, drops off the Hartselle Formation, and flows into Your Cave (Sims 1973). During high discharges, some of the water flows past the Your Cove swallet and flows on down the escarpment into Dry Creek, thus taking a surface route to the Caney Fork River at the base of the escarpment. The two branches of Lost Creek take

Figure 13.6 Lost Creek Falls. After issuing from the mouth of Upper Dodson Cave, the stream flows 75 m perched upon the Hartselle Formation and drops about 25 m into Lost Creek Sink where it flows directly into Lost Creek Cave.

"stair-step" routes, as indicated by a 20 m waterfall in Lost Creek Cave, down through the Monteagle and St. Louis limestones, crossing one resistant layer of limestone, dolomite, shale or chert after another to join with the Caney Fork River Subsurface Cut-Off.

The Caney Fork Subsurface Cut-Off is a water-filled conduit through which much of the Caney Fork River flows after sinking upstream to The Pools. During dry periods, usually throughout the summer and fall months, the entire river flows underground, and its 40 m wide stream bed is dry downstream for the next 6.7 km (Fig. 13.7). The Pools appear to have formed by the collapse of the ceiling of the Caney Fork Subsurface Cut-Off. Except during high discharges, most of the water of the Caney Fork River takes a subsurface route from The Pools to Island Spring.

Figure 13.7 The dry bed of the Caney Fork River. During the dry summer and fall months, the entire Caney Fork River sinks and flows through a subsurface cut-off, resurging at The Pools and again at Island Spring. This results in a dry, surface channel ~40 m wide for over 6.7 km.

The investigation also revealed that some of the water of Lost Creek sinks into a swallet in the northern part of Lost Creek Cove, flows under 5 km of caprock, and resurges from Virgin Falls Cave (Crawford *et al.* 1983). The stream drops off the Hartselle Formation at Virgin Falls (Fig. 13.8) into a large collapse sink in the Monteagle Limestone. It then flows through Virgin Falls Pit Cave into the Caney Fork River Subsurface Cut-Off to resurge with it at The Pools and finally at Island Spring. Island Spring, located directly on the St. Louis–Warsaw contact, is the final resurgence for all Lost Creek water dye traced as of February 1982.

When one compares the subsurface route taken by Lost Creek and other streams with the geologic structure (Fig. 13.5), it is obvious that they are flowing down dip although not down the true dip. The actual course taken appears to correlate well with the local jointing pattern. Joints were mapped in Virgin

Figure 13.8 Virgin Falls. After issuing from the mouth of Virgin Falls Cave, the stream flows 50 m perched upon the Hartselle Formation and falls about 40 m into a large collapse sink in the Monteagle Limestone where it immediately sinks into Virgin Falls Pit Cave. The photo shows only the top half of the 10 m wide waterfall.

Falls Cave and Mill Hole Cave and superimposed upon maps prepared by Sims and others, and by Matthews, Benedict and Craig (Matthews 1971). The caves follow the jointing, switching from one joint set to another causing each cave to resemble the letter "Z". Both caves are in the Bangor Limestone and floored by the resistant Hartselle Formation. The streams responsible for their formation have been controlled to a large degree by jointing while flowing down dip perched upon the Hartselle. Other caves such as Merrybranch and Your Cave have also been greatly influenced by joint orientation.

Source of other cave streams in the Lost Creek Cove area

Not all of the caves in the Lost Creek Cove area were created by subterranean stream invasions. Instead, many caves result from small seeps that flow beneath the talus cover down the slope from the edge of the perched caprock aquifer and into the joints and bedding planes of the Bangor Limestone. These streamlets are believed to be highly aggressive to calcium carbonate because they have not had previous contact with carbonate rock before reaching the Bangor Limestone. Geochemical analysis, specifically the Stenner aggressivity test (Stenner 1971), of such springs issuing from the perched caprock aquifer in the Grassy Cove area of the Cumberland Plateau revealed this to be the case. Therefore, water issuing from the perched sandstone aquifer of the Cumberland Plateau is somewhat comparable to swallet water from streams that flow off the plateau, in terms of aggressivity to calcium carbonate. Of course, caprock streams flowing into swallets would have the advantage of being very aggressive during floods and thus better equipped for dissolving large conduit caves. However, numerous caves appear to have formed along the Cumberland Plateau escarpment due primarily to the action of aggressive water that issues without surfacing from the perched water table of the plateau. These caves do not have swallets, but rather receive diffuse input from vertical shafts.

Many caves of this type have formed just beneath the outcrop of the Hartselle Formation in the underlying Monteagle Limestone. Small cave streams in the Bangor Limestone often drop off the Hartselle Formation into vertical shafts in the underlying Monteagle Limestone without surfacing. Although the Hartselle Formation appears to be virtually impermeable, joints appear to open, near the edge of the escarpment, probably in response to lessening of horizontal confining pressure. Sapping of the underlying limestone, thus decreasing vertical support for the Hartselle, probably also contributes to joint enlargement. Consequently, subsurface streams are often able to penetrate the Hartselle, without surfacing, at enlarged joints sometimes 50 m or more in back of the Hartselle outcrop.

Caves in the underlying Monteagle Limestone often run parallel to the escarpment connecting a series of vertical shafts where water is dropping through the Hartselle Formation. These caves usually follow joints down the dip to a rising where a valley has cut back into the plateau. Therefore, many caves along the Cumberland Plateau escarpment do not have swallets and have not resulted from subterranean stream invasion. However, they are very similar to the caves formed by invasion in that they act as conduits that deliver aggressive water from the perched water table on the plateau to the base of the escarpment.

Conclusions

A comparison of topography, stratigraphy, structure, and surface and subsurface drainage (Fig. 13.5) in the Lost Creek Cove area reveals a very close match with the hypothesized model (Fig. 13.4). If the hypothesized model was to be accepted, the following had to be found in the Lost Creek Cove area: (a) a structural high under Lost Creek Cove, (b) a subsurface drainage system as outlined in the model. The field research, which included geologic mapping, dye tracing, surveying joint orientations and cave investigations, confirms both of these. Figure 13.5 summarizes the results of the investigation by showing the relationship between topography, surface and subsurface hydrology, conduit caves, jointing, structure and stratigraphy in the Lost Creek Cove area.

This research indicates that in the past, Lost Creek flowed upon the caprock, down a structural dome into Dry Creek, thus taking a surface route to the Caney Fork River. After lowering its stream bed to an elevation of ~ 400 m, it intersected the Bangor Limestone at a location in back of the Cumberland Plateau escarpment. Subterranean stream invasion was initiated as aggressive caprock water began to flow into and through the joints and bedding planes of the Bangor Limestone to a resurgence on the Caney Fork River at the base of the escarpment. Since that time, Lost Creek Cove has increased in size and depth, becoming a large karst valley virtually surrounded by non-carbonate rock.

At one time, a surface tributary flowing from northwest to southeast must have joined Lost Creek at about the center of the cove. Unlike the Dog Cove prong of Lost Creek Cove to the east, this prong does not have a stream (Fig. 13.5). This tributary now flows underground as the Hartselle Formation has been breached in three places. A small stream is presently flowing into each of these sinks. It is only a matter of time before Lost Creek itself will cut through the impermeable Hartselle Formation and form a subsurface conduit through the underlying Monteagle Limestone. Thus, in the distant future, it is expected that Lost Creek Cove will become an even larger and deeper karst valley floored by the resistant Warsaw Formation. Even further into the future, Lost Creek Cove will cease to be a karst valley as the retreating Cumberland Plateau escarpment continues to move eastward.

Figure 13.9 Physiographic map of the Cumberland Plateau showing Grassy Cove polje at the head of the Sequatchie Valley.

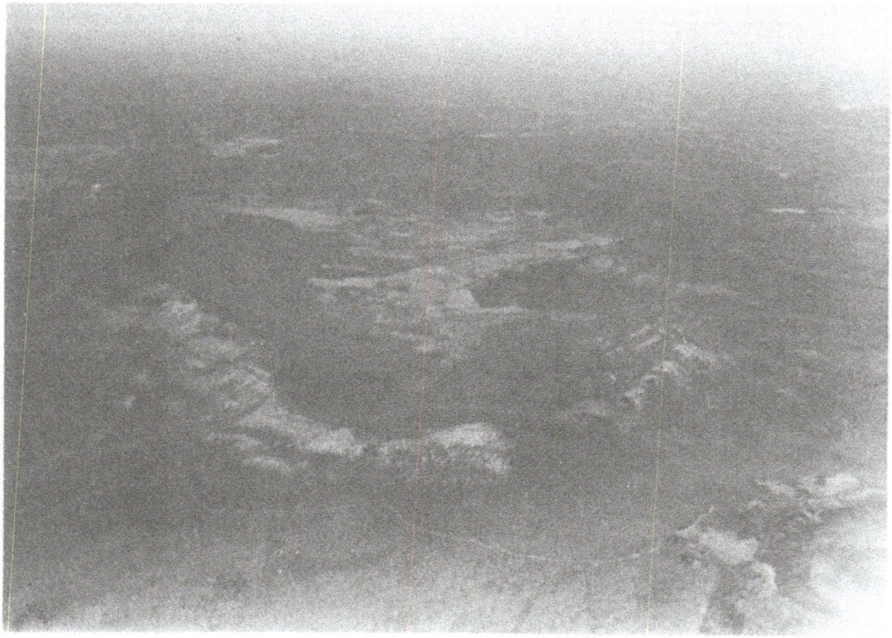

Figure 13.10 Air photo of Grassy Cove. The direction of view is toward the northeast. A light snow outlines the dip slopes of the breached anticlinal mountain.

Grassy Cove area, Cumberland County, Tennessee

Grassy Cove is a massive karst valley, or anticlinal polje, ~13 by 5 km with a flat floor 400 m lower than surrounding mountains. It is the largest karst valley on the Cumberland Plateau (Fig. 13.2 & 13.9) and possibly the largest karst depression in the United States. Grassy Cove is so large that one almost needs to view it by air to perceive that it is indeed a karst depression (Fig. 13.10). The cove is the hollowed-out center of a large anticline; the flanks of the anticline remain as mountains surrounding the cove, protruding 300 m above the surrounding Cumberland Plateau and in places well over 400 m above the floor of the cove.

Grassy Cove is drained by Cove Creek, which flows into Mill Cave on the northwest side of the cove (Fig. 13.11). From here, the stream takes a subsurface route under Brady Mountain to a large spring at the head of the Sequatchie Valley 11 km to the southwest. The spring is the beginning of the Sequatchie River and is appropriately named Head of Sequatchie.

The Sequatchie Valley (Fig. 13.9) is a deep incision ~6–8 km wide with a floor 400–500 m below the Cumberland Plateau. The precipitous walls of the canyon extend in virtually a straight line, cutting into the plateau from the southwest for over 90 km. It is a classic example of an anticlinal valley,

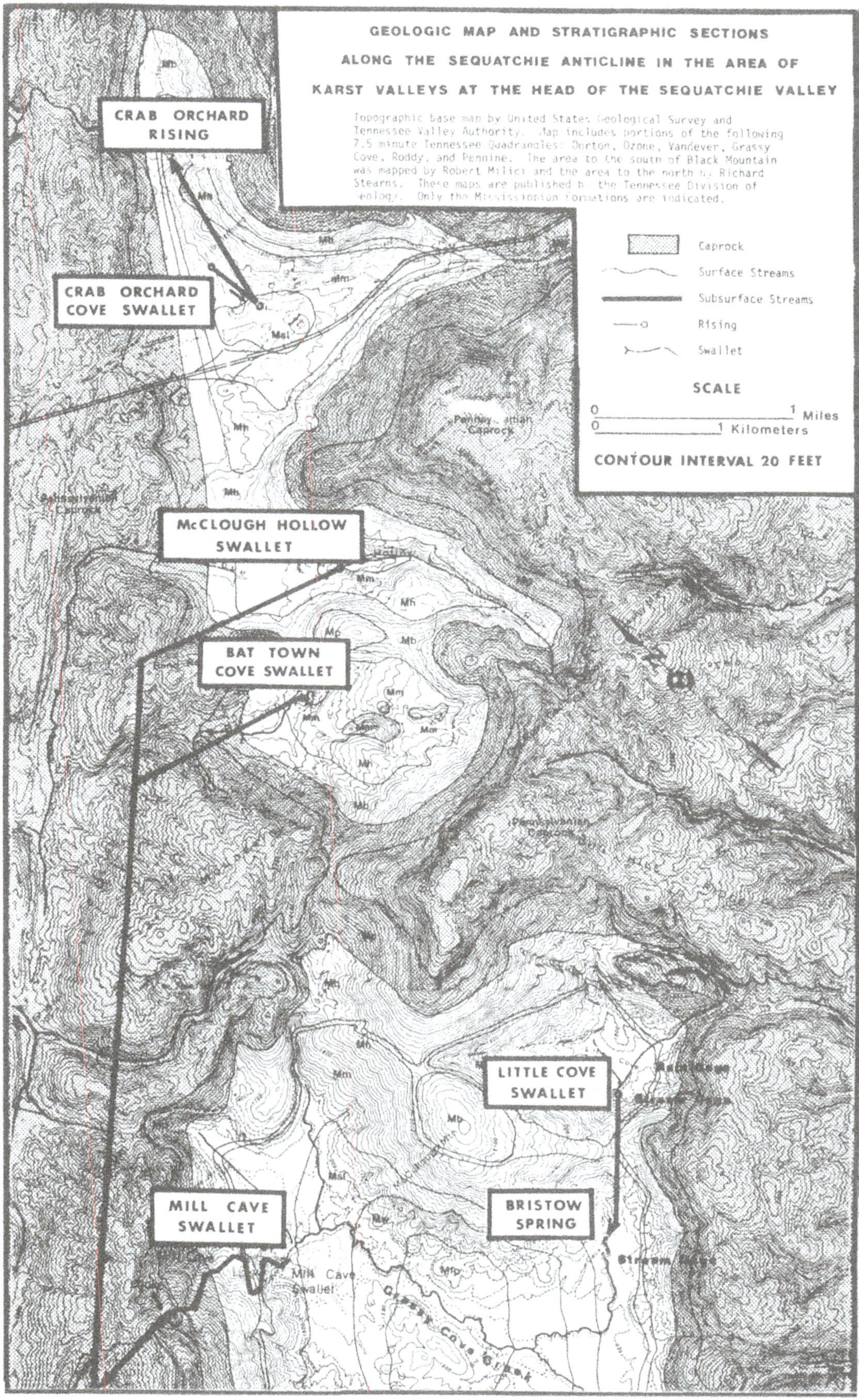
GEOLOGIC MAP AND STRATIGRAPHIC SECTIONS
ALONG THE SEQUATCHIE ANTICLINE IN THE AREA OF
KARST VALLEYS AT THE HEAD OF THE SEQUATCHIE VALLEY
Topographic base map by United States Geological Survey and Tennessee Valley Authority. Map includes portions of the following 7.5 minute Tennessee Quadrangles: Dorton, Ozone, Vandever, Grassy Cove, Roddy, and Pennine. The area to the south of Black Mountain was mapped by Robert Milici and the area to the north by Richard Stearns. These maps are published by the Tennessee Division of Geology. Only the Mississippian Formations are indicated.
Caprock
Surface Streams
Subsurface Streams
Rising
Swallet
SCALE
0
1 Miles
0
1 Kilometers
CONTOUR INTERVAL 20 FEET
CRAB ORCHARD RISING
CRAB ORCHARD COVE SWALLET
McCLOUGH HOLLOW SWALLET
BAT TOWN COVE SWALLET
LITTLE COVE SWALLET
BRISTOW SPRING
MILL CAVE SWALLET
Mill Cave Swallet
Grassy Cove Creek

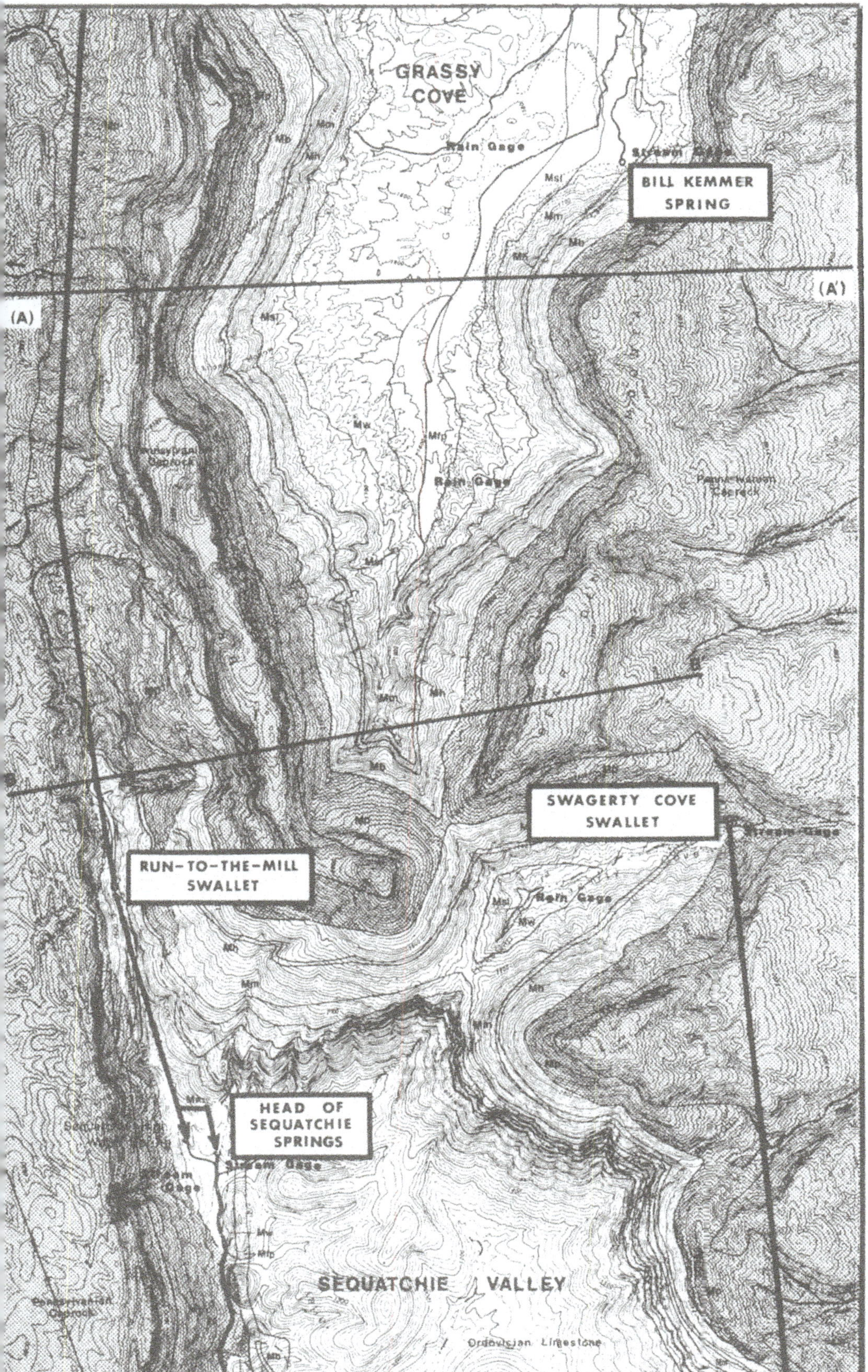
GRASSY COVE
Rain Gage
Stream Gage
BILL KEMMER SPRING
(A)
(A')
Rain Gage
SWAGERTY COVE SWALLET
Stream Gage
RUN-TO-THE-MILL SWALLET
Rain Gage
HEAD OF SEQUATCHIE SPRINGS
Stream Gage
SEQUATCHIE VALLEY
Ordovician Limestone

(b)

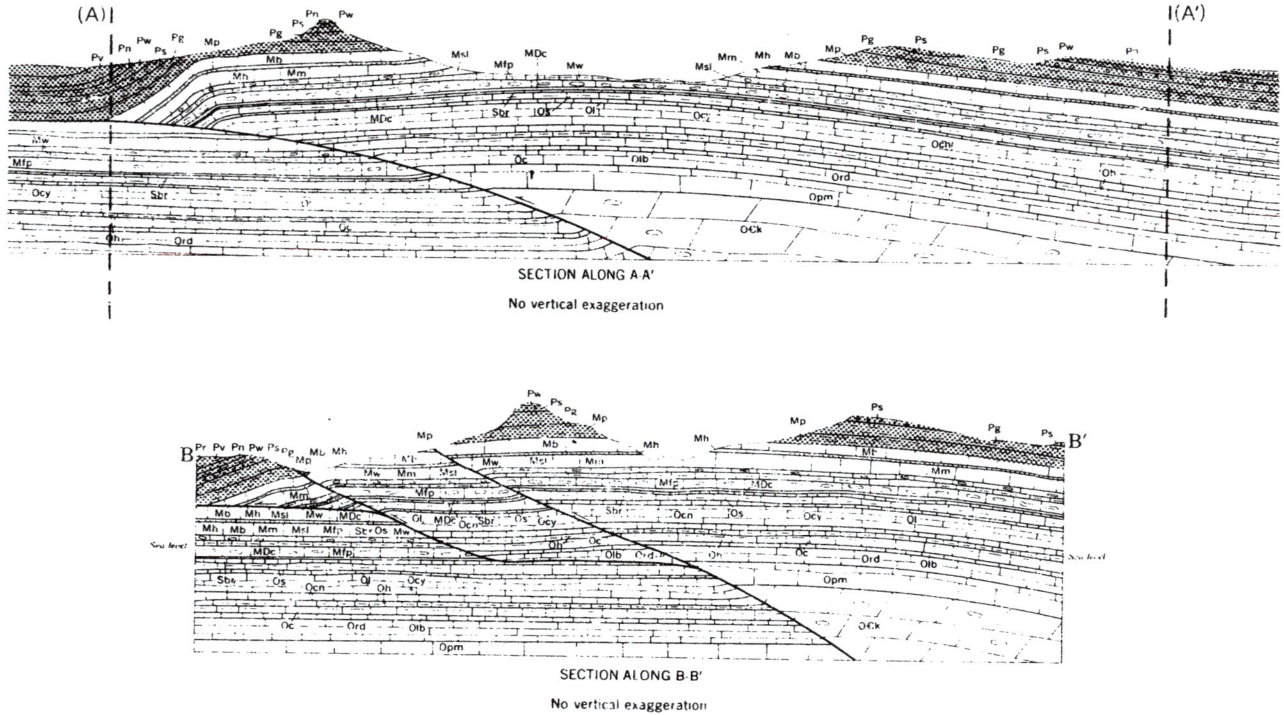

Figure 13.11 (a) Geologic map and (b) stratigraphic sections along the Sequatchie Anticline in the area of karst valleys at the head of the Sequatchie Valley.

having formed along the thrust-faulted Sequatchie Anticline (Fig. 13.12).

The geomorphic history of the Sequatchie Valley is that of anticlinal mountain changed into anticlinal valley. The anticlinal mountain still exists along the northernmost portion of the Sequatchie Anticline, north of Crab Orchard Cove. Between the anticlinal mountain and the head of the Sequatchie Valley is an area ~20 km long that is transitional between mountain and valley (Fig. 13.13). This is a region of karst valleys, locally referred to as coves. Within this area are six examples of karst valley development. From north to south, the major karst valleys are Crab Orchard Cove, McClough Hollow, Bat Town Cove, Little Cove, Grassy Cove and Swagerty Cove.

The author's interest in the relationship between slope retreat and cave processes was originally stirred by reading an article by Milici (1968). Milici implied that there was a relationship between slope retreat and karst processes in the headward erosion of the Sequatchie River and by the solution of limestones and assimilation of uvalas (actually karst valleys) that developed on the crest of the anticline as the Pennsylvanian strata were breached. The same idea had beeen previously stated by Lane (1951).

It is hypothesized by this author that the karst valleys are the result of subterranean stream invasion and that they have developed in a fashion similar to that presented in Figure 13.4. To test this model of the interrelationship between subterranean stream invasion, topography, subsurface hydrology, conduit cavern development, structure and stratigraphy in the Grassy Cove area, it was necessary to determine the major routes taken by subsurface streams. The data were then compared with the topography, structure, and stratigraphy of the area.

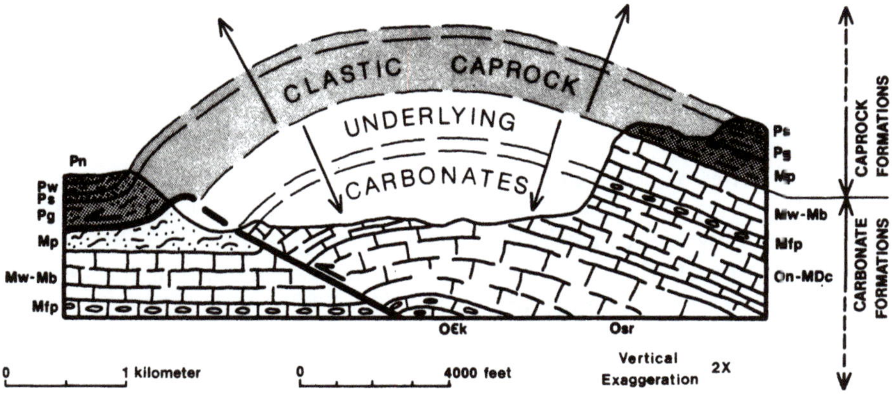

Figure 13.12 Sequatchie Valley (modified from Milici 1968).

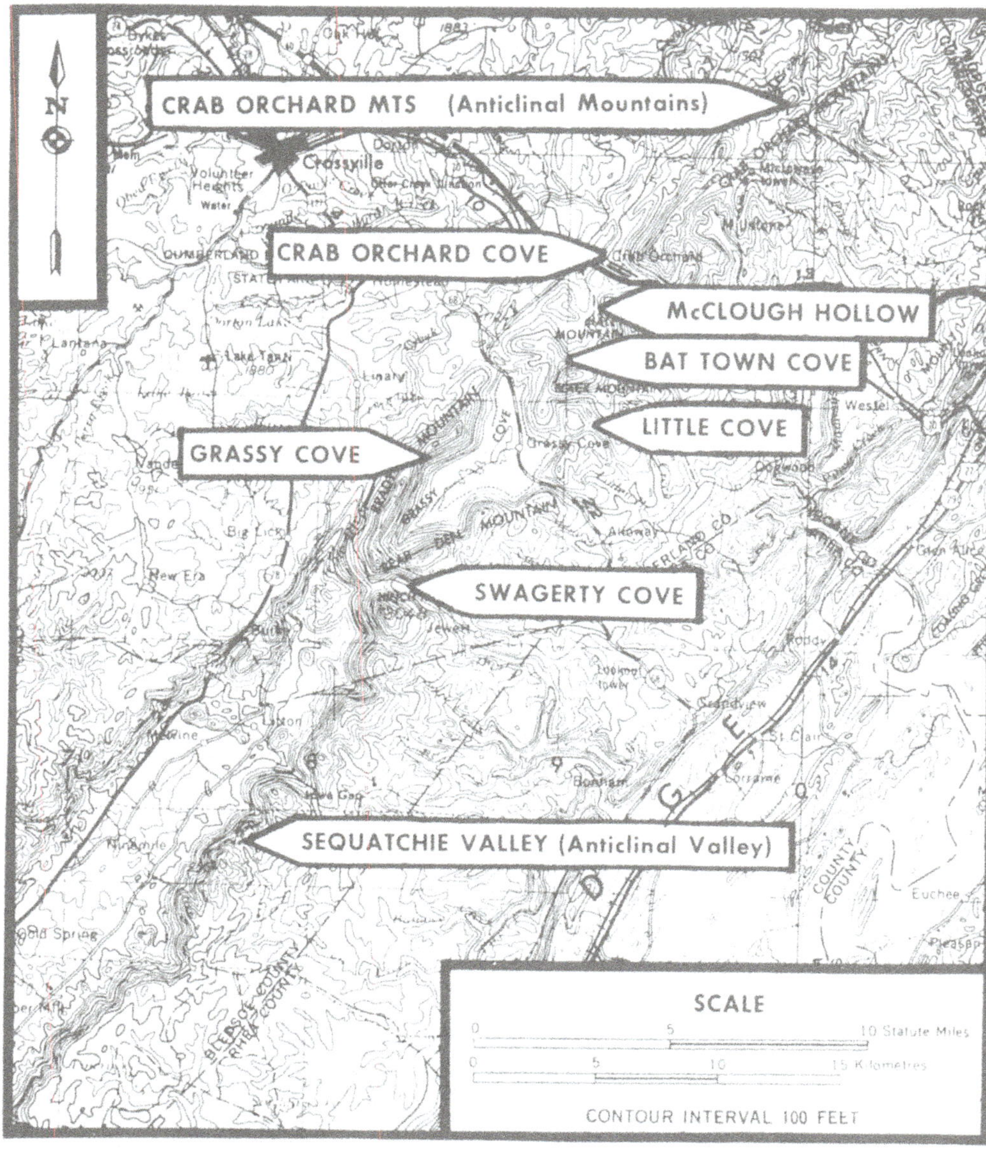

Figure 13.13 Area of karst valley development along the Sequatchie Anticline.

Subsurface flow routes and karst valley development

Figure 13.11 reveals the major subsurface drainage conduits of the Grassy Cove area. Dye traces were performed on the primary swallet of each of the karst valleys except for McClough Hollow. With one exception, all of the streams draining the coves flow south through subsurface conduits to the Sequatchie Valley. Crab Orchard Cove is the exception, draining to the north to a resurgence on Bakers Branch and then flowing into Daddys Creek. The

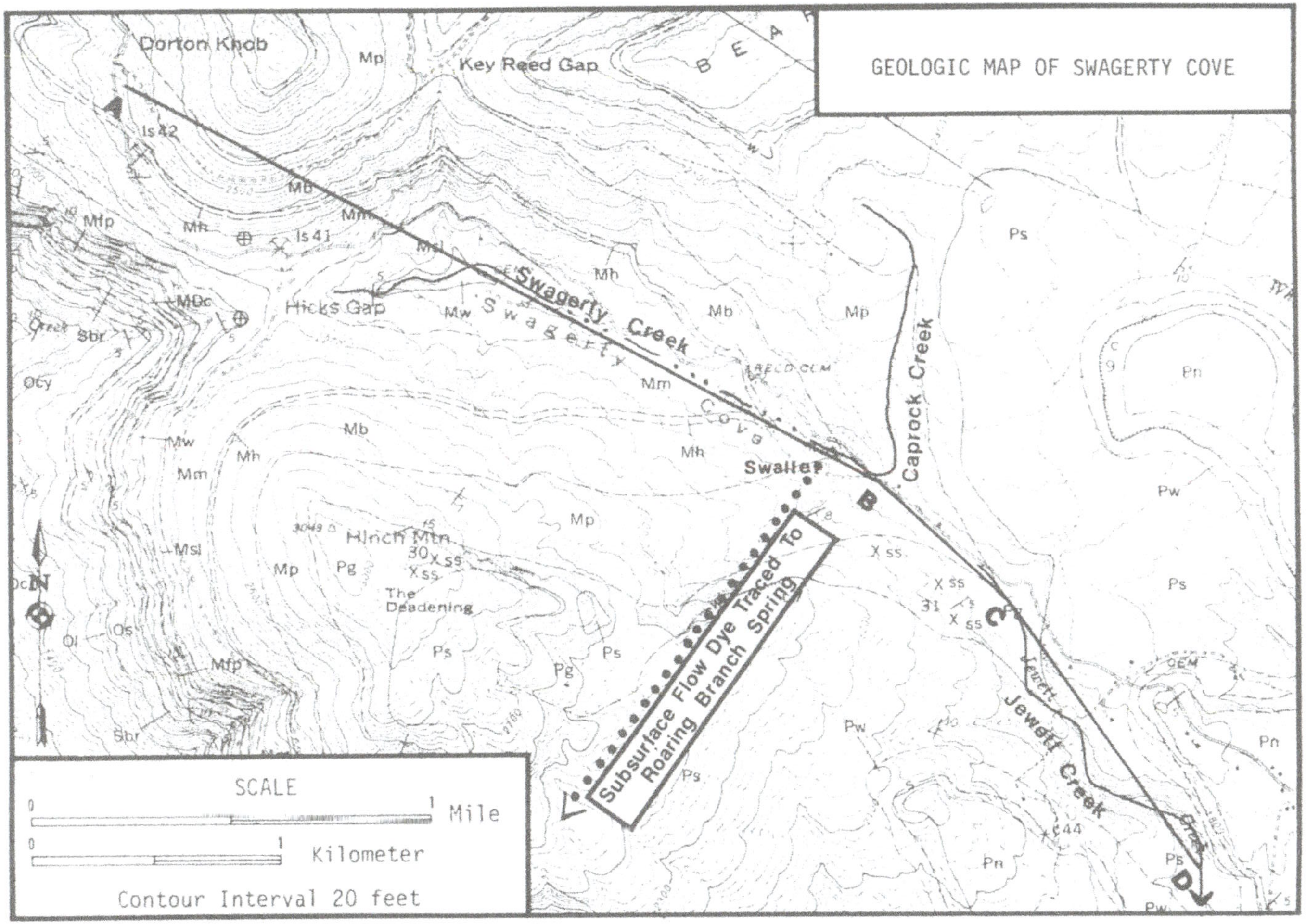

Figure 13.14 Geologic map of Swagerty Cove illustrating the subsurface invasion of upper Jewett Creek.

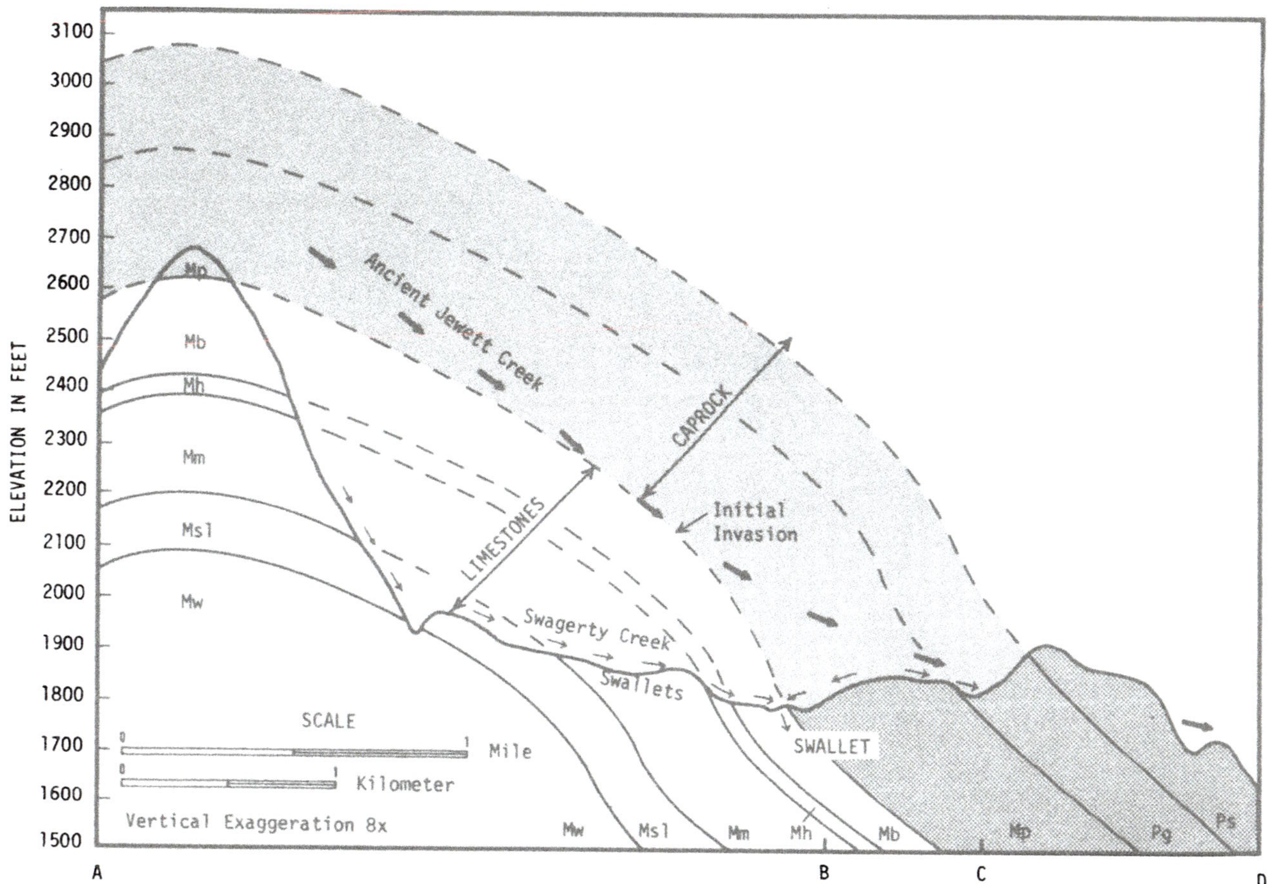

Figure 13.15 Stratigraphic section across Swagerty Cove along line ABCD (Fig. 13.14) illustrating the subsurface invasion of upper Jewett Creek.

drainage from McClough Hollow and Bat Town Cove (Fig. 13.11) flows under two mountains, Flat and Brady, joining the subsurface drainage from Grassy Cove under Brady Mountain, to resurge at Head of Sequatchie Spring. The Little Cove drainage flows under McCullough Mountain to resurge at Bristow Spring in Grassy Cove. The drainage from Swagerty Cove flows around and under Hinch Mountain to resurge at Roaring Branch Spring in the Sequatchie Valley.

When one compares the surface and subsurface drainage with the topography, structure and stratigraphy, the subterranean stream invasion hypothesis does appear to fit quite well. The karst valleys develop as aggressive streams flowing down dip off the anticlinal mountain cut through the caprock sequence and invade the underlying Bangor Limestone thereby forming subsurface drainage conduits.

Swagerty Cove provides a clear example of subterranean stream invasion and associated karst valley development occurring along the flank of the Sequatchie Anticline (Fig. 13.11). Figures 13.14 and 13.15 illustrate invasion that has occurred along upper Jewett Creek in Swagerty Cove. Notice the barbed tributary near the point of invasion. Subterranean stream invasion began to occur as Jewett Creek cut through the caprock into the Bangor Limestone. As Swagerty Creek, the pirated headwaters of Jewett Creek, continued to lower its channel, it cut through the resistant Hartselle Formation into the underlying Monteagle Limestone. Swagerty Creek is now invading the Monteagle as it sinks in several places along its course before flowing into its final swallet in the Bangor. In time, it is expected that Swagerty Creek will enlarge an opening in the Monteagle Limestone sufficient to take all the water, even flood discharges, of Swagerty Creek.

To dye trace the stream that disappears into the swallet in the Bangor Limestone in Swagerty Cove, it was necessary to set up three automatic water samplers, one at Bill Kemmer Spring in Grassy Cove, another at Head of Sequatchie Spring, and a third at Roaring Branch Spring in the Sequatchie Valley (Fig. 13.11). Rhodamine WT dye (20% solution) injected into the stream at the Swagerty Cove swallet came through at Roaring Branch Spring in the Sequatchie Valley. Dye was not detected at Bill Kemmer Spring or Head of Sequatchie Spring.

The East Tennessee Grotto of the National Speleological Society has recently discovered and mapped two extensive cave systems in Swagerty Cove. The cave passages and streams tend to follow the true dip of the strata down the flank of the anticline, influenced by joint orientation in many places. On both sides of the anticline, there is a place where the dip increases (Fig. 13.11). The increased bending of the rock at these places has produced large joints parallel to the anticlinal axis. It is believed that the subsurface Swagerty Creek flows down the true dip until it intersects these joints near the base of the anticline. It then apparently makes a 90° turn to the right to flow along the joints to resurge at Roaring Branch Spring in the Sequatchie Valley. As the subsurface stream flows around and under Hinch Mountain on its way to its

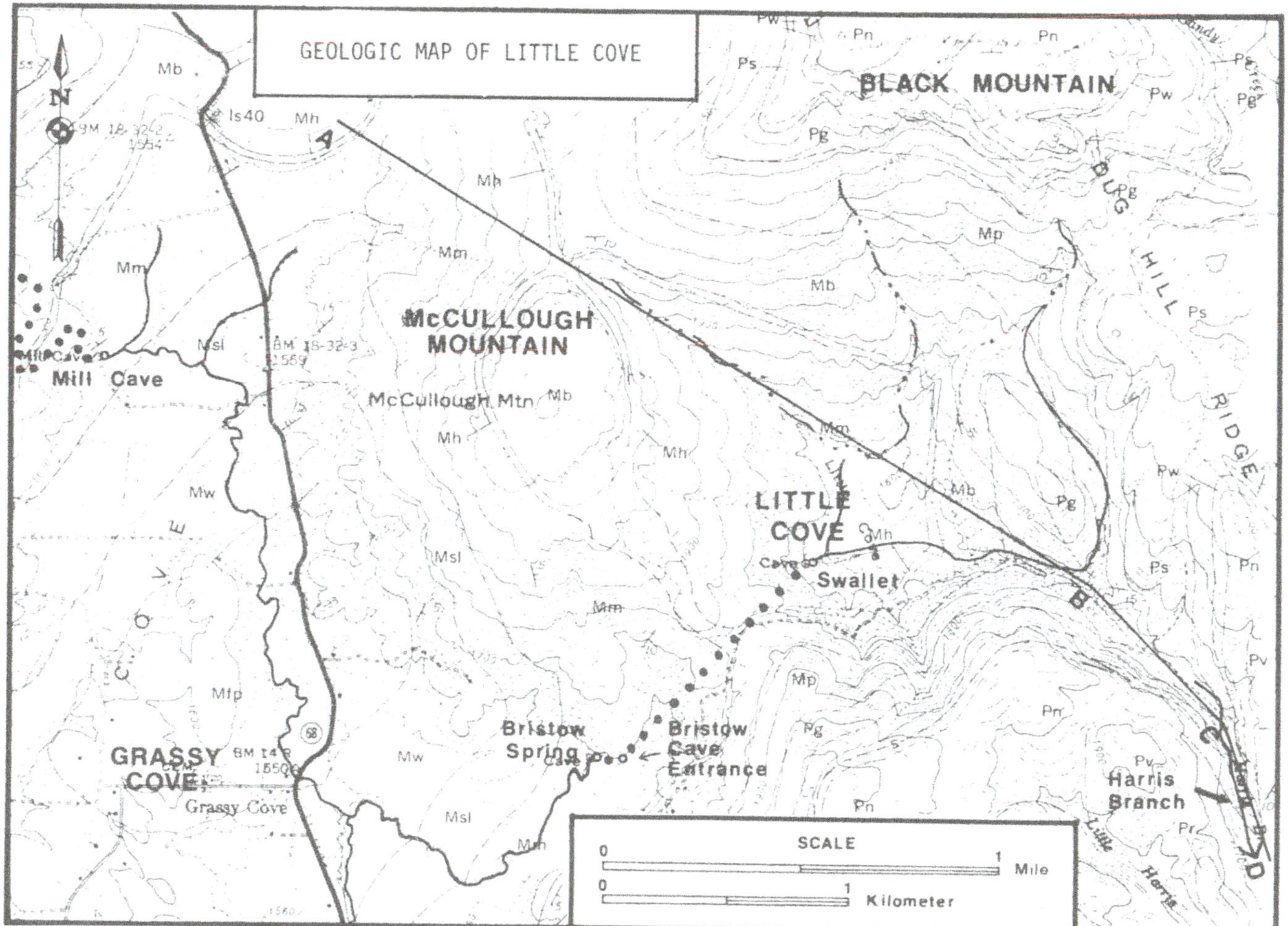

Figure 13.16 Geologic map of Little Cove illustrating the subsurface invasion of upper Harris Branch.

Roaring Branch resurgence, it cuts through the Hartselle, Monteagle and St. Louis formations to resurge at the St. Louis–Warsaw contact, probably breaking through the Hartselle along joints associated with the line of increased dip. The joints, however, do not appear to extend through the resistant, impermeable Warsaw Formation, which acts as a spillover layer. In this case, the subsurface Swagerty Creek appears to resurge at the lowest possible spillover point even though it is up dip slightly from the swallet. In tectonically disturbed areas, such as along the Sequatchie Anticline, joints are more likely to extend through the more resistant and impermeable strata, thus providing flow routes for subsurface streams.

Little Cove, just northeast of Grassy Cove (Fig. 13.11), apparently resulted from the subsurface invasion of Harris Branch (Figs 13.16 & 17). Harris Branch at one time flowed on the caprock, down the dip of the anticline, before cutting through the caprock sequence and invading the Bangor Limestone. From the site of the initial invasion, Little Cove has continued to

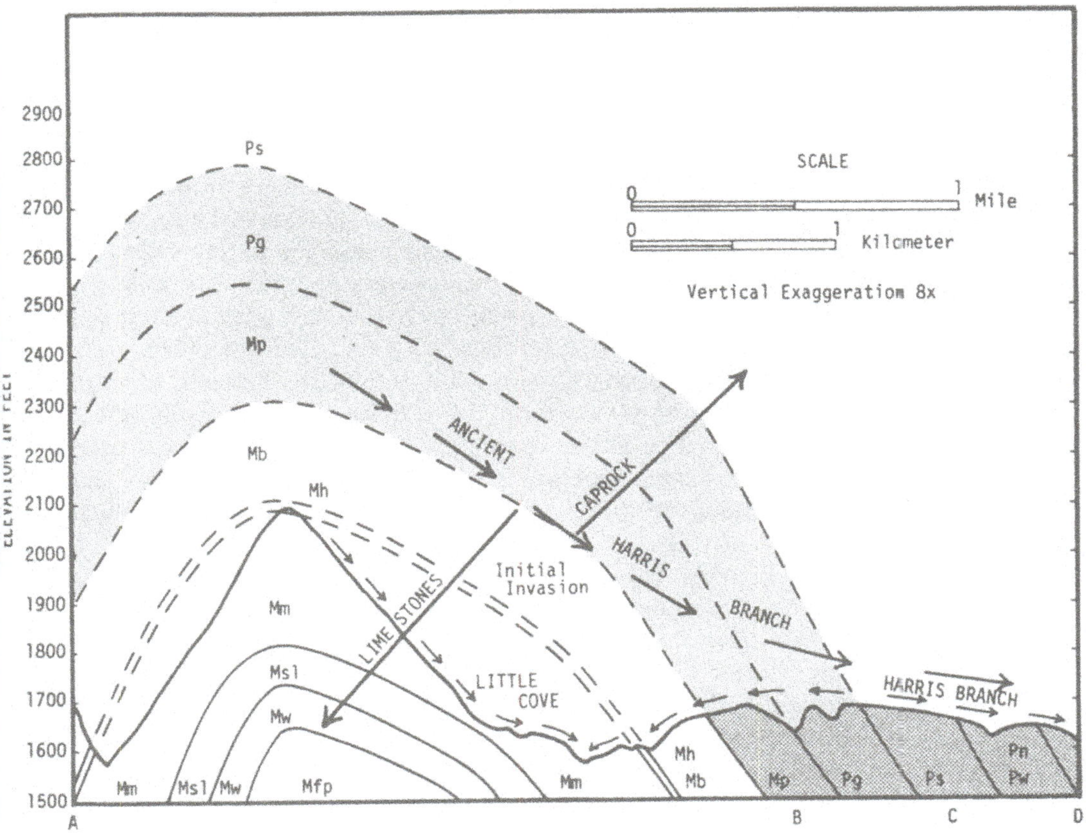

Figure 13.17 Stratigraphic section across Little Cove along line ABCD (Fig. 13.16) illustrating the subsurface invasion of upper Harris Branch.

enlarge and deepen, primarily by sapping of the chemically weak carbonates from under the caprock that surrounds the karst valley. Figure 13.11 shows that the swallet in Little Cove, the rising for Little Cove (Bristow Spring) in Grassy Cove, Bill Kemmer Spring in Grassy Cove, the swallet in Swagerty Cove, and the rising at Roaring Branch Spring all appear to have formed along a line that corresponds with the axis of increased dip along the flank of the anticline.

Grassy Cove, the largest karst valley, has had a similar geomorphic history to that of other coves investigated; it apparently formed by the coalescence of two karst valleys. One developed on the southeast side of the anticline as a perched caprock stream flowing down dip in the area of Stillhouse Gap cut into the underlying carbonates, and the other formed on the northwest side with the surface stream flowing down the dip of the anticline in the area of Low Gap. Stillhouse Gap and Low Gap are the only places where one may enter or exit the cove without crossing high mountains. The cove is drained by Cove Creek, which flows into Mill Cave and under Brady Mountain to resurge at Head of Sequatchie Spring. At 1308 m from the entrance of Mill Cave, there is a massive breakdown room. The collapse appears to extend all the way to the surface, 171 m overhead, as a large collapse sink is evident in the caprock at this location on the topographic map (Fig. 13.11). This breakdown choke backs up the stream after very heavy rains, causing the cave and Grassy Cove to flood.

When Grassy Cove floods, a lake forms that is 1 km wide, 3.5 km long and about 7 m deep at the entrance to Mill Cave (Fig. 13.18). One flood per winter is a common occurrence, often lasting for a week before the water recedes. As long as the floods occur during the winter months, they do little damage. However, floods during the growing season can be very damaging to crops and pasture. All of the karst valleys flood after heavy rains, but the others drain much more quickly than Grassy Cove, usually within 24–48 hours. In Crab Orchard Cove, much of the town of Crab Orchard flooded occasionally before a 1.6 km drainage ditch was excavated in 1976. Because all of the karst valleys in this area are large depressions with steep sides and flat floors and are subject to regular flooding, they may also be considered anticlinal poljes.

The subsurface stream that drains Grassy Cove may be viewed again in the Devilstep Hollow Cave, located ~335 m northwest of Head of Sequatchie Spring. The water then flows to Head of Sequatchie Spring and a very small spring nearby. During high discharge, most of the water resurges at a third spring, Sequatchie High Water Spring, which is located ~60 m directly in front of the cave entrance (Fig. 13.11).

When the author began this research, he believed the Mill Cave to Head of Sequatchie Spring system to be a good subsurface conduit for an input–output investigation of subsurface erosional processes. It appeared to be a single conduit without tributaries (due to the impermeable caprock overhead) for the entire distance. Therefore, recording stream gauges at Mill Cave, Head of

Figure 13.18 Grassy Cove flood lake.

Sequatchie Spring and Sequatchie High Water Spring were installed (Fig. 13.11). It soon became apparent that the system was considerably more complex than originally supposed. After a heavy rain, the discharge at Head of Sequatchie Spring would often show a double peak on the stage recorder, the first peak occurring only a few hours after the rain, the second peak occurring six to nine hours later. This led to the conclusion that the first peak must be due to a swallet on the Sequatchie Valley side of Brady Mountain, the second peak resulting from the discharge from Grassy Cove on the other side of the mountain.

A search was made for a swallet on the Sequatchie Valley side of Brady Mountain, and it was found, as predicted, 2134 m up Devilstep Hollow from Head of Sequatchie Spring (Crawford 1977). Devilstep Hollow Branch enters a swallet in a very large sink, ~ 30 m in diameter and 15 m in depth. It is a mystery why a sink of this size was not on the topographic map. The discovery of Run-to-the-Mill swallet in Devilstep Hollow, plus the discovery that the water from Bat Town Cove and McClough Hollow join the Mill Cave to Head of Sequatchie Spring conduit, greatly increased the complexity of the system, making it a rather poor choice for an input–output study. Also, it appears that small streams in the Gouffre and Grassy Cove Saltpetre caves in Grassy Cove collect water from several vertical shafts along the Hartselle Formation and then flow down dip into Brady Mountain and join the Mill Cave stream under the northwest side of the mountain. All other known drainage from Grassy Cove flows into the Mill Cave swallet.

Conclusions

A comparison of the subsurface hydrology with the surface hydrology, topography, and geologic structure and stratigraphy in the Grassy Cove area appears to support the hypothesized model. Evidence of subterranean stream invasion of streams that once flowed down the anticlinal dip on the virtually impermeable caprock is very clear in Swagerty Cove, Little Cove, Bat Town Cove and McClough Hollow, and somewhat less obvious in the larger karst valleys of Grassy Cove and Crab Orchard Cove. However, there is little doubt that even the largest karst valley, Grassy Cove, once resembled Swagerty Cove in its early evolutionary development. The assimilation of at least two karst valleys, one on each side of the anticline, has occurred to produce Grassy Cove.

Grassy Cove, as well as the other karst valleys, will continue to expand horizontally by slope retreat resulting from the sapping of the surrounding Pennsylvanian caprock by erosion of the underlying Mississippian shales and carbonates. Karst valleys such as Bat Town Cove and McClough Hollow that have only recently breached the Hartselle Formation should lower their floors considerably in the future. Grassy Cove, which is presently floored primarily by the resistant Warsaw and Fort Payne formations, will lower its floor very slowly. However, if it lowers its floor sufficiently to breach the Fort Payne and the thin underlying Chattanooga Shale, there is a high probability that its drainage would be diverted underground once again into the underlying Ordovician limestones. Thus a conduit would be formed in the Ordovician strata from Grassy Cove to the Sequatchie Valley, possibly almost under the present cave system. Of course, this would be dependent on an outlet for the subsurface water since it is conceivable that a passageway through the Ordovician limestones might be blocked by impermeable strata pushed into position by the Sequatchie Thrust Fault which has produced unusually complex structure in the area. If invasion into the Ordovician limestones does occur in the future, Grassy Cove would rapidly lower its floor a hundred meters or more in the vicinity of the invading stream, and the resistant Fort Payne Formation would surround the new depression as a structural terrace.

Devilstep Hollow, the spearhead of the Sequatchie Valley, will advance headward rapidly in the future, assisted somewhat by the assimilation of several small karst valleys that will form due to the subsurface invasion of small streams presently flowing down the northwest side of Brady Mountain. The impermeable Pennington Formation is very thin on the northwest side of the mountain, and it is only a matter of time before several small streams presently flowing down the anticlinal dip into Daddys Creek cut through the Pennington and invade the Bangor Limestone (Fig. 13.11). It is also probable that the northeast flowing Daddys Creek will be captured by Devilstep Hollow Branch, which has a much steeper gradient. Devilstep Hollow Branch will, therefore, someday extend its valley to the northeast, probably even beyond Grassy Cove. Brady Mountain will separate the surface drainage of Grassy Cove from that of Devilstep Hollow until finally erosion will join the two, probably in the vicinity of Low Gap, and the subsurface drainage of Grassy Cove will be diverted to the surface drainage of Devilstep Hollow. Thus, the Sequatchie Valley will have taken a giant step headward as Grassy Cove becomes part of the actual valley itself. As the Sequatchie Valley continues to advance headward, the karst valleys of Bat Town Cove, McClough Hollow, and Crab Orchard Cove will also eventually be assimilated into the valley. By that time, new karst valleys will have formed in the anticlinal Crab Orchard Mountains northeast of Crab Orchard Cove. The first of these new karst valleys will be formed by Terrel Creek. Terrel Creek flowing down the anticlinal dip has already lowered its channel deep into the Pennington Shale, and it is only a matter of time before it cuts through to invade the underlying Bangor Limestone.

By the assimilation of karst valleys, the Sequatchie Valley advances headward up the Sequatchie Anticline from the southwest towards the northeast (Fig. 13.19). The anticlinal mountain is first reduced to karst valleys resulting from subterranean stream invasion, and finally the karst valleys are assimilated into the Sequatchie Valley itself as it advances headward. A very important factor contributing to the formation of karst valleys in a sequential fashion from southwest to northeast in advance of the Sequatchie Valley has been the increased thickness of the Pennsylvanian caprock from southwest to northeast. The Pennsylvanian strata that cap the entire Cumberland Plateau increase in thickness from the southwest to northeast. A similar increase applies to the Sequatchie Anticline because it has the same lithology as the surrounding plateau. Thus, it progressively requires more time for streams flowing down the dip of the anticline to cut through the Pennsylvanian caprock in a southwest to northeast direction. If the caprock along the anticline was the same thickness from southwest to northeast, streams would cut through it at about the same time all along the anticline, and the sequential development of karst valleys from southwest toward the northeast would not occur. This does not appear to have been the case, instead karst valley development has marched right up the anticline from the southwest, followed by the headward advance of the Sequatchie Valley. Karst valley development resulting from subterranean stream invasion, therefore, appears to play a very important role in the geomorphic development of the Sequatchie Valley.

Quantitative investigation of erosional processes associated with cavern development and slope retreat

To quantitatively investigate subsurface erosional processes, two cave systems along the Cumberland Plateau escarpment in the Grassy Cove area were monitored for 15 months. The cave systems were treated as open systems, input of mass and energy occurring at swallets and from diffuse sources, output occurring at risings. An attempt was made to select large but simple, conduit cave systems, those with only one swallet and one rising.

Four recording precipitation gauges and seven stilling wells with recording

Table 13.1 Mean input–output measurements for the Little Cove–Bristow Spring System for 11 occasions when swallet input, diffuse input, and rising output were each sampled on the same day.

	Swallet input Little Cove Swallet			Diffuse input Bristow Cave Drips			Rising output Bristow Spring			
	Mean input	Percent of output	+	Mean	Percent of output	=	Mean measured output	Mean calculated output	Difference between measured and calculated output	Percent difference between measured and calculated output
discharge	0.54105 cms	88.17		0.07257 cms	11.83		0.61362 cms			
total hardness	77.56 ppm	88.17	+	197.11 ppm	11.83	=	89.44 ppm	91.70 ppm	2.26 ppm	2.53
aggressivity	+6.673 ppm	88.17		−16.16 ppm	11.83		+6.345 ppm	+3.972 ppm	+2.373 ppm	37.40

taken during the rise and fall of the stream hydrograph. Certain variables were recorded for each water sample. The following variables were recorded and/or measured in the field: location, date, time, stage height, stage rising or falling, discharge, air temperature, water temperature, pH, specific conductance and drip rate in milliliters per minute (for cave drips only). The following variables were measured within 24 hours in a mobile laboratory: total hardness, calcium hardness, magnesium hardness (by subtraction), aggressiveness towards calcium carbonate (Stenner method), calcium, bicarbonate alkalinity, and carbonate alkalinity. The following elements were measured at the geochemistry laboratory at Vanderbilt University (for selected samples only): potassium, sodium, magnesium, silica, sulfate, chloride, phosphate, nitrate, iron and aluminum.

In addition, suspended-sediment samples were collected at the same time and site as the water samples. A US DH-48 Depth-Integrating Suspended Sediment Sampler was used to collect the samples, and the analysis was performed in both the author's mobile laboratory and the geochemistry laboratory at Vanderbilt University.

The Stenner (1966,1971) method of determining the degree of saturation or aggressiveness to calcium carbonate was used. In discussing the chemistry of cave waters, Picknett (in Ford & Cullingford 1976) mentions several techniques that have been used to quantify aggressivity and maintains that the

Figure 13.20 Stilling well and stage recorder at Mill Cave swallet in Grassy Cove.

Stenner method is simpler and superior to other less direct methods. A detailed description of sampling procedures, data analysis techniques, and the data are published in Crawford (1980).

The quantitative input–output analysis provided considerable evidence in support of the hypothesized model. During the research period:

(a) Caprock streams entering swallets were always aggressive to calcium carbonate.
(b) Water resurging at risings was almost always aggressive to calcium carbonate because of the high percentage of aggressive swallet input (Fig. 13.21). Mixing corrosion may also be partly responsible for the aggressivity at the rising. This is indicated by the 37.10% difference between measured and calculated output at Bristow Spring (Table 13.1).

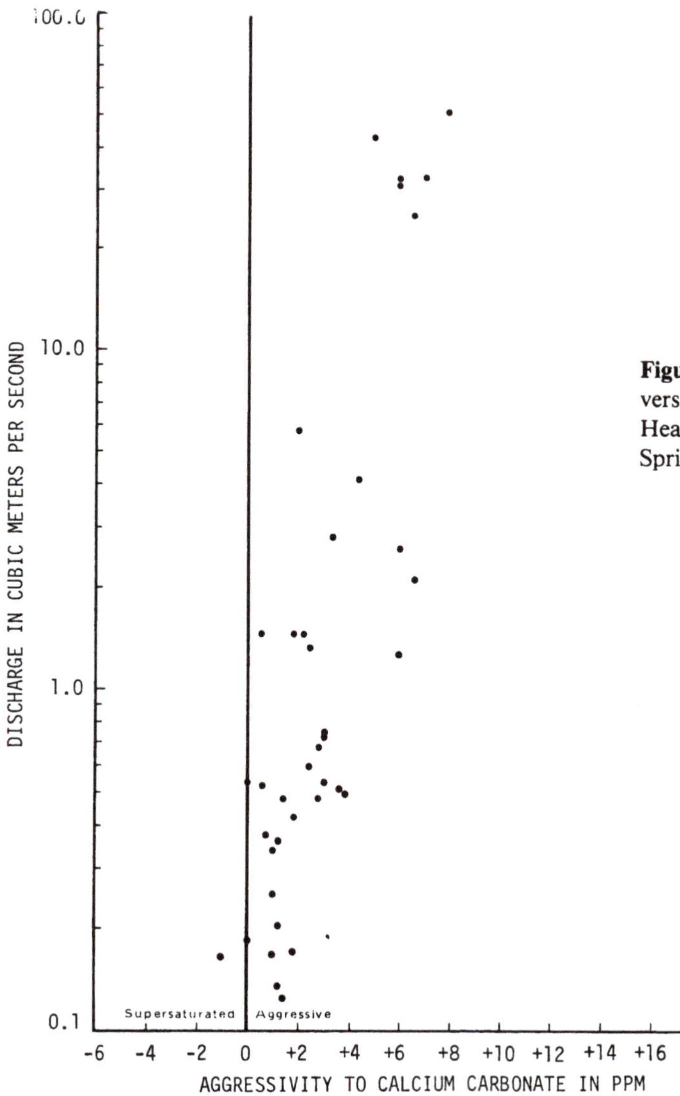

Figure 13.21 Discharge versus aggressivity at Head of Sequatchie Spring.

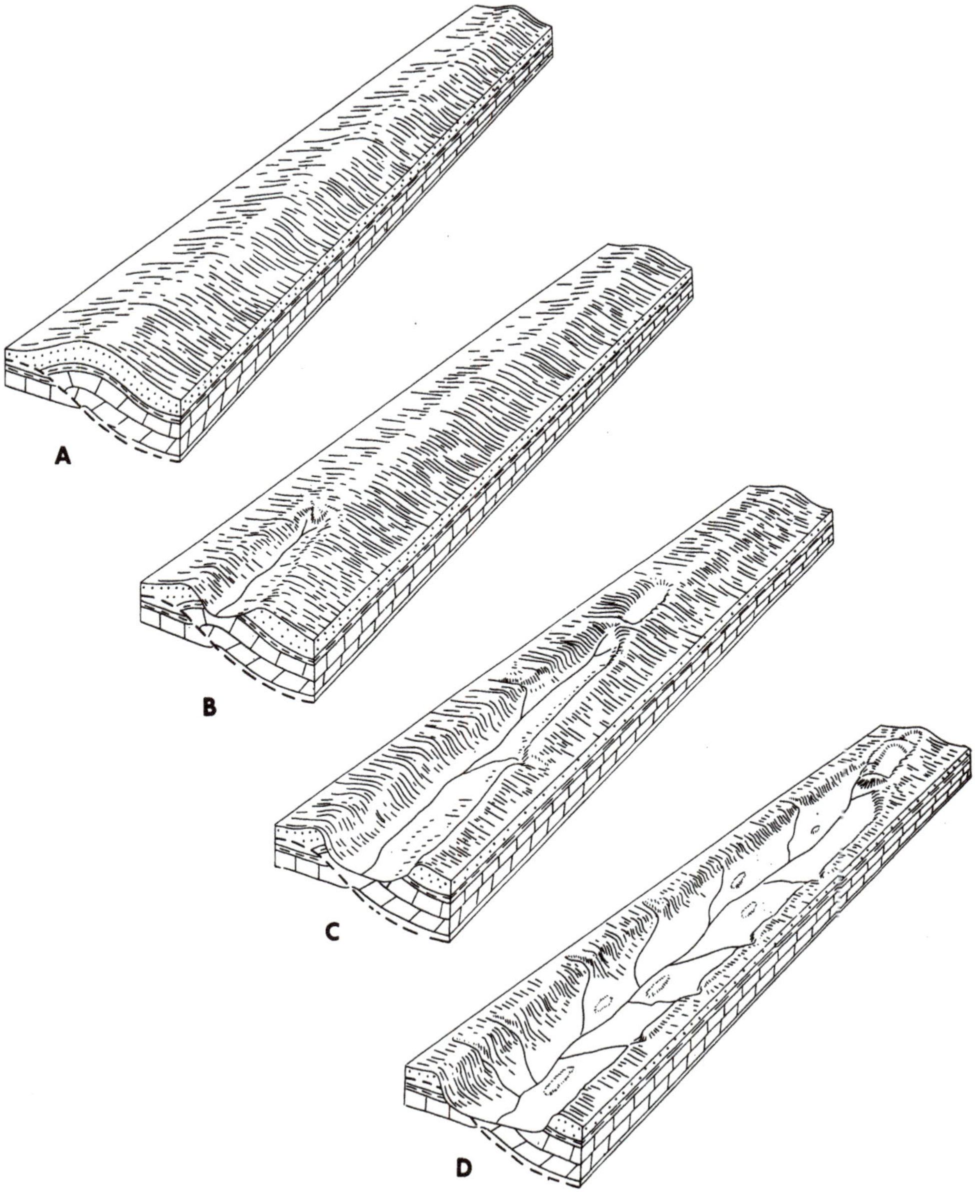

Figure 13.19 Subterranean stream invasion, karst valley development and the headward advance of the Sequatchie Valley (modified from Miller 1974).

stream gauges were installed (Fig. 13.20). Water samples were collected at the swallets and at the risings to investigate changes that occurred as water flowed through the systems. The percolation input was monitored by taking drip samples from within the caves. Precipitation samples were also collected for analysis.

A total of 270 water samples were collected during the 15 month period; intensive sampling occurred during and after heavy rains so that samples were

(c) An inverse correlation existed between total hardness and discharge, and a direct correlation existed between aggressivity and discharge for the swallet input and rising output, indicative of the importance of floods in conduit cavern development (Figs 13.22 & 23).
(d) Due to degassing of carbon dioxide, percolation input was always supersaturated with calcium carbonate on joining the vadose cave systems.
(e) Percolation total hardness and aggressivity were not correlated with drip rate, but instead, with drip water temperature, which reflects soil

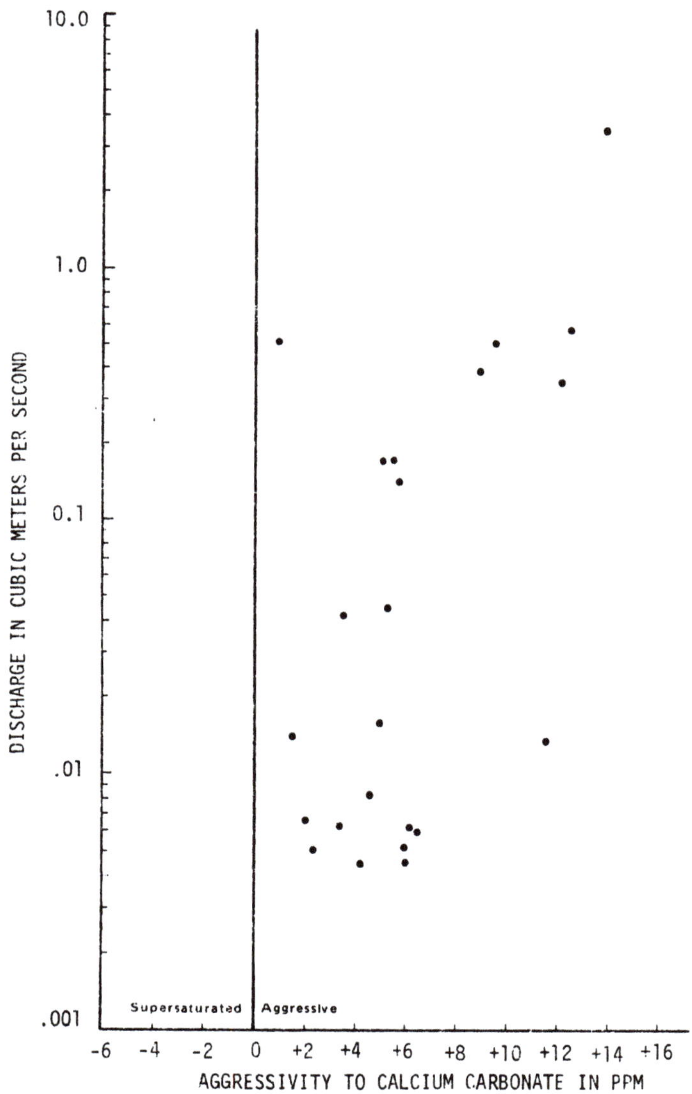

Figure 13.22 Discharge versus aggressivity at Little Cove swallet.

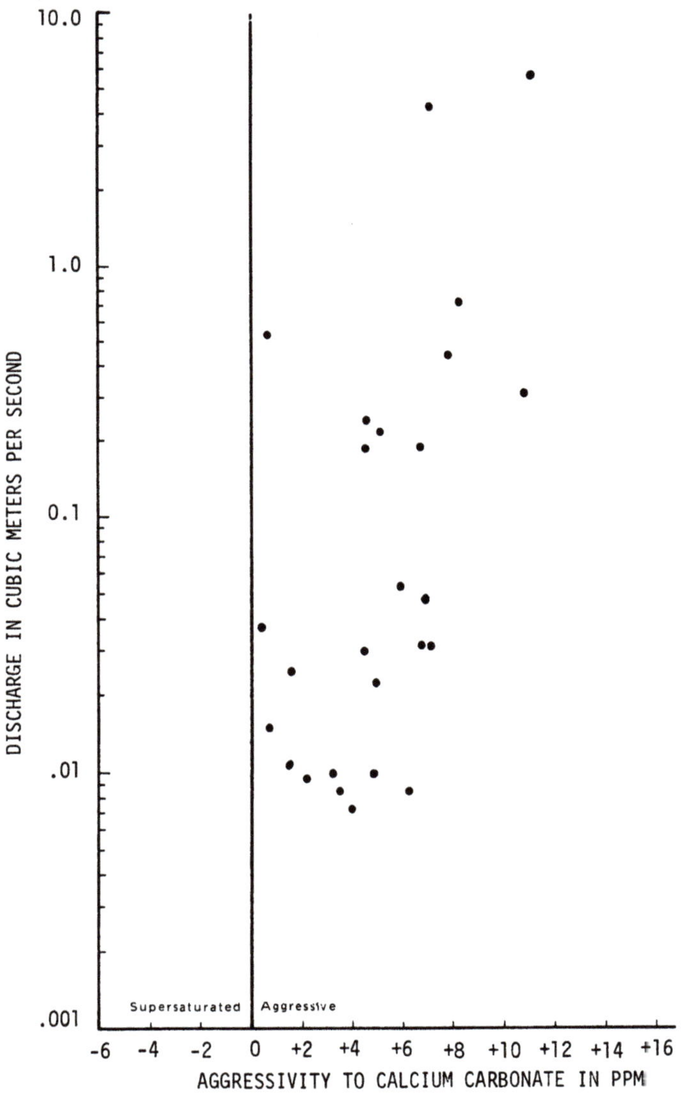

Figure 13.23 Discharge versus aggressivity at Bristow Spring.

temperature and biogenic carbon dioxide content of the soil atomosphere (Figs 13.24 & 25).

(f) All suspended sediment samples were negative for calcite and dolomite, but due to impoundment problems associated with the individual caves investigated, the negative abrasion results are considered inconclusive.

It appears that highly aggressive water entering swallets during high discharges is very important in the corrosion of the conduit floor, walls and ceiling. Because there is a direct correlation between discharge and aggressivity, one would expect a greater increase in milligrams per liter (ppm) total

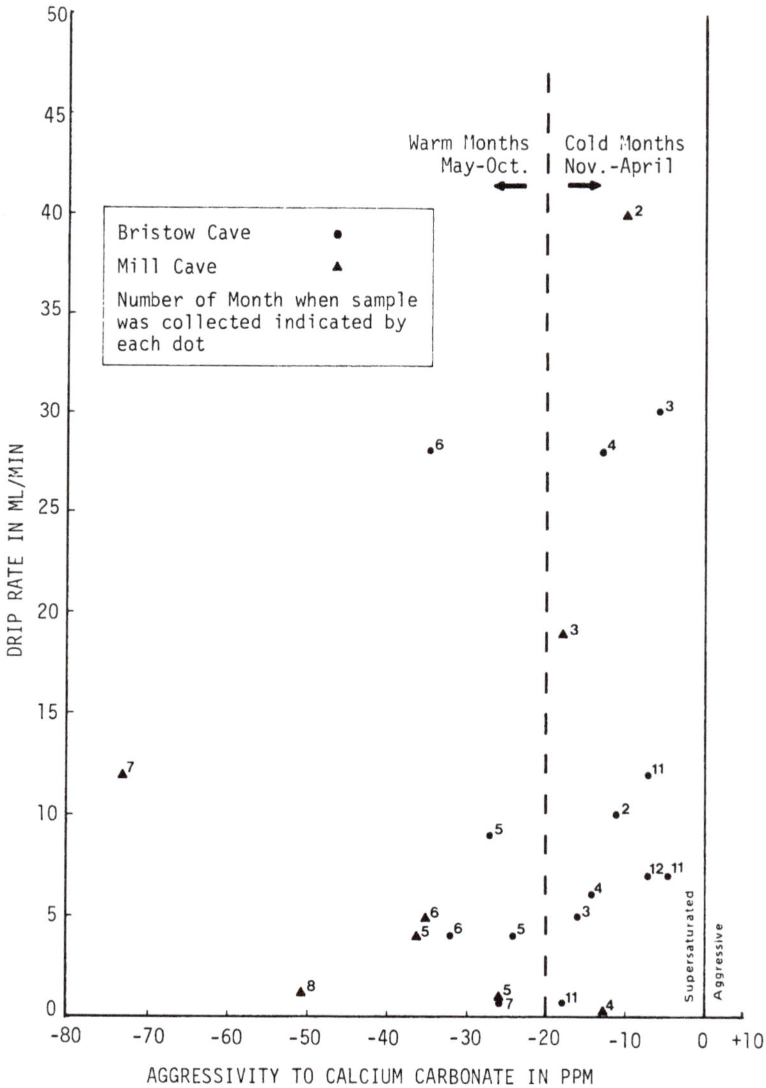

Figure 13.24 Drip rate versus aggressivity for drips from Bristow and Mill caves.

hardness, resulting from the direct corrosion of the conduit during periods of high discharge. There is no question that more dissolved limestone is carried by the stream from the cave during periods of high discharge even though the total hardness is inversely correlated with discharge. However, the input–output analysis tends to indicate that the rising output is approximately equal in total hardness to the percent contributed from the swallet and diffuse input. The increase in total hardness resulting from corrosion of the conduit itself was too small to measure and could not be separated from that which entered the conduit as diffuse and swallet input. Therefore, it was not determined by actual measurement that more calcium carbonate is corroded from the conduit

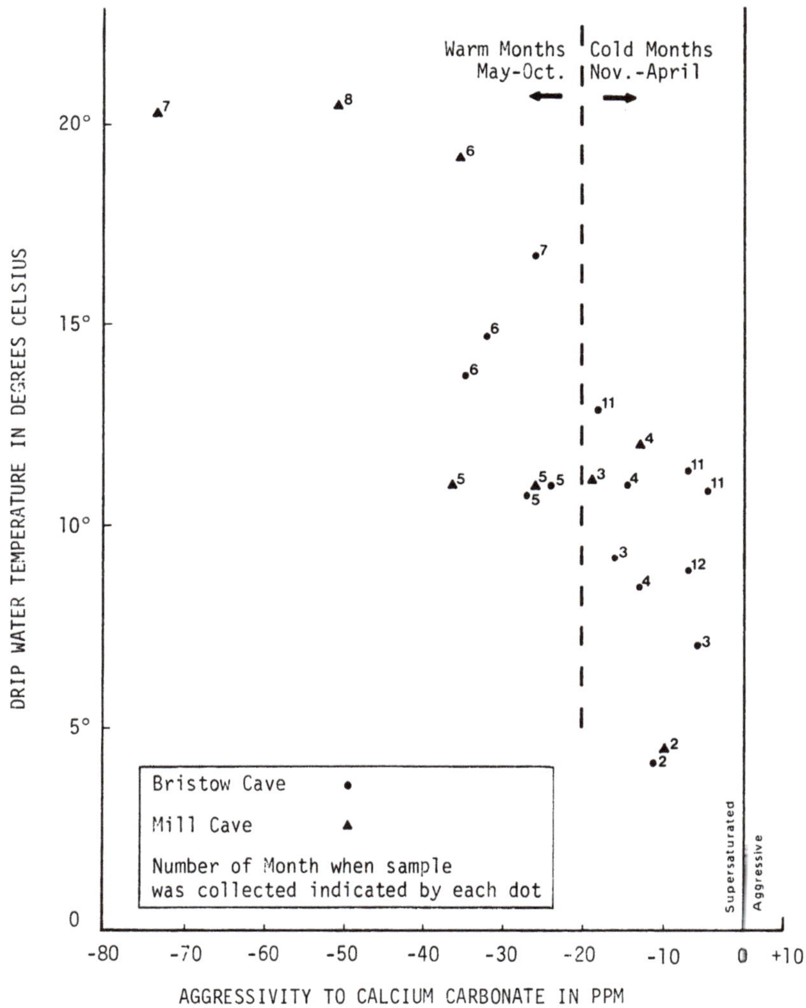

Figure 13.25 Water temperature versus aggressivity for drips from Bristow and Mill caves.

during high discharges. However, because the aggressivity is higher and the quantity of water flowing through the cave is greater, it seems reasonable that the actual quantity of limestone dissolved from the conduit (and thus cavern enlargement) is greater during periods of high discharge.

Although analysis of suspended-sediment samples taken at Little Cove, Mill Cave and Swagerty Cove swallets revealed large quantities of silicious suspended sediment being carried into the conduit caves by invading streams, and although samples taken at Bristow Spring and Head of Sequatchie Spring revealed large quantities of silicious suspended sediment being removed from the caves, this research failed to prove that limestone particles were being removed from the caves in suspension.

Although abrasion by silicious sand-size suspended sediment may not be occurring in the caves investigated, abrasion of limestone by bed load is definitely a factor. Physical evidence in the form of abrasion marks (areas where the gray weathered patina has been removed exposing the lighter limestone) was observed after a large flood in Powder House Sinks Cave, a cave with a bedrock floor. The physical evidence tends to indicate that larger particles, those moved by traction, are exclusively responsible for the abrasion marks.

Even if all abrasion results from bed load and not suspended load, some of the abraded limestone particles should have been carried from the caves in suspension and thus detected in the suspended-sediment samples collected at the resurgences during floods. The fact that all samples collected were negative for calcite and dolomite may be due to one or more of the following reasons:

(a) The very small, abraded limestone particles dissolve rapidly in the aggressive floodwaters and leave the cave in solution rather than in suspension.
(b) The Little Cove to Bristow Spring system and the Mill Cave to Head of Sequatchie Spring system have massive amounts of silicious material passing through them as bed load, thus the cave floors are protected from abrasion by a rather thick covering of non-carbonate sediment.
(c) Only one flood was sufficient to move rocks larger than gravel-size as bed load, and unfortunately, no suspended-sediment samples were taken during this flood event.

For these reasons, the negative results in attempting to quantify the removal of particular limestone in suspension from the cave systems studied are considered inconclusive. Additional research will be necessary before this portion of the model can be accepted or rejected.

Conclusions

This study of groundwater as a geomorphic agent in karst landform development along the Cumberland Plateau escarpment of Tennessee was primarily an investigation of the relationship between subterranean stream invasion, slope retreat, topography, subsurface hydrology, conduit cavern development, structure and stratigraphy. A surface–subsurface erosion model was hypothesized and tested, which includes the following:

(a) Conduit caves are formed by subterranean stream invasion along retreating escarpments that are produced by the more rapid chemical erosion of underlying carbonate rock.
(b) Invasion occurs near the contact between the impermeable caprock and the underlying limestone as surface streams flow down the escarpment.
(c) Invasion by a surface stream is initiated when water highly aggressive to calcium carbonate begins to flow through joints and bedding planes to a resurgence at the base of the escarpment.
(d) After the aggressive water has solutionally enlarged the subterranean stream conduit, the suspended and bed loads of the invading stream are carried through the conduit, thus further enlarging the cave by corrasion.
(e) Both chemical and mechanical erosion are greatly accelerated during floods when aggressive runoff water contributes a much larger percentage of the total discharge entering caves at swallets.
(f) Caves are enlarged almost exclusively by swallet water and vertical shaft water from the clastic caprock aquifer. Of the two types of diffuse input, vertical shaft and percolation, only vertical shaft water is aggressive to calcium carbonate when it enters the cave environment. Vertical shaft water drops directly from the caprock into the underlying limestone through joints that have opened primarily due to release of confining pressure near the edge of the escarpment. This water is from the caprock aquifer and has not had appreciable contact with calcium carbonate. Therefore, it is similar in terms of its aggressivity to water from caprock streams that enter caves at swallets. Subsurface streams fed by vertical shafts form conduit caves in a fashion similar to that of subsurface streams fed by swallets. The importance of floodwater and corrasion, however, are less, and this may account for smaller dimensions in terms of cross-sectional area of most of these caves.
(g) As percolation water passes through the soil, it becomes highly aggressive due to the high percentage of carbon dioxide in the soil atmosphere. It dissolves large quantities of limestone at the regolith–bedrock interface. However, the aggressive water quickly becomes saturated, or nearly saturated, on initial contact with limestone, and therefore, dissolves very

little calcium carbonate as it continues to percolate down through the joints and bedding planes. When the percolation water enters the vadose cave atmosphere where the carbon dioxide content is low, approximately the same as the normal atmosphere, it loses carbon dioxide, thus becoming supersaturated with calcium carbonate. Therefore, percolation water, although capable of dissolving large quantities of limestone at the regolith–bedrock interface and transporting it via joints and bedding planes to cave streams, is inconsequential in the enlargement of the conduit cave systems themselves. It is, however, believed to be the most important process in the general solution of the underlying limestone strata along the escarpment, thereby greatly influencing the retreat of the Cumberland Plateau escarpment.

(h) A subsurface stream takes a stair–step route down the escarpment due to the presence of resistant, almost impermeable layers of shale, sandstone, chert, and even dolomite and limestone.

(i) These resistant impermeable strata tend to act as elevation controls for cave development.

(j) The impermeable strata do not create perched water tables (in the sense of continuous nearly horizontal planes of water), but they do control the elevation of individual subsurface streams.

(k) Normally, subsurface streams flow down dip, floored by resistant, impermeable strata – not always, however, because an impermeable layer can still be a control or "spillover level" for a subsurface stream even when it flows against the dip.

(l) Subsurface streams rarely flow directly down the true dip of the strata; they usually flow somewhere between the true dip and the strike. The actual route taken by the subsurface water down the dip, and consequently cavern development, is greatly influenced by jointing.

(m) Subsurface waterfalls erode vertical shafts where subsurface streams break through impermeable layers.

(n) A sinkhole plain several kilometers wide at the base of the escarpment is a product of caprock removal by slope retreat and follows the retreating escarpment. It is believed that the terra rossa which covers the sinkhole plain has formed from the weathering of detritus from the caprock plateau deposited at the base of the escarpment, combined with residual material, primarily chert, from the weathering of underlying carbonates as the sinkhole plain is lowered.

(o) In areas of folded structure and even in areas of near–horizontal structure where there is a structural high (such as a slight anticline) in back of the retreating escarpment, a special type of subterranean stream invasion may occur. If the gradient of the plateau stream is less than the dip of the structural high, the stream may cut through the caprock into the underlying limestone several kilometers in back of the escarpment, and a large depression called a karst valley, completely surrounded by caprock and often several kilometers in diameter, will be formed.

The hypothesized model was field tested at two locations along the Cumberland Plateau escarpment. The investigation in the Lost Creek Cove area along the western escarpment of the plateau revealed a structural dome under the Lost Creek Cove karst valley, and dye tracing of subsurface streams and investigation of caves indicated that upon invasion, subsurface streams flow down dip, influenced by jointing, and that they are perched upon impermeable strata. In the past, Lost Creek apparently flowed on the caprock, down a structural dome into Dry Creek, thus taking a surface route to the Caney Fork River. After lowering its stream bed to an elevation of ~400 m, it cut into the Bangor Limestone at a location in back of the Cumberland Plateau escarpment. Subterranean stream invasion was initiated as water began to flow into and through the joints and bedding planes of the Bangor Limestone to a resurgence on the Caney Fork River at the base of the escarpment. Since that time, Lost Creek Cove has increased in size and depth, becoming a large karst valley virtually surrounded by non–carbonate rock.

An investigation in the Grassy Cove area at the head of the Sequatchie Valley revealed that caprock streams flowing down dip off the Sequatchie Anticline after breaching the sandstone caprock are diverted underground into the underlying carbonates. The aggressive water of invading caprock streams forms conduit caves as it flows from swallets to risings at the head of the Sequatchie Valley. Slope retreat by sapping proceeds in all directions away from the site of the initial invasion. Large karst valleys result from this subterranean invasion, and it appears that karst valley development plays a major role in changing anticlinal mountain into anticlinal valley, thereby greatly affecting the headward advance of the Sequatchie Valley along the Sequatchie Anticline. The anticlinal mountain is first reduced to karst valleys as surface-flowing streams are diverted underground, and finally the karst valleys are assimilated into the Sequatchie Valley as it advances headward.

The attempt to investigate the processes associated with subterranean stream invasion and conduit cavern development by an input–output analysis was not as successful as had been hoped. The author was not able to quantify the amount of limestone removed from the cave floor, walls and ceiling by corrosion and corrasion. However, the investigation did provide considerable evidence in support of the hypothesized model. The aggressiveness to calcium carbonate of caprock streams entering cave systems as swallet input, the direct correlation between discharge and aggressivity of the swallet input, and the supersaturated state of the percolation input all tend to support the model. The results of the investigation of cavern enlargement by abrasion from suspended, silicious, sand-sized particles was inconclusive, but visual evidence of abrasion by bed load confirms that corrasion is an important process during floods.

All aspects of the hypothesized model have not been tested. An investigation of the origin of the sinkhole plain has only just begun, and additional research concerning the chemical and mechanical processes associated with conduit cavern development is definitely needed. However, it is felt that this investigation produced sufficient evidence that the hypothesized model need not be rejected. As more is learned about the karst hydrogeology of the region, the model will probably need to be modified somewhat, thus evolving with time into a better representation of reality. However, even in its present state, it appears to provide a worthwhile generalization of the complex interrelationships between subterranean stream invasion, slope retreat, surface morphology, subsurface hydrology, conduit cavern development, geologic structure, and stratigraphy along the Cumberland Plateau escarpment of Tennessee, and it may be useful in other areas with similar geologic and hydrologic settings.

Acknowledgements

Support for this research was provided by the Libby Foundation, Clark University, Vanderbilt University, Ford Foundation, US Geological Survey Water Resources Division, Tennessee Division of Geology, Tennessee Division of Water Resources and the Center for Cave and Karst Studies, Western Kentucky University. Illustrations were drafted by Pat Quinlan and Sherri Snell.

References

Crawford, N. C. 1977. The Mill Cave drainage system. *Speleonews* **21**, 72–86.

Crawford, N. C. 1980. *The karst hydrogeology of the Cumberland Plateau escarpment of Tennessee, part IV: erosional processes associated with subterranean stream invasion, conduit cavern development and slope retreat*. Bowling Green, Kent.: Center for Cave and Karst Studies, Western Kentucky University.

Crawford, N. C., J. F. Hoffelt and B. Neff 1983. Karst waterfall development along the Cumberland Plateau escarpment: a hydrogeologic investigation of Virgin Falls and Lost Creek Falls, White County, Tennessee. *Speleonews*, in press.

Ford, T. D. and C. H. D. Cullingford 1976. *The science of speleology*. London: Academic Press.

Hack, J. J. 1966. *Interpretation of Cumberland escarpment and Highland Rim, south-central Tennessee and northeast Alabama*. US Geol Surv. Prof. Pap. 524-C.

Lane, C. F. 1951. *Physiography of the Grassy Cove district, Cumberland County, Tennessee*. PhD dissertation. Northwestern University.

Matthews, L. E. 1971. *Descriptions of Tennessee caves*. Tenn. Div. Geol. Bull. 69.

Milici, R. C. 1968. *The physiography of Sequatchie Valley and adjacent portions of the Cumberland Plateau, Tennessee*. Tenn. Div. Geol. Rep. Invest. 22.

Miller, R. A. 1974. *The geologic history of Tennessee*. Tenn. Div. Geol. Bull. 74.

Olson, C. G., R. V. Ruhe and M. J. Mausbach 1980. The terra rossa limestone contact phenomena in karst, southern Indiana. *Soil Sci. Soc. Am. J.* **44**, 1075–9.

Quinlan, J. F. and R. O. Ewers 1981. Hydrogeology of the Mammoth Cave region, Kentucky. In *Geological Society of America Cincinnati 1981 field trip guidebooks, volume III: geomorphology, hydrogeology, geoarcheology, engineering geology*, T. G. Roberts (ed.), 457–506: Falls Church, Vir.: American Geological Institute.

Ruhe, R. V. and C. G. Olson 1980. The origin of terra rossa in the karst of southern Indiana. In *Field trips 1980 from the Indiana University campus*, Bloomington, R. H. Shaver (ed.), Geol Soc. Am., North-Central Sect., 84–122. Bloomington, Ind.: Geology Department, Indiana University and Indiana Geological Survey.

Sims, J. 1973. The caves of Lost Creek. *Speleonews*, 17, 19–23.

Stenner, R. D. 1969. The measurement of the aggressiveness of water towards calcium carbonate. *Trans Cave Res. Grp G. Br.* **11**(3), 175–200.

Stenner, R. D. 1971. The measurement of the aggressiveness of water to calcium carbonate. *Trans Cave Res. Grp G. Br.* **13**(4), 283–95.

14
Karst groundwater activity and landform genesis in modern permafrost regions of Canada

Derek C. Ford

Introduction

This short chapter addresses two problems. The first is that of establishing the extent of modern karst groundwater circulation and associated karst landform genesis in terrains that are permafrozen today, i.e. may be presumed to have been permafrozen throughout Holocene times. An important part of this problem is determining whether any modern circulation is following, and perhaps expanding, underground routes that were established under differing conditions in earlier times, i.e. whether modern flow is being maintained at given localities in the restrictive permafrost environment because it has chanced to "inherit" groundwater channel systems created in more favorable thermal situations. The second problem is that of determining whether there may be significant development of karst beneath glaciers in regions that will become permafrozen upon deglaciation.

Permafrost is here defined as a thermal condition; beneath a shallow, seasonally thawed, surficial zone (the "active layer"), the ground temperature is permanently below 0°C, (Brown 1970). In Canada, three geographical zones are defined (Fig. 14.1): (a) The southern or **discontinuous** permafrost is limited to depths of a few meters in patches of frost-susceptible materials, which are normally of glacial drift or other detritus. (b) In the **widespread** zone, permafrost becomes more extensive, thickens to tens of meters, and extends into bedrock of all kinds. Discontinuous and widespread permafrost conditions are also present at higher altitude in the southern Rocky Mountains of Canada, generally above 2000 m asl. (c) In the **continuous** permafrost, depth quickly thickens to in excess of 100 m and is 500 m at Lat. 80° N in Ellesmere Island. But it is not present beneath most parts of the shallow seas that surround the many islands of the Arctic region, or beneath the freshwater lakes on the mainland (Brown 1970).

Karst rocks (platform and reefal limestones and dolomites, lagoonal gypsum and minor salt) are abundant in all zones (Fig. 14.1). They are

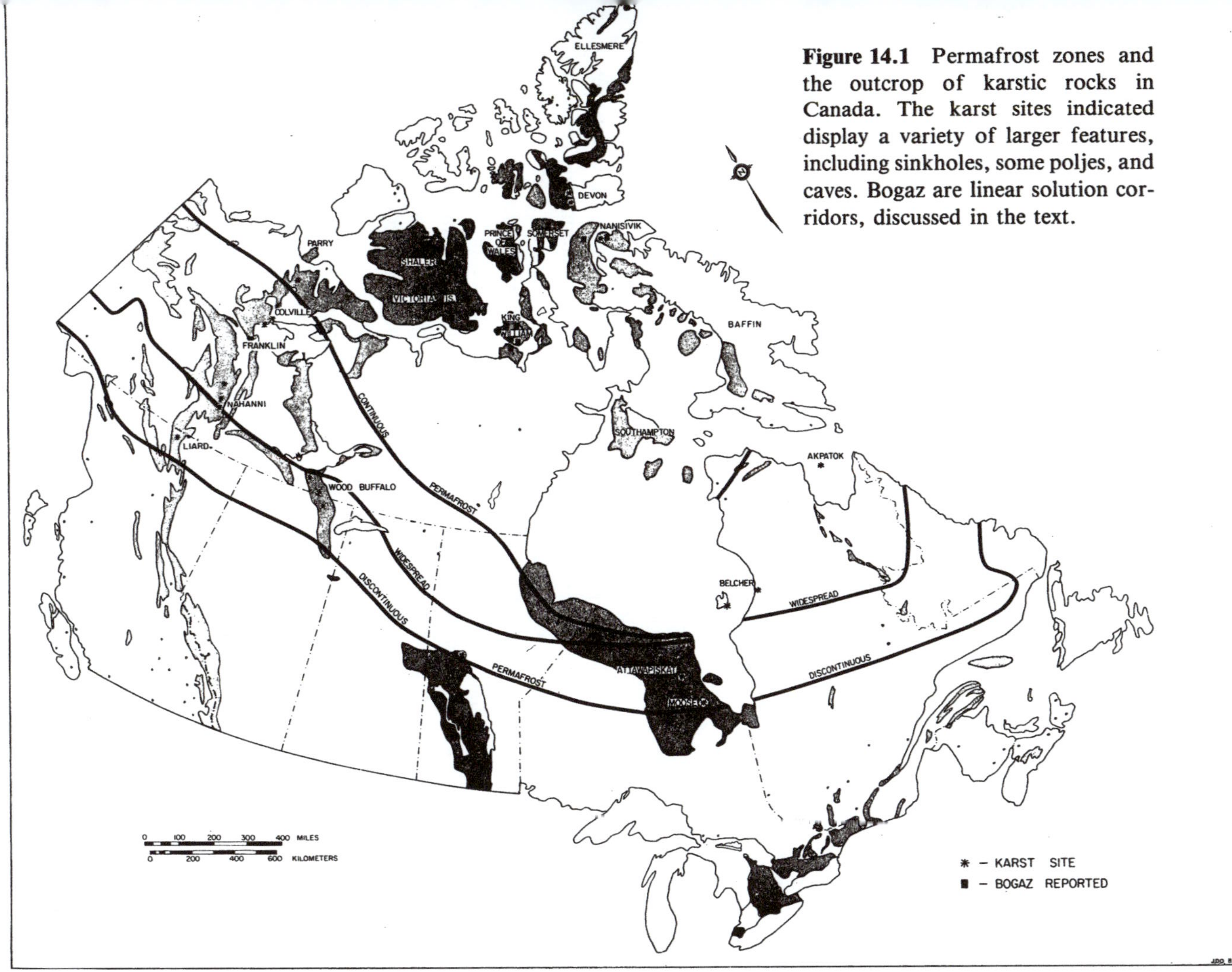

Figure 14.1 Permafrost zones and the outcrop of karstic rocks in Canada. The karst sites indicated display a variety of larger features, including sinkholes, some poljes, and caves. Bogaz are linear solution corridors, discussed in the text.

dominant mountain-forming rocks in the Rockies and in the more northerly Franklin Range, Mackenzie Mountains and Ogilvie Mountains, which extend close to the southern limit of continuous permafrost. A broad lowland belt of carbonates passes through the zones along the western edge of the Shield. The big southern Arctic islands (Victoria, King William, Prince of Wales, Somerset and western Baffin) are dominantly plateaus or lowlands of flatlying carbonates. Other carbonates, deformed in the Innuitian Orogen, extend northwards through Bathurst, Cornwallis and western Devon Island and form central fold belts exposed between the glaciers of Ellesmere Island. There are also diapirs in the remote northwestern islands. These rocks are primarily of Paleozoic age but include major Hadrynian and Helikian formations on Baffin Island. As would be expected in such a great region, there is considerable lithologic variation within the karst rocks. In addition to the massive reefal or well-bedded, formations that tend to host the best karst in any geographical area, there are extensive tracts of thin and interbedded rocks, arenaceous or argillaceous carbonates, not well suited to the development of karst circulation. The less soluble dolomite is generally more abundant than limestone north of Great Bear Lake.

We have one general model for karst landform development in permafrozen terrains. This is the "sub-cutaneous karst" model of Ciry (1962), based upon studies in northern France and similar areas of fossil permafrost phenomena. It asserts that circulation and solution are limited to the seasonally active zone, so favoring the development of spreads of shallow karren ground at the expense of a sinkhole-and-cavern topography. This climamorphic definition should not be confused with the more recent use of a subcutaneous concept by Williams (1983), who refers only to an uppermost zone of greatest solutional fissure and conduit frequency found in most karst situations but best developed, perhaps, where strata are well bedded and jointed and flatlying. From Ciry's application, it is as well, also, to distinguish two categories of permafrost karst: (a) arctic and subarctic regions where permafrost conditions are asserted whenever glacial ice is absent and may also be present beneath the ice (cold-based glaciers), and (b) cool temperate, extra-glacial regions where permafrost conditions may be introduced briefly during episodes of glacial cooling. The latter have received much more study, which is necessarily of a speculative kind because the permafrost condition is absent today.

Modern karst activity in the permafrost zones

Specific karst studies have been limited to date to the discontinuous and widespread permafrost zones in Canada. McMaster University parties have investigated several localities in the Rocky Mountains in detail; these may be taken as valid examples of southerly or marginal mountain permafrost conditions. Brook (1976) has studied the Nahanni karst, an outstanding instance of mountain karst in the center of the widespread zone. Van Everdingen (1981) reports on lowland karst between the Mackenzie River and Great Bear Lake. It is at the northern limit of the widespread zone with some minor extension into the continuous permafrost. North of that there have been limestone solution studies on Somerset, Devon and Ellesmere islands (Smith 1972, Cogley 1972, Woo & Marsh 1977), and Bird (1967) has written an essay on general observations of the limestone landscapes, which stresses the paucity of conventional karst landforms. Some geological mapping reports make passing reference to features in this zone, but without study. As a consequence, the discussion presented here must be considered preliminary and also speculative.

In the Rocky Mountains, many caves that contain permanent ice are now known. In portions of two of them, it is also apparent that the roof and wall rocks are frozen so that groundwaters may not permeate through them. But these are hydrologic fossils perched in mountain spurs. The groundwaters that excavated them are routed elsewhere as a consequence of alpine glacial dissection, and they have negligible catchment today. Our studies suggest that wherever waters may collect into volumes of the size of rivulets or greater, there is no significant permafrost impediment to their passing underground. The crucial test case occurs on the top of a carbonate plateau at Castleguard Mountain, Banff National Park. The altitude is 2300 m asl, the calculated mean annual temperature is $-5\,^{\circ}C$, and subcutaneous frost action has reduced the surface to a felsenmeer of shattered, platy clasts. A mere rivulet from a semi-permanent snowbank keeps open a solutional shaft that becomes impenetrably narrow at a depth of 30 m (Frost Pot – see Ford *et al.* 1976, Smart & Ford 1983). Dye tracing reveals that this water passes to springs 850 m below and 2 km distant in as little as four hours so that the karst groundwater system, despite its primitive enlargement and technical permafrost condition at the top, is very efficient.

The Nahanni karst extends as a narrow belt for some 60 km through the eastern Mackenzie Mountains. It is between Lats. 62–63°N and ranges between 800 m and 1400 m asl. Altitudinal cooling plus a paucity of insulating snow in winter should place its higher parts theoretically in a continuous permafrost condition. Karst landforms are developed in 180–200 m of massive Devonian limestones resting on 1000 m of dolomites. Groundwater circulation is known to pass at least 500 m down into the latter. Central parts of the belt function as regional holokarsts where all drainage is directed to two major systems of springs, each of which has a catchment of hundreds of square kilometers. Included in the karst are poljes that may be inundated by summer rains and which then drain during the fall and winter months (Brook & Ford 1980). At the regional scale, therefore, there is no permafrost impediment to the modern maintenance and development of very efficient karst drainage.

The karst landforms are varieties of sinkholes and solutional corridors. They are often large and occur in dense assemblages. Local relief of some features exceed 150 m. In fact, this is the most rugged karst terrain known in Canada or the United States (Brook & Ford 1978). There are three well-established phases of development: (a) systems of early phreatic conduits, now fossil and exposed in sinkhole and corridor walls, (b) a regional glaciation by a Laurentide ice sheet from the east. From U-series dating of stalagmites in the caves, this occurred before 350 000 years BP, (c) the modern sinkholes, corridors and poljes draining to a lower system of solution conduits that remain largely or entirely phreatic today. The important point here is that the spectacular surface karst features developed after the one known glaciation and, thus, within the Quaternary. It must be presumed that their development took place under permafrost conditions generally as severe as those of today or more severe, for it is known that Laurentide ice sheets approached within a few kilometers on at least two subsequent occasions (Crosbie 1978). There is evidence of one or two phases of milder conditions than the present in the form of large and ornamented speieothem (Harmon *et al.* 1977), but not during the past 250 000 years. So it may be suggested confidently that the conduit systems draining this large and mature karst have enlarged and extended during conditions similar to the modern or more severe, though it is possible that the tiny, but essential, protoconduit nets were generated in milder conditions.

The rugged Nahanni karst contains many isolated, butte-like rock masses with extensive benches below them that are dotted with sinkholes. Where this arrangement occurs, three distinct hydrologic zones may be recognized. The buttes (or high zone) are truly permafrozen, with no groundwater flow taking place. Fossil caves that penetrate them become completely filled with ice or frozen silt within a few meters of the entrances. A low zone is represented by poljes and the bases of all large sinkholes and solutional corridors. It is a zone of unimpeded recharge to the aquifers. The intermediate zone in topographic terms is represented by the benches with sinkholes. It is also intermediate in its permafrost-effect characteristics. It is a zone of a-periodic impedance of the karst recharge. This evidently takes the form of complete ice seals, which build up in horizontal (bedding-plane) conduits at their junctions with vertical-shaft dolines (joint-guided), which they drain. Some of the seals are merely seasonal occurrences; from our observations, others may survive for several years or more. Runoff slowly builds up as standing ponds above them (creating cenote-form sinkholes) until they rupture and drain in the space of a few hours or days. In the summer of 1973, two features that had been brim full ponds the year before were revealed as dry shafts 20 m deep. Nearby, a dry hole of 1972, 25 m in diameter and containing small spruce on its floor, was now flooded. The tops of the trees, still green, could be seen at a depth of 5 m. This feature remained flooded in 1974 but was dry again when visited the following year.

The impedance characteristics of the intermediate zone are of great interest in our context. Karst circulation by slug injection is being maintained today, but it is clear that this occurs through conduit systems that were created under unimpeded (low zone) conditions, i.e. the circulation takes place through inherited systems. Deepening of adjoining major sinkholes and corridors then leaves the benches in an intermediate situation. Catchments of individual sinkholes in the low zone are quite large and include some regular seasonal streams. Catchments of the impeded zone are reduced to a very local scale. There is insufficient supply of warmed, surficial waters to keep the conduits open every summer.

There is a major fluviokarst in the Mackenzie River lowlands to the north-east of the Nahanni region. It extends as a belt for at least 200 km north of Norman Wells (between Lats. 65° and 67°N), crossing thrust structures of the Franklin Mountains and shallow domal structures of the Colville Hills but also including much low-relief plain. The belt is up to 125 km wide. Its southern parts have been examined in detail by Van Everdingen (1981), who maps 1400 sinkhole features, 27 larger compound depressions up to 5 km in length, and at least 63 substantial karst springs. Some of the compound depressions function as poljes. Van Everdingen's mapping extends to the southern limit of continuous permafrost. Reconnaissance reports to the north (Yorath *et al.* 1968, Cook & Aitken 1970) suggest that there may be extensions of the karst at lesser density for some 200 km into the continuous zone.

The karst is developed in Ordovician–Devonian limestones, dolomites and gypsum that dip gently westwards except where locally deformed. It appears that most sinkhole development is to be attributed to the solution of gypsum and collapse of carbonate or clastic cover rocks into the voids created. There is probably some development entirely within limestone or dolomite.

The region was covered by Laurentide ice during the Late Wisconsinan glaciation, and much of it is deeply blanketed with Wisconsinan tills and glaciofluvial and lacustrine deposits. Most sinkholes have the form of shallow saucers in drift, often with fresh secondary collapses towards their centers. This implies that the sinkholes (and therefore the karst system development) are older than the last glaciation, but with Holocene rejuvenation. Purely collapse bedrock sinkholes of evident post-glacial age are rarer but do occur, pointing to the systematic extension of the karst in the prevailing permafrost conditions. Included in the category are examples in the Anderson and Horton River plains, deep within the continuous zone (Yorath *et al.* 1968). Some post-glacial sinkholes may be very impressive. An example near Norman Wells has an elliptical planform measuring 125 × 75 m, with vertical walls of shale descending 40 m to a pond of unknown depth.

Distribution of the sinkholes is very irregular. Greatest densities tend to occur where hydraulic gradients are highest, on the uplifted structures or along the rims of incised river valleys, but many hundreds are found on swampy plains of very low relief. It appears that lithology, structure-gradient effects, and depth of Wisconsinan drift cover are the principal factors determining the location and expansion of elements of the karst system, and that permafrost impedance is ineffective even where hydraulic gradients are very low indeed. It is concluded that, where more soluble rocks such as gypsum are available, unimpeded karst development may extend into southern parts of the continuous zone, but there is certainly an important measure of inheritance in the case in question. It is quite possible that it is an inheritance from pre-Quaternary conditions.

North of this area there have been no systematic inventories of karst features. It is established that there are no rugged karstlands such as Nahanni or extensive sinkhole plains such as the Mackenzie to be found in the northern islands. There appear to be some major springs on dry land; Bird (1967) mentions "icings" (perennial springs freezing at the resurgence in winter) on Southampton Island, Lat. 66°N, and elsewhere. A problem here is that there has been no precise hydrologic budgeting of sufficiently large basins in the islands to establish whether there are surface runoff deficits that might be explained by groundwaters circulating to submarine springs. Corbel (1957) has described such phenomena on Spitzbergen, where the drainage may be inspected in coal mines. There are no deep mines in the Canadian Arctic as yet.

Closed depressions that may be functioning as sinkholes are known on a belt of mixed gypsum–dolomite strata on Devon Island (Lat. 75°N). The features are probably older than the Wisconsinan glaciation of the region. Solution brecciation is noted in gypsums at the northern tip of Ellesmere Island (Lat. 83°N), but it is not established that the process is active today. On limestone and dolomite surfaces, the scant reports suggest that modern karstification is limited to the slow expansion of shallow clint-and-grike karren spreads. In the one detailed quantitative study of relevance, Woo and Marsh (1977) record groundwater circulation to a depth of one meter upon steeply dipping limestones in Ellesmere Island. This precisely fits Ciry's concept of karstification being limited to the seasonally active layer above an impenetrable permafrost. However, we should note that a very large proportion of the carbonate terrains in the continuous permafrost zone of Canada display exactly those characteristics that would favor development of the subcutaneous type of karst assemblage in any climatic region, i.e. they are well bedded and have very low relief and hydraulic gradient. Exacerbating the problem is the abundance of till rich in carbonate clasts that is dumped onto these surfaces in many areas. It has shielded the underlying bedrock from any post-glacial solutional attack; this effect is just as common south of all permafrost in Canada.

In conclusion, it has been shown that karst terrains are well established and developing without significant impediment in the discontinuous and widespread permafrost zones of Canada. Where karst is absent on carbonate rocks in these zones, it appears to be for the non-climatic reasons of unfavorable lithology, inadequate catchment or hydraulic gradient. But in each case, the sufficiently studied karsts are inferred to have been initiated before the Holocene pattern of permafrost zonation and depth was established. Although karst is expanding in the prevailing conditions, it has not been shown categorically that large scale karst circulation can be initiated in them. However, it appears that deeper flow paths were initiated in conditions generally more severe than the present in the Nahanni karst.

Known karst on carbonates in the Arctic island continuous permafrost zone appears to be restricted to the subcutaneous type. But details are scarce, and the structural and topographic settings are of the kind, over much of the area, to favor such development regardless of climatic factors. It is not known whether there is deep groundwater circulation through any of the carbonates. It probably occurs at some sites in the gypsum.

These findings lead one to suggest, tentatively, that Ciry's model is invalid for the fossil permafrost areas in temperate regions to which he has sought to apply it. In such regions, the permafrost will tend to be of the discontinuous type, shallow in depth and brief in duration. It is not proper in logic to suppose that where such conditions are imposed upon karst rocks, these will not previously have developed a karst circulation net, which as an "inheritance," has been shown to permit essentially unimpeded development except in butte situations.

Fossil sub-glacial karst development in modern permafrost zones

Supposing that Ciry's model of karstification restricted to the active layer really does apply in the continuous permafrost zone of Canada today (which may well be true upon the carbonates there), it is of interest to consider whether deeper, more "normal," karstification was possible under the very different thermal conditions that may have prevailed in the upper bedrock when most of the zone was covered by glacier ice. This may strike readers as a curious question because most analyses of the impact of glaciations upon karst terrains (including earlier writings of the author) stress the erasing, deranging or inhibiting effects of glaciers. In net terms, such stress is undoubtedly correct when applied to terrains that become temperate upon deglaciation, such as the alpine karsts of Europe, any glaciated karst in the contiguous United States, etc. This is the class of glaciated terrains that has received most of the analysis to date. But it is not clear that glaciers are destructive of karst in net terms when they spread onto permafrost-impeded ground, as the evidence produced below will suggest.

Glacier ice is a poor thermal conductor. A consequence is that where ice accumulates rapidly to a substantial thickness, the flux of geothermal heat is essentially trapped in its lower zone. This may raise its temperature to the pressure melting point and that of underlying terrain to a somewhat higher value, even in extremely cold regions such as mainland Arctic Canada. The point is discussed in detail in all glaciological textbooks (e.g. Paterson 1969), Shumskii 1964). Pressure meltwater is then generated at the ice base, which permits the ice to slide over the rock, so generating a little additional heat of friction. Sugden (1978) has produced a general model of erosion beneath the full Laurentide ice sheet of Canada, which suggests that the bedrock temperature over much of the present widespread and continuous permafrost zones was raised to the pressure melting point, i.e. the permafrost was eliminated. In more detail, we may suppose that the extent of such thawed permafrost zones will expand and contract with stadial fluctuations during each successive glaciation. Evidences suggest such stadial thawing for most of the karst rocks east of the cordillera and for those exposed in the large southern islands. It is less likely to have occurred in the northern islands such as Cornwallis, Devon and Ellesmere. Thus, water could be made available at the surface of karst rocks that were in a temperate thermal condition in the former areas. It remains to show that the hydraulic gradient could exist to permit the water to circulate deeply, and generate some features of normal karst while doing so.

There are two pieces of evidence (types of karst landforms) that do, indeed, suggest deep Quaternary circulation where there is none today, i.e. sub-glacial circulation. When the karst of Arctic Canada is better known, more may be found. Both evidences occur far within the continuous permafrost zone.

Bogaz (linear erosional corridors) are widespread on platform limestones and dolomites of King William Island and the Adelaide Peninsula (Fraser & Henoch 1959) and on Prince of Wales and Somerset islands (Bird 1967). Smaller patches are seen on eastern Victoria Island, though this terrain is more deeply drift covered, and on the rugged Borden and Brodeur peninsulas of Baffin Island. Individual bogaz extend for several hundred meters and occasionally, up to one kilometer. They are 5–30 m wide. At most sites, they are blanketed by Wisconsinan till or other late-glacial drift. Bedrock walls are not exposed, and the depth to the rock floors is unknown. However, it is clear that these floors lie below the base of the mean modern active layer. Examples in the Borden and Brodeur peninsulas do expose bedrock walls, which are being degraded by frost shatter. Some of these are 10–12 m deep to floors of scree. From the descriptions, they are very similar in form and scale to many solution corridors in the Nahanni karst.

The features are joint-aligned. In a given area, there may be preferential extension along a particular joint direction coinciding with the direction of glacier flow, but there are always strong minor orientations and cases of intersection are common. These relationships demonstrate that the features are not products of simple glacial scour. They pre-date emplacement of the latest till, are fossil today and are limited to massive limestones and dolomites. There is little doubt that they are karst landforms. They developed when water could circulate to depths of at least 10–15 m in the rock, with flow paths having lengths of some kilometers. Their development was arrested by growth of post-glacial permafrost. It is concluded that they were created by sub-glacial meltwaters. It is stressed that they are best developed on the low-relief plains where today, hydraulic gradients are minimal. This indicates that lack of bedrock topographic gradient need be no impediment to sub-glacial karst development.

The second evidence of sub-glacial karst circulation is a single feature. It occurs in the only Canadian Arctic mine that penetrates carbonate rocks to a significant extent. This is the Nanisivik zinc-lead mine at latitude 73°N on the Borden Peninsula, Baffin Island. The ore extends as a horizontal body for 3 km through a horst formed in steeply dipping, hard and impermeable, dolomites of Helikian age. Immediately to the south, an adjoining graben has been scoured to form a typical glaciated valley. The area was completely glaciated during the Late Wisconsinan but is now free of ice.

The Nanisivik mine passes through the ore body. It has an ambient temperature of −11°C, implying that permafrost temperatures extend 100+ m below. Small bodies of ground ice are common in vugs, and there is no flowing water.

For a distance of 200 m along the southwest wall of the mine (i.e. against the adjoining graben valley), the ore is in a state of arrested collapse into some deeper void. The arresting agent is a firm cement of ground ice formed by freezing downwards. Individual ice bodies between blocks of broken sulphides attain volumes of many cubic meters. They are supporting the blocks. Two boundary normal faults hading at 70° can be seen at the western limit of the collapse. Mining has not yet revealed its full extent to the east. The collapse must extend at least 50 m below the mine, where it is known by drilling that there are dolomites with minor sulphide bodies.

The collapse is interpreted as a product of solution by waters injected from the graben valley floor, which lies to the south and 10–15 m higher than the exposure in the mine. This injection is unlikely to have occurred in some pre-Quaternary, pre-permafrost situation because the collapse was unstable and still settling when arrested. It is suggested that the waters were injected sub-glacially during one or more glaciations. The injections extended to a depth of at least 60 m in bedrock and effected considerable solution.

In conclusion, it has been shown here that at some sites, lowland and mountainous, in the continuous permafrost zone where field visits have revealed karst processes and circulation to be limited to an active layer ~1 m in depth today, there was sub-glacial circulation to depths of 10–15 m over areas of many square kilometers, and to a depth of 60+ m at the one site that mining has opened for inspection.

In the widespread permafrost zone and at the southern limit of the continuous permafrost, it was shown in the previous section of this chapter that there is unimpeded karstic circulation and expansion, but with fissure or conduit networks probably inherited from previous conditions. In this section, it appears that sub-glacially generated karst systems cannot be passed on as a viable inheritance when the terrain reverts to deep, continuous permafrost conditions upon deglaciation.

References

Bird, J. B. 1967. *The physiography of Arctic Canada with special reference to the area south of Parry Channel*. Baltimore: Johns Hopkins Press.

Brook, G. A. 1976. *Geomorphology of the North Karst, South Nahanni River region, N.W.T., Canada*. PhD dissertation. McMaster University.

Brook, G. A. and D. C. Ford 1978. The nature of labyrinth karst and its implications for climaspecific models of tower karst. *Nature* **280**, 383–5.

Brook, G. A. and D. C. Ford 1980. Hydrology of the Nahanni Karst, northern Canada, and the importance of extreme summer storms. *J. Hydrol.* **46**, 103–21.

Brown, R. J. E. 1970. *Permafrost in Canada*. Toronto: University of Toronto Press.

Ciry, R. 1962. Le role du droid dans la spéléogenèse. *Spelunca Mem.* **2**(4), 29–34.

Cogley, J. G. 1972. *Processes of solution in an Arctic limestone terrain*. Inst. Br. Geogs Spec. Publn 4, 201–11.

Cooke, D. C. and J. D. Aitken 1970. *Geology, Colville Lake map-area and part of Coppermine map-area, Northwest Territories*. Geol Surv. Can. Pap. 70–12.

Corbel, J. 1957. *Les karsts du N.-O. de l'Europe et de quelques regions de comparaison*. Univ. Lyon Mem. Doc. 12.

Crosbie, M. L. 1978. *The nature and origin of deposits in Ram Plateau canyons, Mackenzie Mountains, N.W.T.* MS thesis. McMaster University.

Ford, D. C., R. S. Harmon, H. P. Schwarcz, T. M. L. Wigley and P. Thompson 1976. Geohydrologic and thermometric observations in the vicinity of the Columbia Icefield, Alberta and British Columbia, Canada. *J. Glaciol.* **16**(74), 219–30.

Fraser, J. K. and W. S. Henoch 1959. *Notes on the glaciation of King William Island and Adelaide Peninsula, N.W.T.*, Dept Mines Tech. Surv. Geog. Branch Pap. 22.

Harmon, R. S., D. C. Ford and H. P. Schwarcz 1977. Interglacial chronology of the Rocky and Mackenzie mountains based upon $^{230}TH/^{234}U$ dating of calcite speleothems. *Can. J. Earth Sci.* **14**(11), 1730–8.

Paterson, W. S. B. 1969. *The physics of glaciers*. New York: Pergamon Press.

Shumskii, P. A. 1964. *Principles of structural glaciology*. New York: Dover.

Smart, C. C. and D. C. Ford 1983. Hydrogeology of the Castleguard karst, Main ranges, Rocky Mountains of Canada. *J. Hydrol.* **61**, 193–200.

Smith, D. I. 1972. *The solution of limestone in an Arctic environment*. Inst. Br. Geogs Spec. Publn 4, 187–99.

Sugden, D. E. 1978. Glacial erosion by the Laurentide ice sheet. *J. Glaciol.* **20**(83), 367–92.

Van Everdingen, R. O. 1981. *Morphology, hydrology and hydrochemistry of karst in permafrost near Great Bear Lake, Northwest Territories*. Nat. Hydrol. Res. Inst. Pap. 11.

Williams, P. W. 1983. The rôle of the subcutaneous zone in karst hydrology. *J. Hydrol.* **61**, 45–68.

Woo, M-K., and P. E. Marsh 1977. Effect of vegetation on limestone solution in a small High Arctic basin. *Can. J. Earth Sci.* **14**(4), 571–81.

Yorath, C. J., H. R. Balkwill and R. W. Klassen 1968. *Geology of the eastern part of the Northern Interior and Arctic Coastal Plains, Northwest Territories*. Geol Surv. Can. Pap. 68–27.

15
Hydrogeomorphic evolution of karsted plateaus in response to regional tectonism

Ernst H. Kastning, Jr

Introduction

The origin of solutional caves has been the subject of considerable controversy for some time. In its infancy, karst science was highly descriptive and many early concepts attributed cavern development to catastrophic events or scouring of rock by subsurface streams (Davies 1966, 1968, Quinlan 1968, 1978, Shaw 1979). Little attention was paid to the details of geologic and hydrologic settings as controlling factors until the late 19th century. Systematic studies of cave origin began to flourish in the mid-20th century when several classic papers on the subject appeared in major geological journals (Davis 1930, Swinnerton 1932, Gardner 1935, Malott, 1937, Bretz 1942, Woodward 1961). Differences among the theories presented in these accounts centered largely around the vertical position within the groundwater zone at which solutional excavation takes place. Do caves develop above the water table (vadose zone) or below it (phreatic zone)? Each paper was based on fieldwork confined to relatively few caves in specific hydrogeologic settings rather than in diverse settings in many locales. (See White 1959, Halliday 1960, Warwick 1962, Palmer 1962, DeSaussure 1963, Kastning 1975, Powell 1975b, and Sweeting 1981 for reviews of the classic papers.) It soon became apparent that no single theory was sufficient to explain the origin of all solutional caves, and recent studies focus in detail on factors that control or influence cavern development. Greater emphasis is now placed on the roles of geologic structure (Davies 1960, Ford 1965, 1971, Ford & Ewers 1978) and the hydrodynamics of groundwater in various structural settings (A. N. Palmer 1972, 1975, 1977, 1981, M. V. Palmer 1976, White 1969, 1977).

Together, geologic structure, lithic character of the bedrock, topographic conditions, and hydraulic gradient within the aquifer determine the final configurations of solutional conduits. Under different settings, any or all of these factors may dominate; however, structural framework determines initial avenues of groundwater circulation, whether these are bedding-plane partings, joints or faults. Moreover, the attitude of strata (strike and dip) in conjunction with the position of the water table, will determine whether passages will develop horizontally along the strike in the shallow-phreatic zone or will be inclined down the dip in the vadose zone (Palmer 1972, 1977). In steeply dipping strata, the frequency of fractures may determine whether groundwater flows along deep phreatic loops or moves within the shallow-phreatic zone (Ford 1971, Ford & Ewers 1978). Geologic structure, in part, controls collapse of passage ceilings, a process that may block active conduits and lead to backflooding, development of diversion routes, and excavation of maze passages (Palmer 1972, 1975).

Karst features as aids to interpreting structure

The structure underlying a landscape, particularly one of relatively low relief, may be difficult to study. This is particularly true if deformation has been slight or the area under study is covered by thick soil or glacial deposits. The problem of mapping structure may be lessened considerably in karst terrane where geologic structure promotes enlargement of fractures and guides development of caves or surface karst landforms.

Because caves provide access to the subsurface, careful stratigraphic mapping with precise vertical control along cave passages can often divulge subtle changes in attitude or thickness of strata (Palmer 1981). In areas of extensive cavern development, composite geologic maps of caves can provide working structural maps of the region. Orientations of linear cave-passage segments suggest significant fracture trends that can be used to identify structural elements and to interpret tectonic stresses. Alignments of surface karst landforms, such as dolines or solutionally enlarged fissures in limestone pavements, may provide similar information. This technique of structural mapping is often enhanced by use of remote sensing, especially through analysis of aerial photographs, or by noting alignments on topographic maps.

Avenues for groundwater flow, such as partings and fractures, are generally created or modified when a region undergoes tectonic stress and deformation. Fracture patterns, including fault zones and joint sets, commonly reflect distinct episodes of tectonism. Complex patterns, consisting of several superimposed fracture sets, may represent two or more sequential or synchronous structural events.

Influence of tectonism on karst is most easily assessed where trends can be mapped and compared on a regional basis. This study focuses on three extensive karsted plateaus in the United States (Fig. 15.1): (a) the Helderberg Plateau, east-central New York, (b) the Mississippian Plateau, western Kentucky, and (c) the Edwards Plateau, central Texas. Although these regions differ stratigraphically and in climatic conditions that prevailed during karstification, each one represents an upland surface developed on a gently dip-

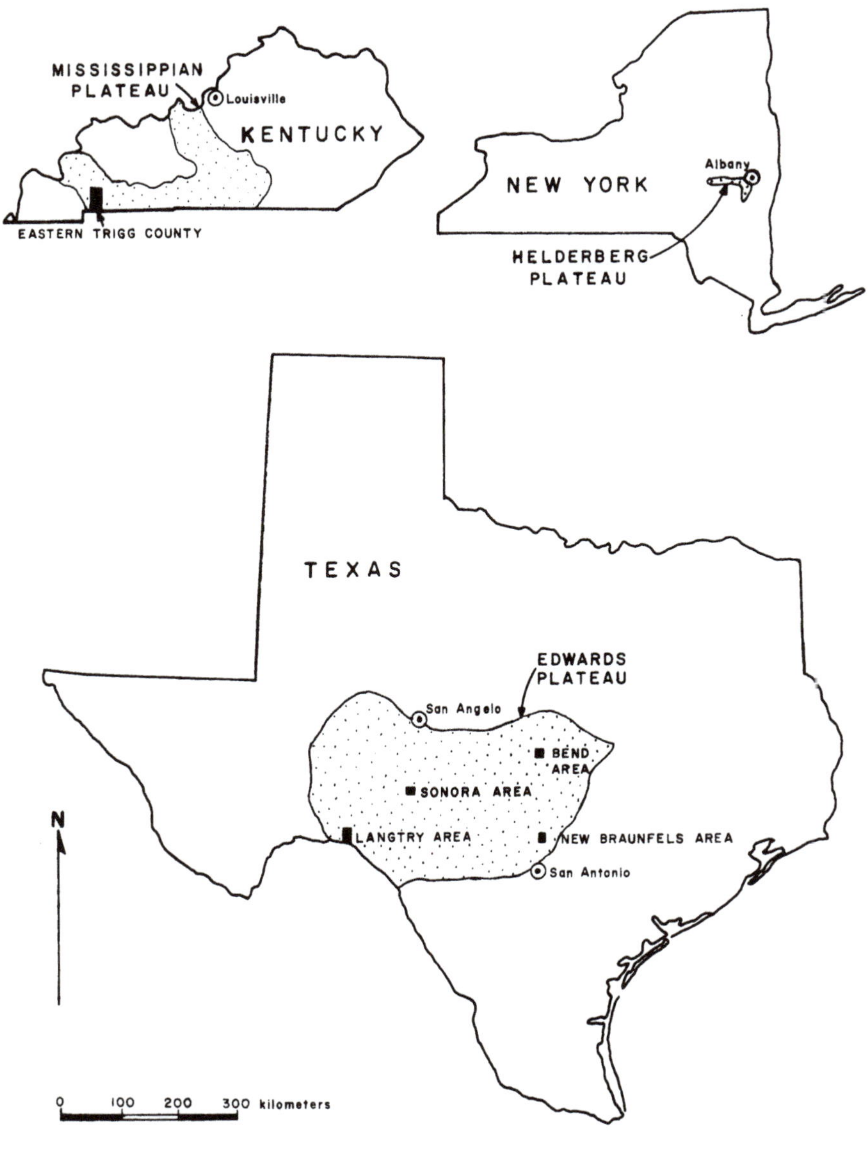

KMK 11-12-81

Figure 15.1 Location map showing the Helderberg Plateau of east-central New York, the Mississippian Plateau of western Kentucky, and the Edwards Plateau of central Texas. Sub-areas discussed in the text are indicated by black rectangles.

ping sequence of carbonate rocks, and each is characterized by systematic fractures that have guided development of karst landforms.

Similar studies have been done in other areas, including Alabama (Wilson 1977), Arkansas (Barlow & Ogden 1982), Indiana (Powell 1975a, 1977), Kentucky (Hine 1970), Missouri (Barnholtz 1961), Pennsylvania (Lattman & Parizek 1964, Deike 1969), Texas (Rich 1928, Wermund & Cepeda 1977, Wermund *et al.* 1978) and West Virginia (Ogden 1974).

Helderberg Plateau, east-central New York

The Helderberg Plateau in Schoharie and Albany counties, New York, is underlain by a sequence of Silurian and Devonian limestone, sandstone and shale that crops out as a narrow band up to 8 km in width. Carbonate beds of the Helderberg Group and Onondaga Formation dip between 1.5° and 2.0° to the SSW and form topographic benches. These surfaces are uplands with respect to low-lying areas below the Helderberg Escarpment to the north and east. The escarpment represents the northern boundary of the glaciated Allegheny Plateau of the Appalachian Physiographic Province.

Carbonate rocks of the region consist principally of dense crystalline limestone and dolostone within two stratigraphic groups (Fig. 15.2). Units of the Helderberg Group vary significantly in composition and thickness of beds. Limestone units high in calcite content (80–90% $CaCO_3$) include the Thacher Member (Manlius Formation), Ravena Member (Coeymans Formation) and the upper Becraft Formation. Remaining units, the Cobleskill, Chrysler, Kalkberg, and New Scotland beds, are high in silica, magnesium, and insoluble material (up to 53%). Bedding thicknesses vary from a few centimeters up to 25 cm within most units; however, the upper Thacher and Ravena members are consistently thick bedded (12–50 cm). In general, the thin-bedded units contain numerous shale partings whereas the massive units are relatively free of argillaceous beds.

Calcite contents of rock of the Onondaga Formation vary between 79% and 98%. The Edgecliff and Moorehouse limestone members are relatively pure in calcite (84–98%), but the Schoharie Formation and Nedrow Member are high in silica and insoluble material (up to 17%). Strata of the Onondaga are consistently thin (1–12 cm), but locally contain abundant bedded chert up to 20 cm thick.

These upper Silurian and Devonian beds were deformed during the Acadian and Allegheny orogenies in the late Paleozoic (Rodgers 1970). Carbonate rocks of eastern Albany County experienced locally intense folding, faulting and fracturing during the late Acadian Orogeny. Mild warping, which resulted in the present-day regional dip, is ascribed to the Allegheny phase.

Joint and other fracture patterns are of particular interest in interpreting the area's karst features. Parker (1942) identified three regional joint sets in east-

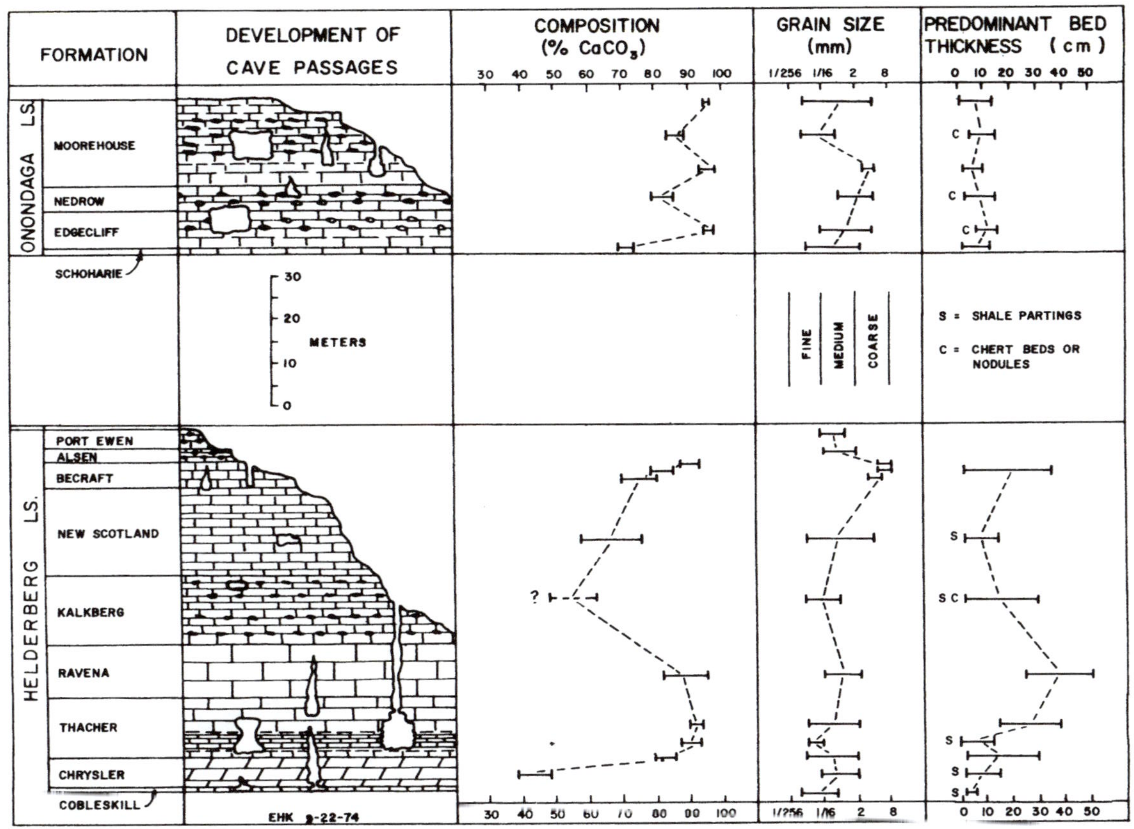

Figure 15.2 Relationship of cave-passage development to limestone composition, grain size and bedding thickness, Albany County, New York. Based on data from Johnsen (1958), Rickard (1962, 1975) and Kastning (1975).

central New York (Fig. 15.3). Joints of the principal set (Set I) are remarkably planar, vertical and smooth. They generally strike between N2°E and N30°E. Their extents range from a few centimeters to hundreds of meters horizontally and vertically, and they are numerous in both competent and incompetent rocks. Shear fractures of this set commonly intersect at acute dihedral angles and suggest simultaneous compression and tension at the time of fracturing (Parker 1942, Muhlberger 1961). Joints of Set II have irregular, curved surfaces and are generally confined to impure shaly or sandy beds. However, they are scarce in thick-bedded units. They strike between N45°W and N85°W. The rough nature of the joint surfaces in inhomogeneous rocks suggests that these fractures are the product of tensional forces acting perpendicular to the joint planes (Parker 1942). Joints of Set III are horizontally long and vertical and many are curved. They strike between N55°E and N65°E and occur primarily in weak homogeneous beds. Set III joints account for less than 15% of the total joints of the area, and their extents are dependent on the lithologic properties of the rock. The joints rarely extend upwards or downward into thicker-bedded units.

Parker (1942) found a regional variation of systematic jointing throughout the slightly deformed Paleozoic rocks of New York. Strike directions of Set I joints vary slightly from west to east, forming a series of radiating linears sweeping across the state. Set II joints, which are generally orthogonal to those of Set I, form a concentric arcuate pattern. These configurations extend from New York to Pennsylvania and Ohio (Nickelsen & Hough 1967, 1969, Parker 1969).

In New York, orientation of systematic joints is independent of other structural elements, such as regional dip, folds and faults. By application of crosscutting relationships, Parker (1942) found that jointing predates other struc-

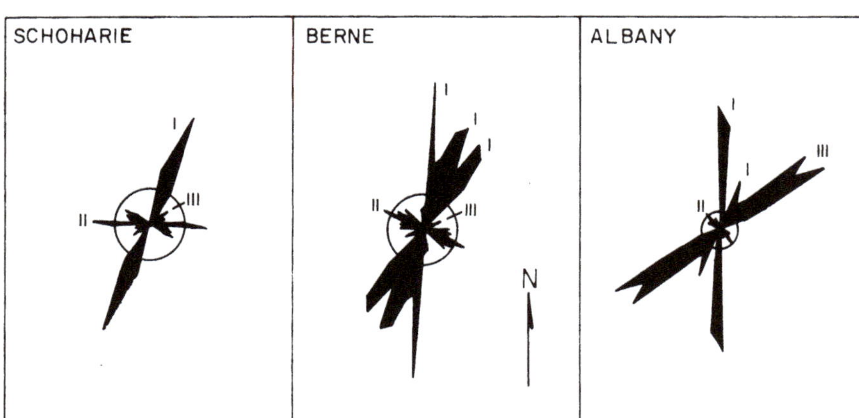

Figure 15.3 Orientation diagrams for joint sets of the Schoharie, Berne and Albany 15-minute quadrangles, east-central New York. Radii of circles represent 10 joints. Roman numerals indicate joint sets. Modified from Parker (1942).

tures and can be attributed to stresses incurred during regional warping and topographic uplifting that occurred prior to the onset of the Appalachian orogenies. Therefore, systematic joints may be related to basement tectonics.

Joints of Set I and Set II are probably conjugate as suggested by their mutual orthogonality and similarity in regional variation. Those of Set III have a fairly constant strike throughout the state, indicating stresses independent of those that formed joints of sets I and II. Set III joints may be of a different age.

Numerous local structures, such as folds, faults and other fractures, are superimposed on regional joint patterns. Folds increase in amplitude and decrease in wavelength toward the east side of the Helderberg Plateau, in proximity to the Appalachian foldbelt. Here limbs of folds may dip as much as 30°, and in general, fold axes are aligned in a northeast–southwest direction, parallel to the axis of the Appalachian deformation.

High-angle normal faults and low-angle reverse (thrust) faults are present throughout the Helderberg region, but decrease in abundance to the west, away from the foldbelt. Many faults are shown on geologic maps of the area (Ruedemann 1930, Johnsen 1958, Fisher *et al.* 1970–1971, Isachsen & McKendree 1977): however, several previously unreported faults were located during speleological studies (Palmer 1972, Gregg 1974, Kastning 1975, 1977, Mylroie 1977). Movements along normal faults are generally <0.3 m, but along thrust

Table 15.1 Character of joints in the carbonate units of the Helderberg Plateau, modified from Palmer (1962) and Kastning (1975).

Formation	Character of joints
Onondaga Limestone	Vertical joints up to 30m long with 0.6–3.0m spacing. Lower Moorehouse subunit has joints dipping up to 75°.
Becraft Limestone	Joints up to several hundred meters long with spacings from 1.5 to 6m. Vertical, well-defined joints.
New Scotland and Kalkberg limestones	Generally poorly defined joints. Occasional well-defined joints extend upward from the Ravena Limestone. Joint spacing is 0.9–1.5m. Joints are generally vertical.
Ravena Limestone	Many well-defined, vertical joints spaced ~ 1.5–7.5m apart. Joints often several hundred meters long.
Thacher Limestone	Many well-defined, vertical joints spaced 0.2–2.0m apart. Joints up to several hundred meters long.
Chrysler Dolomite	Many poorly defined joints. Very few well-defined joints. Joints are vertical and spaced 0.9–3.0m apart.
Cobleskill Dolomite	Many short joints spaced 0.01–0.2m apart.

faults, movements of tens of meters have been recorded (Darton 1893, 1894, Kastning 1975, 1977).

The three regional joint sets described above account for most observed fractures of the Helderberg Plateau, but joints associated with folds are locally abundant in areas of deformation. Moreover, joint swarms occur as fracture zones associated with faults and commonly extend for distances several times the length of single joints within the swarms (Fig. 15.4). It is probable that small joints also formed during expansion of bedrock in response to removal of overburden by fluvial or glacial erosion, or to subsidence and spalling along escarpments.

Lithologic character has affected spacing and horizontal and vertical extents of joints and fracture zones. Characteristics of joints in the cave-forming rock units are summarized in Table 15.1. Generally, thin-bedded units (e.g. the upper Cobleskill, Thacher, and Becraft limestones) have small joint spacing. Conversely, medium-bedded to massive units (e.g. the Chrysler, Ravena, and Onondaga limestones) have larger joint spacings. Additionally, horizontally and vertically extensive joints developed in thick-bedded, relatively homogeneous units (e.g. the Thacher, Ravena, Kalkberg, New Scotland, and Onondaga limestones). Joint surfaces are smooth in homogeneous units (e.g. the Thacher and Ravena limestones), but are noticeably rougher in units with an abundance of fossils, chert or irregular bedding planes (e.g. the Chrysler, Kalkberg, New Scotland, and Onondaga limestones).

Joints have exerted the most obvious structural control on cave-passage configurations in the Helderberg Plateau. Many caves of the region consist of linear, vertically oriented, parallel passages aligned along principal joint sets (Sets I, II or III of Parker 1942) or along joints related either to the Appalachian tectonic events or to unloading. Cave-joint associations were recognized by early investigators of caves and groundwater resources of the Helderberg region (Grabau 1906, Cook 1907, Goldring 1935, Berdan 1948, 1950), and detailed mapping of caves during the last three decades has substantiated the prevalence of joint control (Anderson 1961, Palmer 1962, Egemeier 1969, Kastning 1975, Baker 1976, Mylroie 1977).

Other structures influenced cave development on a local scale. Onesquethaw Cave, Albany County, formed in part along the flanks of an anticline (Palmer 1972). Thrust faults guided passage orientations in McFails Cave (Palmer 1976), Barytes Cave (Mylroie 1977), Spider Web Cave (Mylroie 1977), Shelter Cave (Mylroie 1977), and Van Vliets Cave (Mylroie 1977) in Schoharie County and Clarksville Cave (Kastning 1975, 1977) in Albany County. Vertical, normal faults controlled passage development in Tufa Cave (Mylroie 1977) and Schoharie Cave (Fig. 15.4) (Kastning 1975,1977) in Schoharie County and in Onesquethaw Cave (Palmer 1972) in Albany County. Faults do not always behave as zones of weakness along which permeability is enhanced (Kastning 1977). In some cases, mapped faults have exerted little or no influence on cavern development, e.g. Howe Caverns (Gregg 1974) and Secret-Bensons Cave (Mylroie 1977) in Schoharie County. In others, thrust faulting has

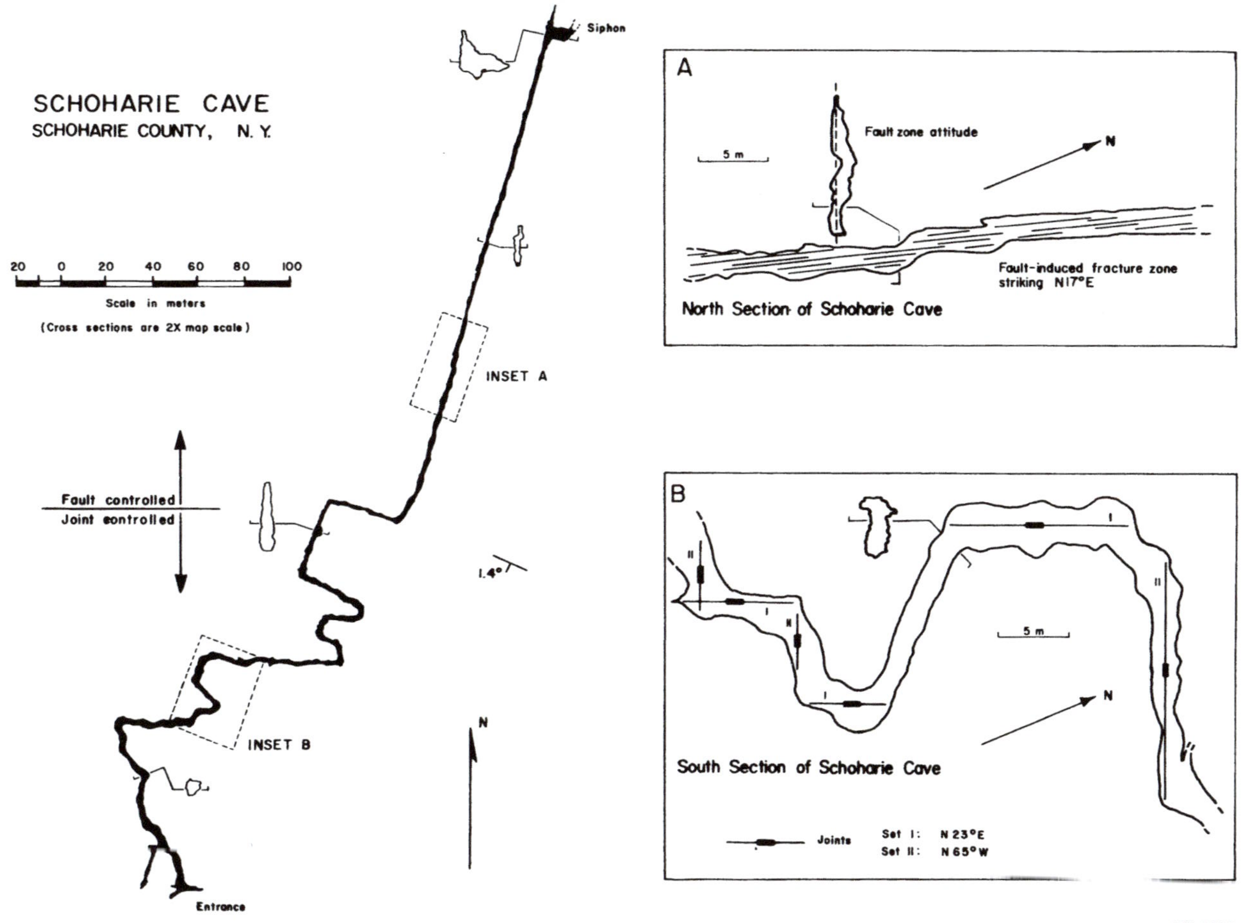

Figure 15.4 Planimetric map of Schoharie Cave, Schoharie County, New York, showing orientations of linear passage segments along (a) fracture zone produced by tensional faulting and (b) fractures of sets I and II of Parker (1942). Modified from Kastning (1975, 1977).

recrystallized limestone along fault planes, forming an impermeable surface on which groundwater becomes perched, e.g. Clarksville Cave in Albany County (Kastning 1975, 1977).

The greatest cavern development in the Helderberg region is generally confined to thinly bedded, well-jointed strata that are rich in calcite, such as the Thacher and Onondaga limestone units (Fig. 15.2). The abundance of bedding-plane partings augmented horizontal permeability within these rocks, leading to relatively large discharge and dissolution rates, particularly along joint/bedding-plane intersections. The Ravena limestone, a massive, high-calcite unit with extensive joints spaced farther apart, (Table 15.1), is also a good cave former. However, passages within this member usually extend downward into the Thacher beds. The Thacher and Ravena limestones behave as a single hydrostratigraphic unit, and the most extensive caves have passages in both members (Palmer 1962, Baker 1973, Kastning 1975).

During cavern development in the Helderberg Plateau, meteoric water, which entered the upland surfaces as recharge through joints, was guided down the dip to local base-level discharge points. Lateral flow occurred within those limestone beds of most favorable lithology and along the most open joints, along bedding-plane partings, or both. Angulate patterns of caves in plan view (Fig. 15.4) reflect the alternate use of fracture sets as water moved down dip.

Complex sections of caves, where passages form mazes along orthogonal joints, e.g. Skull Cave and the Camp Allen Cave System in Albany County (Kastning 1975), are products of cavern development in the floodwater zone (Palmer 1975). Under these conditions, fractures of all degrees of openness are excavated by turbulent, chemically aggressive water under steep hydraulic gradients.

In summary, karstic groundwater flow and excavation of most caves in the Helderberg Plateau occurred in a down-dip direction, following the most advantageous initial avenues (bedding-plane partings and extensive, open joints). Cave-passage configurations reflect regional systematic jointing produced prior to the Appalachian Revolution, regional dip originating from tectonism during the Allegheny Orogeny, and fracturing incurred during the Acadian Orogeny.

Mississippian Plateau, western Kentucky

The non-glaciated Mississippian Plateau formed on a broad, arcuate belt of Mississippian formations, predominantly limestones, along the west flank of the Cincinnati Arch. The plateau is divided by the 60 m high Dripping Springs Escarpment into (a) the Pennyroyal Plain, below and to the south of the escarpment and (b) the Mammoth Cave Plateau, above and to the north (Sauer 1927). The Pennyroyal Plain is principally underlain by cavernous St., Louis and Ste. Genevieve limestones. The Mammoth Cave Plateau developed on a cap of Cypress Sandstone lying above the Ste. Genevieve (McFarlan 1943).

A part of the Mississippian Plateau, encompassing nearly all of eastern Trigg County in western Kentucky (Fig. 15.1), was investigated to assess the role fractures play in karst formation. Drainage of the Pennyroyal Plain varies from deeply entrenched streams to extensive upland areas that drain internally through numerous dolines and caves. The topography is gently rolling. Where stream incision has occurred, however, bluffs line some valley sides (Fig. 15.5), and in some areas, local relief exceeds 70 m.

The St. Louis Limestone is, in part, a fine- to medium-grained limestone with alternating massive bioclastic beds, 0.6–1.6 m thick, and thin argillaceous beds. In places, the unit consists of very finely crystalline, silty dolomitic beds. Nodular, bedded and irregular chert is present in much of the unit, and gypsum occurs in some beds (Ulrich & Klemic 1966).

The lower part of the Ste. Genevieve Limestone consists of interlayered beds that vary texturally from finely crystalline and partly dolomitic to coarse-grained clastic and contain detrital fossil fragments. A fine-grained oölitic limestone commonly occurs at the base, and the contact with the underlying St. Louis Limestone is well defined and conformable. Nodular, spherical and tabular chert comprises from a few percent up to 50% of some beds. The top of the zone of abundant nodular chert is distinct and serves as a marker bed (Ulrich & Klemic 1966). The lower Ste. Genevieve Limestone, as mapped in part of the area (Ulrich & Klemic 1966, Klemic & Ulrich 1967, Klemic *et al.* 1968), is correlative with the upper member of the St. Louis Limestone, as mapped elsewhere (Fox 1965, Nelson & Seeland 1968, Seeland 1968).

The upper part of the Ste. Genevieve Limestone is a fine- to very coarse-grained, oölitic, bioclastic, commonly crossbedded limestone, with local thin beds of fine-grained dolomitic limestone. A few thin lenses of chert are present in upper beds, and bedded chert is locally abundant in lower beds (Ulrich & Klemic 1966).

In the southwestern part of the study area, Mississippian formations are unconformably overlain by the Tuscaloosa Formation (Upper Cretaceous), consisting of gravel, sand, silt and clay (Roberts 1929). Large clasts are typically well-rounded ellipsoidal or discoidal pebbles or cobbles of quartzite and sandstone. In places, well-rounded to angular chert derived from the Mississippian limestones are mixed with the gravels. Locally, Lower Cretaceous colluvial deposits underlie the Tuscaloosa and consist of sandstone rubble composed of blocks and slabs of sandstone slumped together with blocks of limestone (Klemic *et al.* 1968). Quaternary alluvial deposits occur in valley bottoms of major streams and within valleys of abandoned meanders. Paired terraces composed of this material are common along the streams.

Regional uplift of the western Kentucky region followed deposition of Mississippian sediments (Meramec and Chester series). During uplift, structural doming dominated to the northeast, southeast (Cincinnati Arch), and northwest, while structural subsidence to the north formed the Illinois (or

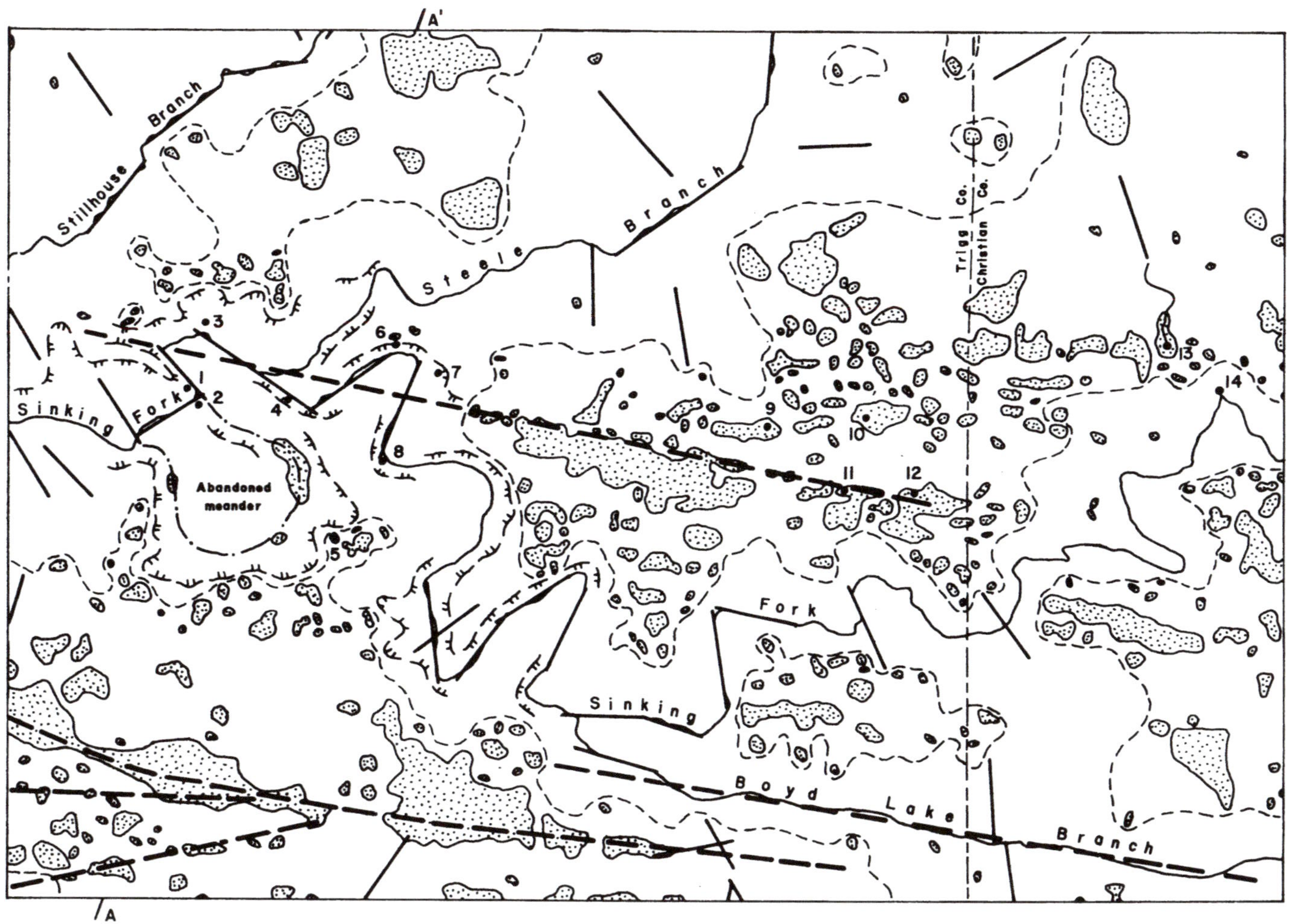
A'
A
Stillhouse Branch
Steele Branch
Sinking Fork
Sinking Fork
Abandoned meander
Boyd Lake Branch
Trigg Co.
Christian Co.
1
2
3
4
5
6
7
8
9
10
11
12
13
14

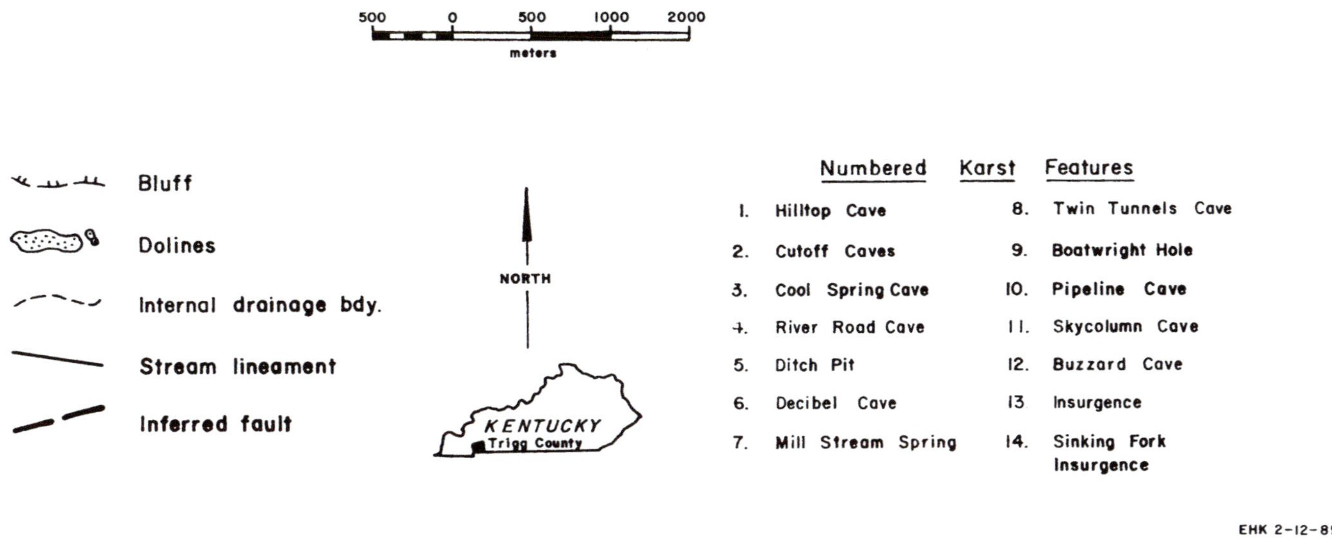

Figure 15.5 Geomorphic map of a part of the Caledonia quadrangle, Mississippian Plateau, eastern Trigg County and western Christian County, Kentucky. Dolines, areas of internal drainage, caves, and other karst features are indicated. Stream lineaments and inferred faults were identified from topographic maps and aerial photography. A–A′ is a part of the line of section in Figure 15.6. Modified from Kastning and Kastning (1981).

Western Coal) Basin. Surf erosion progressed and increased in intensity as upwarping continued from the Pennsylvanian to Tertiary (Dicken & Brown 1938, McFarlan 1943, Brown & Lambert 1963). The Nashville Basin, a truncated structural dome along the axis of the Cincinnati Arch, lies southeast of the area. Extensive normal faulting apparently occurred between the Middle Permian and Middle Cretaceous to recent (Trace 1974).

Regional dip in eastern Trigg County is $<1.0°$ toward the Illinois Basin to the north (Dicken & Brown 1938). Structure contours on 1 : 24000 scale geologic maps (Fox 1965, Ulrich & Klemic 1966, Klemic & Ulrich 1967, Seeland 1968, Nelson & Seeland 1968, Klemic *et al.* 1968) indicate local variations in magnitude and direction of regional dip, and in some cases, dips are reversed. Most local changes in bedding attitude are associated with block faulting.

Many faults have been identified and mapped in the western Mississippian Plateau. However, mapping has been somewhat inconsistent, with greater detail present in maps of the fluorspar district at the western end of the plateau in Kentucky. Here, faulting has been intense, and the economic importance of fluorite mineralization along fault planes has led to exhaustive structural mapping (Weller & Sutton 1951, Heyl & Brock 1961, Heyl 1972, 1974, Hook 1974). To the southeast in eastern Trigg County, fewer faults have been mapped. A large graben extends ENE across two 7.5 minute quadrangles (Seeland 1968, Nelson & Seeland 1968), but few faults have been mapped to the southeast of this feature.

Mapping during the course of this study indicates that many additional faults are likely in the area (Kastning & Kastning 1981). Fracture patterns (presumably representing both faults and joints) have been recognized and mapped by the following methods: (a) measurement of linear stream segments on topographic maps, (b) analysis of linear passage segments on published cave maps, (c) mapping of photolineaments from black-and-white (1 : 20000) and color (1 : 100000) aerial photographs, (d) interpretation of alignments of dolines and closed depressions on topographic maps, and (e) confirmation of fractures in caves and on the surface.

There is strong evidence for structural control of surface and subsurface drainage throughout eastern Trigg County. (A part of this area is shown in the geomorphic map, Figure 15.5, and in the geologic section, Figure 15.6.) Four dominant directions were observed for linear stream segments over 0.5 km in length: N70°–90°W, N10°–30°W, N10°–30°E, and N50°–60°E. These coincide with joints measured at surface outcrops and with orientations of linear passage segments on published cave maps (Mylroie 1978, 1979, 1980, Moore & Mylroie 1979).

Many dolines and closed depressions shown on topographic maps are highly aligned (Fig. 15.5); however, few of these alignments are visible on aerial photographs. A comparison of doline alignments with mapped faults shows that many dolines lie along fault traces, and others lie along hypothetical extensions of the faults. Doline alignments are typically extensive; in some cases,

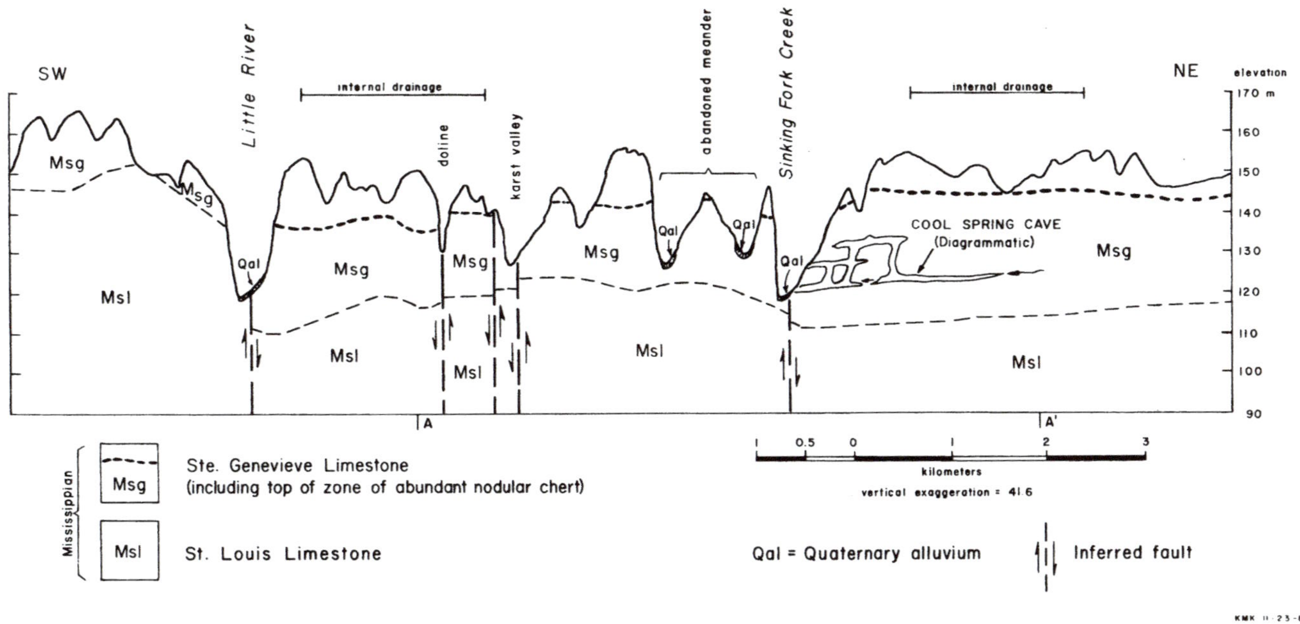

Figure 15.6 Geologic section of eastern Trigg County, Kentucky, showing geologic units and structure. A–A′ indicates extent of Figure 15.5 along line of section. Modified from section by K. Kastning, based on geologic map by Ulrich and Klemic (1966) and field and map interpretations.

they can be traced for distances of as much as 8–10 km. Several major alignments do not correspond to previously mapped faults; however, they are parallel or sub-parallel to nearby known faults and thus suggest additional faults in the subsurface. Composite maps and sections of known and supposed faults (e.g. Figs 15.5 & 6) show that their patterns represent a system that is consistent with regional faulting in western Kentucky, particularly in the fluorspar district. It appears possible to map "faults" by simply determining major alignments of closed depressions. Because the Mississippian limestone bedrock is blanketed by pervasive residual soils (Dicken & Brown 1938) and is not exposed except along bluffs and in steep-walled dolines, many faults have presumably gone unmapped. It appears that the determination of major alignments of closed depressions is a suitable means of detecting additional faults in the western Pennyroyal Plain.

Subsurface drainage paths, as determined by tracing of dyes injected at points of recharge (Moore & Mylroie 1979), agree with lineaments and supposed fractures mapped from doline alignments and stream segments (Fig. 15.5). Groundwater, infiltrating through dolines or as streams sinking into swallow holes, travel along dominant open fractures, enlarging these through dissolution. As conduits enlarge, they grow headward, capture additional surface drainage, and become integrated into dendritic networks consisting of master channels and tributaries. Such flow systems presently drain extensive interfluvial uplands where nearly all surface water enters dolines.

After regional uplift, erosion stripped Pennsylvanian and upper Mississippian sedimentary rocks from the area, exposing the highly fractured Ste. Genevieve and St. Louis Limestone beds. Lithic character of the bedrock and degree of fracturing have influenced the form and position of karst features that prevail today. Maximum degradation has occurred in the vicinity of fault zones, and reaches of some base-level streams that drain the limestone plateau are commonly located along major faults (Figs 15.5 & 6).

Depth of dolines is apparently stratigraphically controlled; those developed in large internally drained areas seldom extend below the top of the nodular chert beds of the Ste. Genevieve Limestone. Moreover, extensive cave passages have developed in the oölitic, relatively chert-free beds below, indicating that vertical infiltration in the uplands is transmitted along fractures to the oölitic zone where water is then conducted horizontally to base-level springs along the bluffs of major stream valleys. Cave passages are generally horizontal and at grade to the surface drainage. In some cases (e.g. Cool Spring Cave, Fig. 15.6), upper cave levels appear to be relict conduits that once communicated with earlier and higher positions of surface streams. This suggests that denudation and stream incision in eastern Trigg County may have been episodic and closely related to the Quaternary history of the Ohio River and its major tributaries (the Cumberland River in this case).

In summary, doline alignments and stream alignments in the western Mississippian Plateau have been influenced by regional structure, and in particular, by faults radiating from the Fluorspar District to the northwest. Most caves and proven groundwater flow paths have been guided by fractures associated with the faults. Karst features aid in recognizing and mapping faults that may otherwise go undetected where a thick mantle of soil conceals the bedrock.

Edwards Plateau, central Texas

The Edwards Plateau of central Texas is an upland surface underlain by a thick sequence of Early Cretaceous limestone and dolostone in the west and south, and by Paleozoic carbonate and clastic rocks in the northeast. With an area of 82 900 km^2, it is one of the largest contiguous karst regions of the United States (Fig. 15.1). Because of the economic importance of karst aquifers as water supplies for over one million people, the hydrogeology of the plateau has been extensively studied by the United States Geological Survey and the Texas Department of Water Resources (Maclay & Small 1976, 1983, Walker 1979); yet until recently (Kastning 1983b), extensive study of karst landforms, including caves, has been minimal.

Four sub-areas have been selected within the plateau to assess influence of tectonism on karst in a variety of structural settings (Fig. 15.7): (a) the Sonora

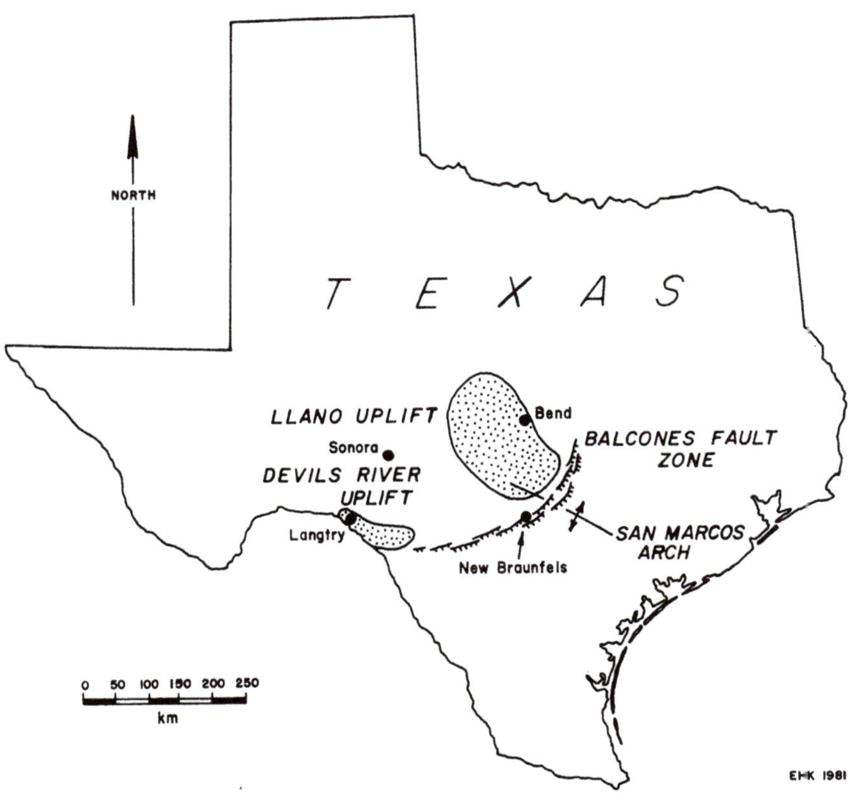

Figure 15.7 Structural elements of the Edwards Plateau, including the Devils River and Llano uplifts, San Marcos Arch and Balcones Fault Zone.

area in the west, which lies within the interior of the plateau and is only slightly deformed, (b) the Langtry area in the southwest, which sits astride the Devils River Uplift, (c) the New Braunfels area in the southeast, which lies within the Balcones Fault Zone and along a flank of the San Marcos Arch, and (d) the Bend area in the northeast, which is on the northeast margin of the Llano Uplift.

The interior topography of the Edwards Plateau is undulating to slightly rolling and is moderately dissected by drainage systems flowing southeast to the Gulf of Mexico. However, the Balcones Escarpment along the south and east plateau margins is deeply dissected by streams, and slopes are correspondingly steeper there than in the plateau interior.

The area contains a remarkable diversity of bedrock types, ranging from sedimentary units in the plateau proper to metamorphic and igneous rocks in the Llano Uplift. Paleogeographic and structural elements have influenced deposition, resulting in gradations in the lithic character and thickness of carbonate units throughout the plateau. This is particularly true for the three major cavernous stratigraphic groups under consideration here: (a) the Ellenburger Group (Ordovician), (b) the Trinity Group (Lower Cretaceous) and (c) the Edwards Group (Lower Cretaceous). Discussions of regional aspects of the character and depositional history of these deposits are found in Cloud and Barnes (1946), Stricklin *et al.* (1971) and Rose (1972).

The solubility of carbonate units of the plateau is quite variable, and throughout the plateau, caves are generally confined to relatively pure limestone strata of highly porous dolostone beds. In some situations, cave passages developed just above beds of reduced solubility or are capped by resistant units (Kastning 1983a).

Cavern morphology is strongly influenced by the nearby topographic setting. Caves in upland areas are generally isolated and relict features, consisting of one or more irregularly shaped chambers connected by short passages. They are poorly integrated, and complex or maze caves typically exhibit spongework patterns. In contrast, where surface streams have incised deeply, particularly along the Balcones Escarpment, caves near base level are well integrated, consist of long, horizontal corridors of rather uniform cross section, and often contain perennial streams. Of over 2000 documented caves on the plateau, most are clustered along the deeply dissected Balcones Escarpment (Fieseler *et al.* 1978, p. 20), suggesting that cavern development has been accelerated by topographic dissection and steepening of hydraulic gradients (Kastning 1978, 1981, Woodruff & Abbott 1979).

The structural framework underlying the Edwards Plateau (Fig. 15.7) significantly affected cavern development. Where caves are confined to specific stratigraphic horizons, groundwater generally moved down the dip, resulting in a predominance of dip-oriented passages. In most cases, caves are excavated along favorable (open) joints giving cave passages a slightly angulate pattern, wherein linear passage segments are aligned parallel or subparallel to the dip direction. However, where beds are nearly horizontal and where fractures of tensional origin are numerous (high frequency of open fractures), passages are commonly guided by dominant joint sets rather than by attitudinal control. These effects are illustrated by examining cavern development in areas where fractures are associated with major structural elements.

Sonora area

The Sonora area in Sutton County lies within the relatively undeformed interior of the Edwards Plateau (Fig. 15.7). Beds of the Segovia Formation of the Edwards Group, which contain the largest caves, lie generally parallel to the pre-Cretaceous surface and are nearly horizontal, with dips of 0.1°–0.2° to the southwest (Cartwright 1932, Rall & Rall 1958, Brown *et al.* 1965, Blank *et al.* 1966, Walker 1979).

The topography of the Sonora area is largely developed on the upper and middle parts of the Segovia Formation, which consist of sequences of dolomitic limestone and marl. The uppermost beds are medium- to coarse-grained, cherty, medium-bedded to massive, and contain abundant mollusc fragments, spar and scattered rudists. In this area, several thick limestone beds form prominent resistant ledges that provide a caprock over underlying beds of the Segovia (Fig. 15.8). The middle beds are dolomitized micrite and marl, are thin- to thick-bedded, soft and porous, and tend to weather recessively in contrast to the upper Segovia beds (Kastning 1983a).

Most passages of large caves of the area (e.g. Caverns of Sonora, Felton Cave, Silky Cave) are conformable with the Segovia beds; the largest conduits are excavated in part along highly burrowed, massive beds of marl and limestone, consisting of finely crystallized micrite and fine, skeletal biomicrite or biosparite. Intervening units that are typically more dolomitic, nodular and cherty are less cavernous.

The Segovia Formation was fractured during three distinct events (Kastning 1983a): (a) differential compaction over a north–south trending structural high (Cartwright 1932), (b) upward extension of pre-existing fractures related to the Chadbourne fault system of Pennsylvanian age (Conselman 1958, Rall & Rall 1958), and (c) unloading through erosion of overlying rocks. The first two events produced fractures dominantly oriented north–south or slightly to the northeast–southwest. The latter event may explain fractures of other orientations.

Vertical extents and orientations of fractures vary among lithic units, suggesting (a) differences in competency and brittleness among strata, (b) differences in initial openness of fractures among strata prior to dissolutional enlargement, (c) fracturing of lower beds prior to deposition of upper beds, or (d) changes in direction in hydraulic gradients and groundwater flow as cave levels evolved in sequence. From evidence in Caverns of Sonora, Felton Cave and Silky Cave, upper Segovia beds apparently have fractures of a northeast orientation, whereas fractures in lower beds are oriented NNE. Moreover, upper Segovia fractures are more numerous and because of their orientation,

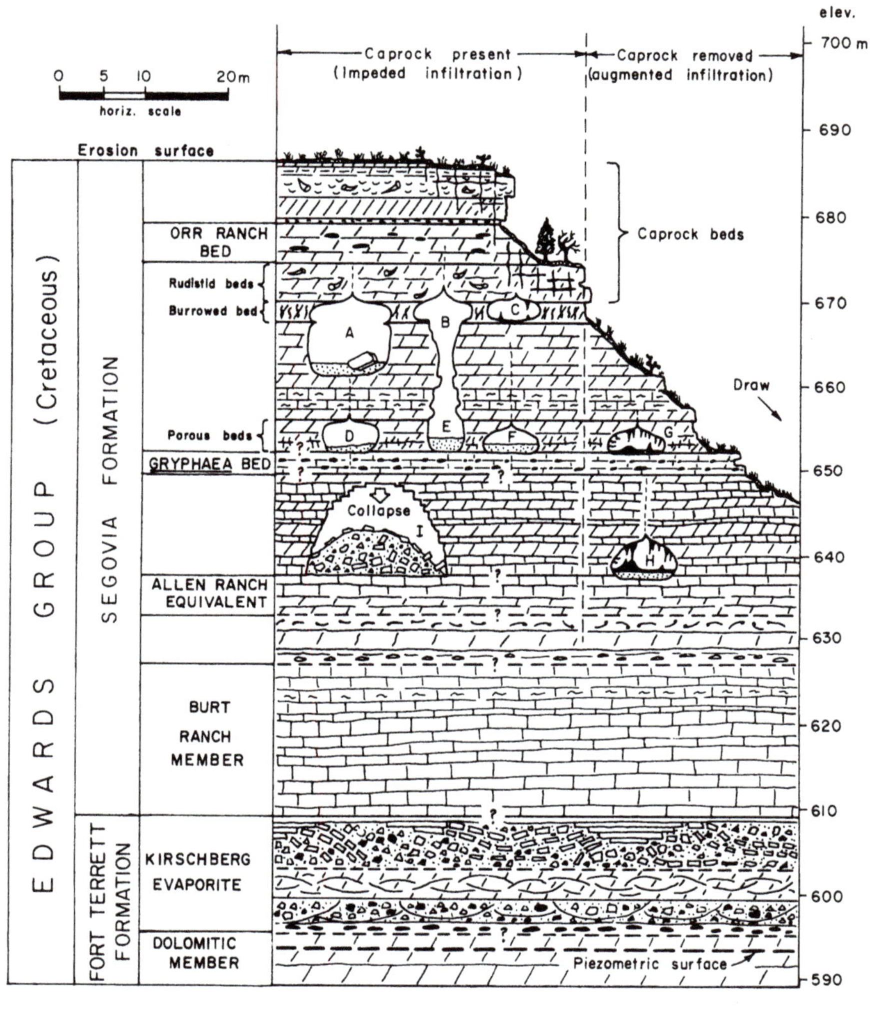

Figure 15.8 Generalized geologic section of the Sonora area, central Texas, showing stratigraphy and lithic character (after Rose 1972), topographic weathering profile, piezometric surface and horizons of cavern development. A through I are typical cross sections of cave passages. From Kastning (1983a), used with permission of Elsevier Publishing Company, Amsterdam.

may reflect differential compaction of Cretaceous strata into pre-Cretaceous valleys that flank a north–south oriented pre-Cretaceous topographic high (Cartwright 1932). Conjugate fractures of short horizontal extent in the Sonora area are generally oriented at right angles to the north and northeast sets in the Caverns of Sonora and Felton Cave and the east set in Silky Cave.

Orientations of lineaments mapped from aerial photographs and those of cave passages agree very well with fracture orientation. In the Sonora area, mapped photolineaments and the orientations of elongate playas and lake basins trend predominantly NNE (Pool 1977) as do major cave passages (Fig. 15.9). Some short passage segments have been excavated along conjugate sets, but are limited in length either because these fractures may not have experienced tension during tectonism and remained relatively closed, or because they are not aligned parallel to the dip, in the direction of groundwater flow.

Morphology of cave passages in the Sonora area suggests that initial development in the shallow-phreatic zone led to later canyon cutting into passage floors by vadose streams and modification by periodic reflooding of the caves. At the onset of cavern development, groundwater traveled down the gentle southwest regional dip, excavating conduits of tubular cross section along beds favorable to dissolution. Maximum enlargement occurred just beneath the water table. As degradation and dissection of the Edwards Plateau

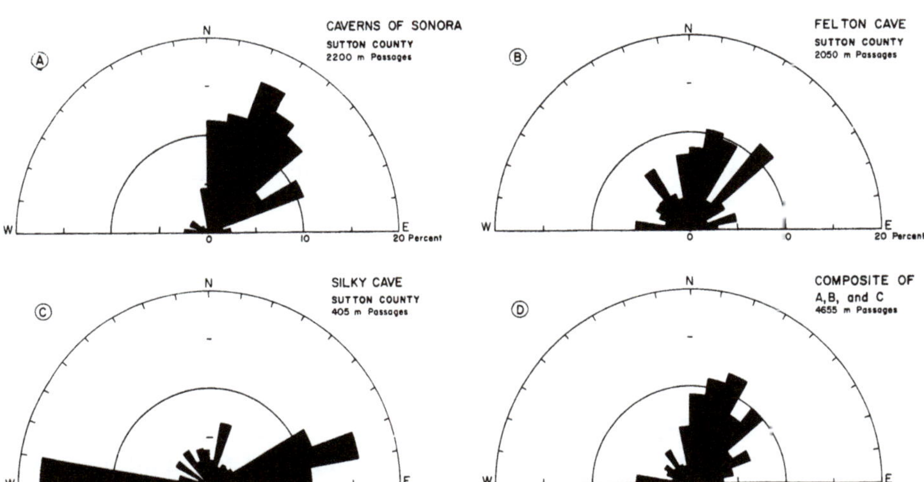

Figure 15.9 Orientation diagrams of linear cave-passage segments in (A) Caverns of Sonora, (B) Felton Cave, (C) Silky Cave and (D) the three caves combined. From Kastning (1983a), used with permission of Elsevier Scientific Publishing Company, Amsterdam.

progressed, local water tables dropped in response to lower base levels. When water tables dropped beneath the conduits, free-surface streams continued solutional enlargement by excavation of passage floors, and this produced high, relatively narrow, canyon passages. Localized collapse of ceiling beds and partial infilling of caves with sediment have since modified passage configurations, producing irregular floor profiles (Kastning 1983a). In places, debris has blocked the flow of vadose streams, and in such cases, ponded floodwater has subsequently excavated maze-like passages and diversion routes.

In general, groundwater that produced the master conduits in the Sonora area flowed along the hydraulic gradient and down the dip, following joints that were left most open by tectonism.

Langtry area

The Langtry area lies within the Devils River Uplift (Fig. 15.7), an anticlinal structure developed on high-standing Precambrian rocks of the Ouachita System (Flawn 1959). Uplift occurred during the Paleozoic (Webster 1980) and may have continued during the Cretaceous. Present attitudes of beds and fractures are, in part, related to pre-Cretaceous events because basement structure influenced deposition of Cretaceous limestone. However, later tectonism, including regional faulting and uplift that occurred during the Miocene, also deformed the rocks.

The lowest exposed bedrock unit in the Langtry area is the Devils River Formation (Lower Cretaceous), a thin- to thick-bedded limestone that is finely crystalline and very fossiliferous. The lithic character of the unit is relatively uniform over its 260 m thickness, and the formation is nearly pure in calcite (96.0–99.7%; Rodda *et al.* 1966). The top 60 m are exposed by erosion in the deeply dissected region adjacent to the Rio Grande. The Devils River is overlain by the Buda Limestone, which is a thick- to very thick-bedded, dense, porcelaneous limestone in its upper part, and a thin- to thick-bedded and clayey limestone with nodular weathering in its middle part. The Buda is 24–27 m thick and is exposed along valley slopes of the Rio Grande and the Pecos River. The Boquillas Flags overlie the Buda Limestone and consist of four distinguishable units of clastic, clayey to silty limestone. These beds are generally thinly laminated calcarenite and limestone, and the unit is ~47–55 m thick.

The Langtry area exhibits a systematic fracture pattern (Fig. 15.10) consisting of three or four principal fracture sets (Leonard 1977). Northwest-trending fractures coincide with the northern boundary of the uplift and are related to compressional forces from late Paleozoic deformation. A set of northeast-trending lineaments mapped by Freeman (1964) and Leonard (1977) are orthogonal to the northwest set, suggesting that they are also related to the Devils River Uplift. Most northeast-trending fractures apparently resulted from contemporaneous tensional forces applied during extension and faulting

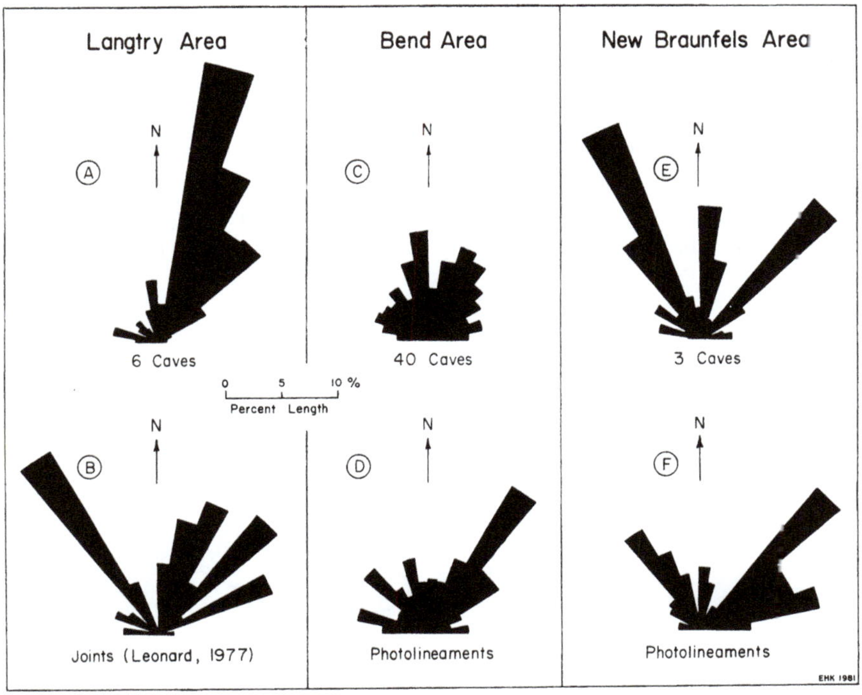

Figure 15.10 Orientation diagrams indicating relationships between linear cave-passage segments and mapped fractures and photolineaments for the Langtry, Bend and New Braunfels areas, central Texas. From Kastning (1981).

along the Balcones Fault Zone and San Marcos Arch during the late Cretaceous and early Tertiary epochs (Fowler 1956, Leonard 1977).

Caves in the Langtry area (Kunath & Smith 1968) are generally deep (>80 m). This is explained by (a) the massive and uniform character of the highly calcitic Devils River Formation, (b) deep incision of the Rio Grande drainage through 90 m of the stratigraphic section, creating steep hydraulic gradients, (c) regional dip of 0.3°–1.0° to the southwest towards the Rio Grande, and (d) vertically extensive joints carrying recharge to great depths along bathyphreatic loops (Kastning 1983a). Horizontal cave passages are relatively short, and lie along distinct stratigraphic horizons. These segments are connected by steeply sloping or vertical conduits.

Nearly all cave passages in the Langtry area are predominantly oriented parallel to the northeast fracture sets (Fig. 15.10a & b). Note that considerably fewer passage segments are aligned along the dominant northwest fracture set. This difference is attributable to initial openness of fractures. Those fractures formed by extension (i.e. the northeast sets) separated under deformation;

however, fractures formed by compression remained more closed (Fig. 15.11). Open fractures promoted circulation of groundwater and dissolutional enlargement whereas closed fractures inhibited this process. Analysis of photolineaments observable on aerial imagery confirms these relationships (Freeman 1968, Leonard 1977). Open fractures (the northeast sets) are readily visible as lineaments because (a) vegetation (which is sparse in the region) takes root in those fractures with the ability to accumulate moisture, and (b) many lineaments in this region consist solely of vegetal alignments (Kastning 1981).

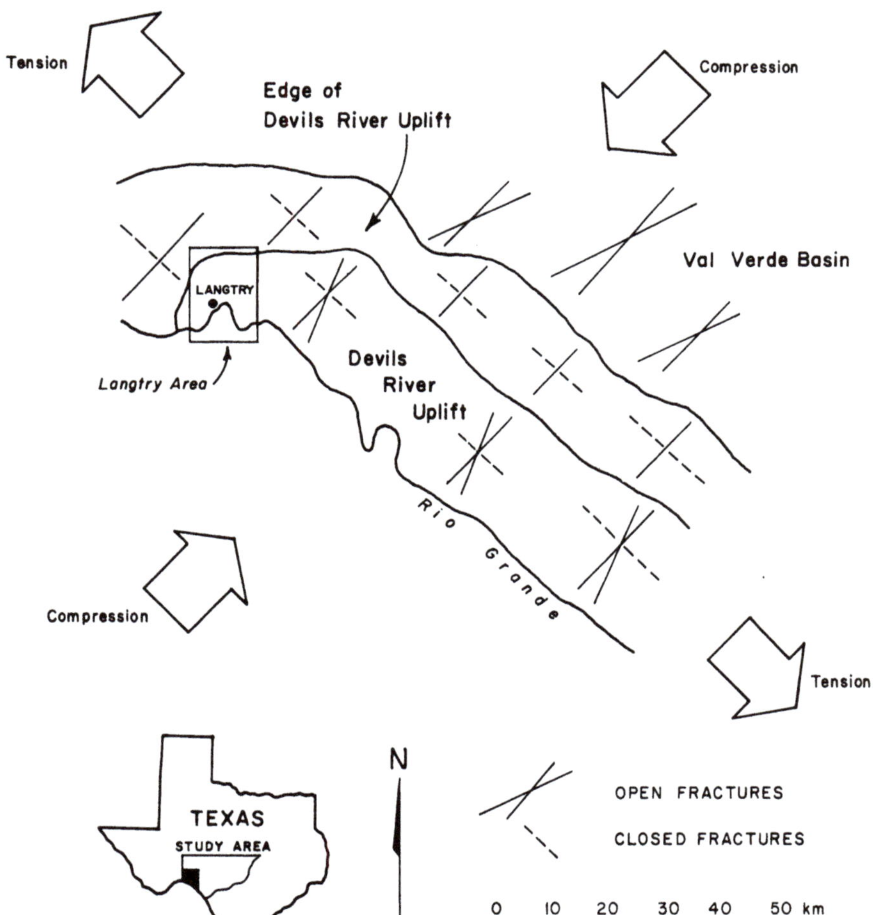

Figure 15.11 Devils River Uplift structural trend showing compressional and tensional forces acting during uplift of the Edwards Plateau, and general configuration of open and closed fractures. From Kastning (1983a) as modified from Leonard (1977), used with permission of Elsevier Scientific Publishing Company, Amsterdam.

New Braunfels area

The New Braunfels area lies astride the Balcones Fault Zone and on the southwest flank of the San Marcos Arch (Fig. 15.7). The arch was raised along its northwest-trending axis during the close of the Early Cretaceous (Woodruff & Abbott 1979). The Balcones Fault Zone consists of a series of *en échelon* normal faults extending from the southwest part of the Edwards Plateau to San Antonio, thence northeast to Austin. Faulting occurred during the Miocene as the plateau region was uplifted, and created areas of high local relief in the vicinity of the Balcones Escarpment.

The cavernous region of the New Braunfels area is underlain by rocks of the Trinity and Edwards groups. The Glen Rose Formation of the Trinity Group is the lowest cavernous unit in the sequence. It consists of alternating resistant and recessive beds of limestone, dolostone and marl. The limestone is aphanitic to fine grained, hard to soft and marly. The dolostone is fine grained and porous. The upper part of the Glen Rose is thin bedded, more dolomitic and less fossiliferous than the lower part. The Edwards Limestone, which unconformably overlies the Glen Rose, is generally fine- to coarse-grained with abundant chert and fossils. Solution zones and collapse breccia are common.

The deformation that produced the San Marcos Arch created a primary set of extensional fractures oriented northwest–southeast and parallel to the arch axis, and two conjugate sets (Fig. 15.10f). Fractures of the primary set are long in comparison to those of the conjugate sets. Later deformation during Balcones faulting produced northeast–southwest extensional fractures, parallel to individual faults, and conjugate fractures of shorter length (Fig. 15.10f).

Cavern development in the New Braunfels area progressed in response to two distinct tectonic episodes (Woodruff & Abbott 1979). First, following uplift of the San Marcos Arch, Lower Cretaceous rocks were subjected to subaerial exposure and heightened topographic relief that promoted deep circulation of groundwater through fractures produced during uplift. Primary porosity in the Glen Rose Formation was solutionally enhanced as cavities enlarged, but remained poorly integrated. During the Late Cretaceous, the region was covered episodically by shallow seas that deposited calcareous and clastic sediments above the Glen Rose Formation. The second disturbance, Balcones faulting during the Miocene, produced high local relief in the immediate area, which resulted in steep hydraulic gradients. Groundwater moving along the gentle southeast dip enlarged pre-existing northwest–southeast fractures as well as many of those produced by the faulting. As a result, major caves of the area, such as the Natural Bridge Caverns, consist of well-integrated conduits of large cross section that are angulate in plan view in response to flow along open fractures favorably aligned to flow paths (Kastning 1980).

Incision of streams draining the Edwards Plateau was rapid in the vicinity of the Balcones Escarpment. Groundwater levels declined in response to valley cutting, and cave-passage levels developed in a descending sequence. Passages were enlarged into canyons by subsequent vadose streams and later abandoned as water circulated at greater depths. In some places, ceilings collapsed, blocking major active conduits and promoting development of diversion routes.

In general, flow was to the southeast, and northwest–southeast trending fractures account for a large percentage of the overall orientation of cave pasages (Fig, 15.10e). However, many northeast–southwest oriented fractures and some conjugate fractures provided cross-over paths between adjacent master fractures and thereby account for a significant number of cave passages as well (Fig. 15.10e).

Bend area

The Bend area lies on the northeast flank of the Llano Uplift (Fig. 15.7). Ordovician rocks are cut by a series of *en échelon* normal faults striking NNE, formed by tensional epeirogenic stresses during the Pennsylvanian (Cloud & Barnes 1946).

Exposed rocks are primarily of the Gorman and Honeycut formations of the Ellenburger Group (Ordovician). The Gorman consists of thick- to thin-bedded, cherty limestone and dolostone, which contain thin arenaceous beds. It is subdivided into a lower dolomitic facies and an upper calcitic facies (Cloud & Barnes 1946). In the Bend area, the dolomitic facies is 25–66 m thick, but only the upper few meters are exposed along the lower valley walls of the Colorado River, which passes through the area. This unit is microgranular, fine grained and contains few fossils. The overlying calcite facies, which is exposed over most of the uplands, is 76–117 m thick, and consists of upper and lower units of aphanitic, thick- to thin-bedded limestone and a middle unit of microgranular to fine-grained dolostone.

The Honeycut Formation, overlying the Gorman, is exposed in the northwest and northeast parts of the area and consists of thick- to thin-bedded, cherty, fossiliferous limestone and dolostone. Locally, the Honeycut forms a caprock over the Gorman Formation.

Regionally, beds of the Ellenburger Group dip 1.0°–1.4° toward the Colorado River to the northeast. However, *en échelon* faults striking NNE have displaced large blocks of the Ellenburger, and in places, the dip is 0°–4° to the northwest. Faulting has compartmentalized groundwater flow toward the river.

The Ellenburger units are highly fractured, and joints are typically vertical and spaced a few meters apart. Dominant fracture trends are N30°–60°E, N10°–20°W, N30°–50°W, and N70°–90°W and agree with mapped photolineaments (Fig. 15.10d). Many fractures parallel major faults, but others are apparently conjugate sets related to faulting, or are the result of unloading in response to topographic dissection and denudation in the vicinity of the Colorado River (Kastning 1981).

Orientations of cave-passage segments (from maps in Reddell 1973) agree moderately well with those of mapped fractures (Fig. 15.10c & d). Long caves have been selectively excavated along open extensional fractures oriented along the northeast dip, toward the river. In areas where dips are to the northwest, caves also follow favorable fractures, but caves and their individual passage segments are generally much shorter.

Caves in the Bend area are typically confined to highly calcitic limestone beds of the calcitic facies of the Gorman Formation. Bedding-plane partings have provided little permeability and few avenues for groundwater circulation in comparison with those provided by vertically oriented fractures. As a result, long and well-integrated flow systems are arranged down dip along northeast-trending fractures, but within calcite-rich units. Many caves are angulate in pattern, suggesting that groundwater has followed those fractures initially most open and able to conduct flow most efficiently along prevailing hydraulic gradients to base-level springs at river level.

Conclusions

In moderately deformed areas, exemplified by the foregoing regional studies, fracturing is generally uniform and systematic. Flexures are broad with respect to their amplitude, dips of strata are gentle ($<5°$), and faulting is normal and extensional in origin. Extensive plateaus underlain by carbonate rocks typically exhibit distinct fracture patterns that are related to tectonic episodes.

Most systematic sets of fractures correspond to major structural elements; for example, (a) most fractures of the western Mississippian Plateau, Kentucky, relate to the Cincinnati Arch, Illinois Basin, and to extensive faults in the area, and (b) many fractures of the Edwards Plateau, Texas, are clearly related to the Devils River and Llano uplifts, the San Marcos Arch and the Balcones Fault Zone. In some cases, fracture sets are related to basement tectonism where the underlying structure is now deeply buried; this is apparently true for the Helderberg Plateau, New York.

Cave passages are usually excavated along bedding-plane partings or fractures that are the most open initially, or are favorably oriented along prevailing hydraulic gradients. These openings will enlarge at the greatest rates and become the master conduits of well-integrated cave systems. Consequently, cave passages may accentuate the existence of major fractures as well as indicate past groundwater flow paths. In addition, surface karst landforms, such as dolines and fissured pavements, represent zones of recharge and tend to become aligned along vertically oriented fractures that serve as avenues for infiltration.

A precise understanding of the origin of particular caves or cave systems requires careful mapping of geologic structure and its relation to topographic factors, lithic character of the rock, and hydrogeologic framework of the karst terrane. Conversely, maps showing geometry, geographic position, and orientations of cave passages are often highly useful in recognizing structural trends, particularly where surface features are obscured by a mantle of soil or glacial deposits, or by dense vegetal cover.

Acknowledgements

I thank my wife, Karen, for contributions to this paper. Much of the data and interpretation on the western Mississippian Plateau is from her research project currently in progress. She assisted me in fieldwork in Texas, and discussed and helped develop the ideas presented in this paper. Robert F. Black of the University of Connecticut and Victor R. Baker of the University of Arizona reviewed earlier drafts of the Helderberg and Edwards plateaus studies, respectively, and many of their suggestions are incorporated herein. Financial support was provided by a Cave Research Foundation Fellowship, the 1977 National Speleological Society Ralph W. Stone Research Award, and two Ronald K. Deford scholarships from the Department of Geological Sciences, University of Texas at Austin.

References

Anderson, R. 1961. Geology of Barton Hill. *Nat. Speleol Soc. Bull.* **23**, 11–4.

Baker, V. R. 1973. Geomorphology and hydrology of karst drainage basins and cave channel networks in east central New York. *Water Resour. Res.* **9**, 695–706.

Baker, V. R. 1976. Hydrogeology of a cavernous limestone terrane and the hydrochemical mechanisms of its formation, Mohawk River basin, New York. *Empire St. Geogram* **12**, 2–65.

Barlow, C. A. and A. E. Ogden 1982. A statistical comparison of joint, straight cave segment, and photo-lineament orientations. *Nat. Speleol Soc. Bull.* **44**, 107–10.

Barnholtz, S. 1961. Influence of jointing on cavern development in Miller County, Missouri. *Mo. Speleol.* **3**, 66–73.

Berdan, J. M. 1948. Hydrology of limestone terrane in Schoharie County, New York. *Trans Am. Geophys. Un.* **29**, 251–3.

Berdan, J. M. 1950. *The ground-water resources of Schoharie County, New York.* St. NY Water Power Cont. Com. Bull. GW-22.

Blank, H. R., W. G. Knisel, Jr. and R. W. Baird 1966. *Geology and groundwater studies in part of the Edwards Plateau of Texas, including Sutton and adjacent counties.* US Dept Ag., Agricul Res. Serv. Rep. ARS 41-103.

Bretz, J. H. 1942. Vadose and phreatic features of limestone caverns. *J. Geol.* **50**, 675–811.

Brown, J. B., L. T. Rogers and B. B. Baker 1965. *Reconnaissance investigation of the groundwater resources of the middle Rio Grande basin, Texas.* Texas Water Com. Bull. 6502, M1–M80.

Brown, R. F. and T. W. Lambert 1963. *Reconnaissance of ground-water resources in the Mississippian Plateau region, Kentucky.* US Geol Surv. Water-Supply Pap. 1603.

Cartwright, L. D., Jr. 1932. Regional structure of Cretaceous on Edwards Plateau of southwest Texas. Bull. Am. Assoc. Petrolm Geols **16**, 691–700.

Cloud, P. E., Jr. and V. E. Barnes 1946. *The Ellenburger Group of central Texas.* Univ. Texas Publn 4621.

Conselman, F. B. 1958. Chronology of movements along Fort Chadbourne fault system in west-central Texas: discussion. *Bull. Am. Assoc. Petrolm Geols* **42**, 2783–5.

Cook, J. H. 1907. Limetone caverns of eastern New York. In *Third report of the Director of the Science Division, 1906*, J. M. Clarke (ed.), 32–51. Albany: New York State Museum.

Darton, N. H. 1893. On two overthrusts in eastern New York. *Geol Soc. Am. Bull.* **4**, 436–9.

Darton, N. H. 1894. Preliminary report on the geology of Albany County. In *Thirteenth annual report of the State Geologist for the year 1893*, J. Hall (ed.), 229–61. Albany: Geological Survey of New York.

Davies, W. E. 1960. Origin of caves in folded limestone. *Nat. Speleol Soc. Bull.* **22**, 5–18.
Davies, W. E. 1966. The earth sciences and speleology. *Nat. Speleol Soc. Bull.* **28**, 1–14.
Davies, W. E. 1968. The earth sciences and speleology: reply. *Nat. Speleol Soc. Bull.* **30**, 93.
Davis, W. M. 1930. Origin of limestone caverns. *Geol Soc. Am. Bull.* **41**, 475–628.
Deike, R. G. 1969. Relations of jointing to orientation of solutional cavities in limestones of central Pennsylvania. *Am. J. Sci.* **267**, 1230–48.
DeSaussure, R. 1963. The general formation and development of limestone caves. *Cave Notes* **5**, 1–5.
Dicken, S. N. and H. B. Brown 1938. *Soil erosion in the karst lands of Kentucky.* US Dept Ag. Cir. 490.
Egemeier, S. J. 1969. Origin of caves in eastern New York as related to unconfined groundwater flow. *Nat. Speleol Soc. Bull.* **31**, 97–110.
Fieseler, R. G., J. F. Jasek and M. Jasek 1978. *An introduction to the caves of Texas.* Nat. Speleol Soc. Guide. 19.
Fisher, D. W., Y. W. Isachsen and L. V. Rickard 1970–71. *Geologic map of New York.* NY St. Mus. Sci. Serv. Map Chart Ser. 15.
Flawn, P. T., 1959. Devils River Uplift. In *Geology of the Val Verde Basin and field trip guidebook, November 5, 6, 7, 8, 1959*, R. L. Cannon, R. T. Hazzard, A. Young and K. P. Young (eds), 74–8. West Texas Geological Society.
Ford, D. C. 1965. The origin of limestone caverns: a model from the central Mendip Hills, England. *Nat. Speleol Soc. Bull.* **27**, 109–32.
Ford, D. C. 1971. Geologic structure and a new explanation of limestone cavern genesis. *Trans Cave Res. Grp G. Br.* **13**, 81–94.
Ford, D. C. and R. O. Ewers 1978. The development of limestone cave systems in the dimensions of length and depth. *Can. J. Earth Sci.* **15**, 1783–98.
Fowler, P. 1956. Faults and folds of south-central Texas. *Trans Gulf Coast Assoc. Geol Soc.* **6**, 37–42.
Fox, K. F., Jr. 1965. *Geology of the Cadiz quadrangle, Trigg County, Kentucky.* US Geol. Surv. Geol Quad. Map GQ-412.
Freeman, V. L. 1964. *Geologic map of the Langtry quadrangle, Val Verde County, Texas.* US Geol Surv. Misc. Geol. Invest. Map I-422.
Freeman, V. L. 1968. *Geology of the Comstock–Indian Wells area, Val Verde, Terrell, and Brewster counties, Texas.* US Geol Surv. Prof. Pap. 594–K.
Gardner, J. H. 1935. Origin and development of limestone caverns. *Geol Soc. Am. Bull.* **46**, 1255–74.
Goldring, W. 1935. *Geology of the Berne quadrangle.* NY St. Mus. Bull. 303.
Grabau, A. W. 1906. *Guide to the geology and paleontology of the Schoharie Valley in eastern New York.* NY St. Mus. Bull. 92, 77–386.
Gregg, W. J. 1974. Structural control of cavern development in Howe Caverns, Schoharie County, New York. *Nat. Speleol Soc. Bull.* **36**, 1–6.
Halliday, W. R. 1960. Changing concepts of speleogenesis. *Nat. Speleol Soc. Bull.* **22**, 23–9.
Heyl, A. V., Jr. 1972. The 38th parallel lineament and its relationship to ore deposits. *Econ. Geol.* **67**, 879–94.
Heyl, A. V., Jr. 1974. Some fluorite–barite deposits in the Mississippi Valley in relation to major structures and zonation. In *A symposium on the geology of fluorspar*, D. W. Hutcheson (ed.), 55–7. Kent. Geol Surv. Ser. X Spec. Publn 22.
Heyl, A. V., Jr. and M. R. Brock 1961. *Structural framework of the Illinois–Kentucky mining district and its relation to mineral deposits.* US Geol Surv. Prof. Pap. 424-D, D3–D6.
Hine, G. T. 1970. *Relation of fracture traces, joints, and ground-water occurrence in the area of the Bryantsville quadrangle, central Kentucky.* Kent. Geol Surv. Ser. X, Thesis Ser. 3.
Hook, J. W. 1974. Structure of the fault systems in the Illinois–Kentucky fluorspar district. In *A symposium on the geology of fluorspar*, D. W. Hutcheson (ed.), 77–86. Kent. Geol Surv. Ser. X Spec. Publn 22.
Isachsen, Y. W. and W. McKendree 1977. *Preliminary brittle structures map of New York.* NY

St. Mus. Sci. Serv. Map Chart Ser. 31.

Johnsen, J. H. 1958. *Preliminary report on the limestones of Albany County, N.Y.* NY St. Mus. Sci. Serv. Unnum. Rep.

Kastning, E. H. 1975. *Cavern development in the Helderberg Plateau, east-central New York.* NY Cave Surv. Bull. 1.

Kastning, E. H. 1977. Faults as positive and negative influences on ground-water flow and conduit enlargement. In *Hydrologic problems in karst regions*, R. R. Dilamarter and S. C. Csallany (eds), 193–201. Bowling Green: Western Kentucky University.

Kastning, E. H. 1978. *Caves and karst hydrogeology of the southeastern Edwards Plateau, central Texas: guidebook, geology field excursion, National Speleological Society annual convention, New Braunfels, Texas, June 18–23, 1978.* Nat. Speleol Soc. Guide. 19A.

Kastning, E. H. 1980. Structural, lithologic, and topographic controls on the origin of Natural Bridge Caverns, Comal County, Texas (Abs.). *Nat. Speleol Soc. Bull.* **42**, 32.

Kastning, E. H. 1981. Tectonism, fractures, and speleogenesis in the Edwards Plateau, central Texas, U.S.A. In *Proceedings of the Eighth International Congress of Speleology, Bowling Green, Kentucky, July 18 to 24, 1981*, B. F. Beck (ed.), 692–5. Hunstville: National Speleological Society.

Kastning, E. H. 1983a. *Geomorphology and hydrogeology of the Edwards Plateau karst, central Texas.* PhD dissertation. University of Texas at Austin.

Kastning, E. H. 1983b. Relict caves as evidence of landscape and aquifer evolution in a deeply dissected carbonate terrain: southwest Edwards Plateau, Texas, U.S.A. In *V. T. Stringfield Symposium: processes in karst hydrology*, W. Back and P. E. LaMoreaux (guest eds), *J. Hydrol.* **61**, 89–112.

Kastning, K. M. and E. H. Kastning 1981. Fracture control of dolines, caves, and surface drainage: Mississippian Plateay, western Kentucky, U.S.A. In *Proceedings of the Eighth International Congress of Speleology, Bowling Green, Kentucky, July 18 to 24, 1981*, D. F. Beck (ed.), 696–8. Huntsville: National Speleological Society.

Klemic, H. and G. E. Ulrich 1967. *Geologic map of the Roaring Spring quadrangle, Kentucky-Tennessee.* US Geol Surv. Geol. Quad. Map GQ-658.

Klemic, H., G. E. Ulrich and S. L. Moore 1968. *Geologic map of part of the Johnson Hollow quadrangle, Trigg County, Kentucky.* US Geol Surv. Geol. Quad Map GQ-722.

Kunath, C. E. and A. R. Smith (eds), 1968. *The caves of the Stockton Plateau.* Texas Speleol Surv. **3**(2).

Lattman, L. H. and R. R. Parizek 1964. Relationship between fracture traces and the occurrence of ground water in carbonate rocks. *J. Hydrol.* **2**, 73–91.

Leonard, R. C. 1977. *An analysis of surface fracturing in Val Verde County, Texas.* MA thesis. University of Texas at Austin.

Maclay, R. W. and T. A. Small 1976. *Progress report on geology of the Edwards aquifer, San Antonio area, Texas, and preliminary interpretation of borehole geophysical and laboratory data on carbonate rocks.* US Geol Surv. Open-File Rep. 76-627.

Maclay, R. W. and T. A. Small 1983. Hydrostratigraphic subdivisions and fault barriers of the Edwards aquifer, south-central Texas, U.S.A. In *V. T. Stringfield Symposium: processes in karst hydrology*, W. Back and P. E. LaMoreaux (guest eds), *J. Hydrol.* **61**. 127–46.

Malott, C. A. 1937. Invasion theory of cavern development. *Proc. Geol Soc. Am. 1936*, 323.

McFarlan, A. C. 1943. *Geology of Kentucky.* Lexington: University of Kentucky.

Moore, F. M. and J. E. Mylroie 1979. Influence of master stream incision on cave development, Trigg County, Kentucky. In *Western Kentucky Speleological Survey annual report, 1979*, J. E. Mylroie (ed.), 47–68. Murray: College of Environmental Sciences, Murray State University.

Muehlberger, W. R. 1961. Conjugate joint sets of small dihedral angle. *J. Geol.* **69**, 211–9.

Mylroie, J. E. 1977. *Speleogenesis and karst geomorphology of the Helderberg Plateau, Schoharie County, New York.* NY Cave Surv. Bull. 2.

Mylroie, J. E. 1978. Caves and karst features of western Kentucky as compiled May 1, 1978. In *Western Kentucky Speleological Survey annual report, 1978*, J. E. Mylroie (ed.), 11–55. Murray: College of Environmental Sciences, Murray State University.

Mylroie, J. E. 1979. Western Kentucky cave listing and description update. In *Western Kentucky Speleological Survey annual report, 1979*, J. E. Mylroie (ed.), 69–84. Murray: College of Environmental Sciences, Murray State University.

Mylroie, J. E. 1980. Cave list and update. In *Western Kentucky Speleological Survey annual report, 1980*, J. E. Mylroie (ed.), 49–77. Murray: College of Environmental Sciences, Murray State University.

Nelson, W. H. and D. A. Seeland 1968. *Geologic map of the Gracey quadrangle, Trigg and Christian counties, Kentucky*. US Geol Surv. Geol. Quad. Map GQ-753.

Nickelson, R. P. and V. D. Hough 1967. Jointing in the Appalachian Plateau of Pennsylvania. *Geol Soc. Am. Bull.* **78**, 609–30.

Nickelson, R. P. and V. D. Hough 1969. Jointing in south-central New York: reply. *Geol Soc. Am. Bull.* **80**, 923–6.

Ogden, A. E. 1974. The relationship of cave passages to lineaments and stratigraphic strike in central Monroe County, West Virginia. In *Proceedings of the Fourth Conference on Karst Geology and Hydrology*, H. W. Rauch and E. Werner (eds), 29–32. Morgantown: West Virginia Geological & Economic Survey.

Palmer, A. N. 1962. *Geology of the Knox Cave System, Albany County, New York*. BA honors thesis. Williams College.

Palmer, A. N. 1972. Dynamics of a sinking stream system: Onesquethaw Cave, New York. *Nat. Speleol Soc. Bull.* **34**, 89–110.

Palmer, A. N. 1975. The origin of maze caves, *Nat. Speleol Soc. Bull.* **37**, 57–76.

Palmer, A. N. 1977. Influence of geologic structure on groundwater flow and cave development in Mammoth Cave National Park, U.S.A. In *Karst hydrogeology: proceedings of the 12th Congress of the International Association of Hydrogeologists, Huntsville, Alabama*, J. S. Tolson and F. L. Doyle (eds). *Int. Assoc. Hydrogeols Mem.* **12**, 405–14.

Palmer, A. N. 1981. A geological guide to Mammoth Cave National Park. Teaneck: Zephyrus Press.

Palmer, M. V. 1976. *Ground-water flow patterns in limestone solution conduits*. MA thesis. State University of New York, Oneonta.

Parker, J. M., III 1942. Regional systematic jointing in slightly deformed sedimentary rocks. *Geol Soc. Am. Bull.* **53**, 381–408.

Parker, J. M., III 1969. Jointing in south-central New York: discussion. *Geol Soc. Am. Bull.* **80**, 919–22.

Pool, J. R. 1977. *Morphology and recharge potential of certain playa lakes of the Edwards Plateau of Texas*. Baylor Geol Studies Bull. 32.

Powell, R. L. 1975a. Joints in carbonate rocks in south-central Indiana. *Proc. Ind. Acad. Sci.* **84**, 343–54.

Powell, R. L. 1975b. Theories of the development of karst topography. In *Theories of landform development*, W. N. Melhorn and R. C. Flemal (eds), 217–42. London: George Allen & Unwin (1981 re-issue).

Powell, R. L. 1977. Joint patterns and solution channel evolution in Indiana. In *Karst hydrogeology: proceedings of the 12th Congress of the International Association of Hydrogeologists, Huntsville, Alabama*, J. S. Tolson and F. L. Doyle (eds), *Int. Assoc. Hydrogeols Mem.* **12**, 255–69.

Quinlan, J. F., Jr. 1968. The earth sciences and speleology: discussion. *Nat. Speleol Soc. Bull.* **30**, 87–92.

Quinlan, J. F., Jr. 1978. *Types of karst, with emphasis on cover beds in their classification and development*. PhD dissertation. University of Texas at Austin.

Rall, R. W. and E. P. Rall 1958. Pennsylvanian subsurface geology of Sutton and Schleicher counties, Texas. *Bull. Am. Assoc. Petrolm Geols* **42**, 839–70.

Reddell, J. (ed.) 1973. *The caves of San Saba County*, 2nd edn. Texas Speleol Surv. **3**(7–8).

Rich, J. L. 1928. Jointing in limestones as seen from the air. *Bull. Am. Assoc. Petrolm Geols* **12**, 861–2.

Rickard, L. V. 1962. *Late Cayugan (Upper Silurian) and Helderbergian (Lower Devonian) stratigraphy in New York*. NY St. Mus. Sci. Serv. Bull. 386.

Rickard, L. V. 1975. *Correlation of Silurian and Devonian rocks in New York State*. NY St. Mus. Sci. Serv. Map Chart Ser. 24.

Roberts, J. K. 1929. The Cretaceous deposits of Trigg, Lyon, and Livingston counties, Kentucky. In *Pleistocene of northern Kentucky and other papers*, 281–326. Kent. Geol Surv. Ser. VI **31**.

Rodda, P. U., W. L. Fisher, W. R. Payne and D. A. Schofield 1966. Limestone and dolomite resources, lower Cretaceous rocks, Texas Univ. Texas, Austin, Bur. Econ. Geol. Rep. Invest. 56.

Rodgers, J. 1970. *The tectonics of the Appalachians*. New York: Wiley.

Rose, P. R. 1972. *Edwards Group, surface and subsurface, central Texas*. Univ. Texas, Austin, Bur. Econ. Geol. Rep. Invest. 74.

Ruedemann, R. 1930. *Geology of the Capitol district (Albany, Cohoes, Troy, and Schenectady quadrangles)*. NY St. Mus. Bull. 285.

Sauer. C. O. 1927. *Geography of the Pennyroyal (region, Kentucky)*. Kent. Geol Surv. Ser. VI **25**.

Seeland, D. A. 1968. *Geologic map of the Cobb quadrangle, Trigg and Caldwell counties, Kentucky*. US Geol Surv. Geol. Quad. Map GQ-710.

Shaw, T. R. 1979. *History of cave science: the scientific investigations of limestone caves, to 1900*. Crymych: Anne Oldham.

Stricklin, F. L., Jr., C. I. Smith and F. E. Lozo 1971. *Stratigraphy of lower Cretaceous Trinity deposits of central Texas*. Univ. Texas, Austin, Bur. Econ. Geol. Rep. Invest. 71.

Sweeting, M. M. (ed.) 1981. Karst geomorphology. Benchmark papers in geology, Vol. 59. Stroudsburg: Hutchinson Ross.

Swinnerton, A. C. 1932. Origin of limestone caves. *Geol Soc. Am. Bull.* **43**, 663–93.

Trace, R. D. 1974. Illinois–Kentucky fluorspar district. In *A symposium on the geology of fluorspar*, D. W. Hutcheson (ed.), 58–76. Kent. Geol Surv. Ser. X Spec. Publn 22.

Ulrich, G. E. and H. Klemic 1966. *Geologic map of the Caledonia quadrangle, Trigg and Christian counties, Kentucky*. US Geol Surv. Geol. Quad. Map GQ-604.

Walker, L. E. 1979. *Occurrence, availability, and chemical quality of ground water in the Edwards Plateau region of Texas*. Texas Dept Water Resour. Rep. 235.

Warwick, G. T. 1962. The origin of limestone caverns. In *British Caving*, 2nd edn, C. H. D. Cullingford (ed.), 55–85. London: Routledge & Kegan Paul.

Webster, R. E. 1980. Structural analysis of Devils River Uplift: Southern Val Verde Basin, Southwest Texas. *Bull. Am. Assoc. Petrolm Geols* **64**, 221–41.

Weller, S. and A. A. Sutton 1951. *Geologic map of the western Kentucky fluorspar district*. US Geol Surv. Min. Invest. Field Stud. Map MF-2.

Wermund, E. G. and J. C. Cepeda 1977. Regional relation of fracture zones to the Edwards Limestone aquifer, Texas. In *Karst hydrogeology: proceedings of the 12th Congress of the International Association of Hydrogeologists, Huntsville, Alabama*, J. S. Tolson and F. L. Doyle (eds). *Int. Assoc. Hydrogeols Mem.* **12**, 239–53.

Wermund, E. G., J. C. Cepeda and P. E. Luttrell 1978. *Regional distribution of fractures in the southern Edwards Plateau and their relationship to tectonics and caves*. Univ. Texas, Austin, Bur. Econ. Geol. Cir. 78-2.

White, W. B. 1959. Speleogenesis. In *Speleo Digest – 1959*, J. R. Dunn and A. D. McGrady (eds), 2.1–2.34. Pittsburgh: National Speleological Society.

White, W. B. 1969. Conceptual models for carbonate aquifers. *Ground Water* **7**, 15–21.

White, W. B. 1977. Conceptual models for carbonate aquifers: revisited. In *Hydrologic problems in karst regions*, R. R. Dilamarter and S. C. Csallany (eds), 176–87. Bowling Green: Western Kentucky University.

Wilson, J. R. 1977. Lineaments and the origin of caves in the Cumberland Plateau of Alabama. *Nat. Speleol Soc. Bull.* **39**, 9–12.

Woodruff, C. M., Jr. and P. L. Abbott 1979. Drainage-basin evolution and aquifer development in a karstic limestone terrain, south central Texas, U.S.A. *Earth Surf. Proces.* **4**, 319–34.

Woodward, H. P. 1961. A stream piracy theory of cave formation. *Nat. Speleol Soc. Bull.* **23**, 39–58.

Index

For Product Safety Concerns and Information please contact our EU
representative GPSR@taylorandfrancis.com
Taylor & Francis Verlag GmbH, Kaufingerstraße 24, 80331 München, Germany

www.ingramcontent.com/pod-product-compliance
Ingram Content Group UK Ltd.
Pitfield, Milton Keynes, MK11 3LW, UK
UKHW052214190926
475702UK00006B/1007
9780367464479